Managing AutoCAD® in the Design Firm

A Manual for Architects and Interior Designers

Karen A. Vagts

Addison-Wesley Publishing Company

READING, MASSACHUSETTS • MENLO PARK, CALIFORNIA • NEW YORK
DON MILLS, ONTARIO • WOKINGHAM, ENGLAND • AMSTERDAM
BONN • SYDNEY • SINGAPORE • TOKYO • MADRID • SAN JUAN
PARIS • SEOUL • MILAN • MEXICO CITY • TAIPEI

Many of the designations used by manufacturers and sellers to distinguish their products are claimed as trademarks. Where those designations appear in this book, and Addison-Wesley was aware of a trademark claim, the designations have been printed in initial capital letters or all capital letters.

The author and publisher have taken care in preparation of this book, but make no expressed or implied warranty of any kind and assume no responsibility for errors or omissions. No liability is assumed for incidental or consequential damages in connection with or arising out of the use of the information or programs contained herein.

Screen shots are actual screen captures of AutoCAD® R12 and AutoCAD® R13 software and other Autodesk products and are reprinted with permission from and under the copyright of Autodesk, Inc.

This reproduction is provided with RESTRICTED RIGHTS. Use, duplication, or other disclosure by the U.S. Government is subject to restrictions set forth in FAR 52.227-19 (Commercial Computer-Software-Restricted Rights) and DFAR 252.227-7013(c)(1)(ii) (rights in Technical Data and Computer Software), as applicable.

Library of Congress Cataloging-in-Publication Data

Vagts, Karen A.
    Managing AutoCAD in the design firm : a manual for architects and interior designers / Karen A. Vagts
        p. cm.
    Includes bibliographical references and index.
    ISBN 0-201-48960-0
    1. Architecture—Data processing. 2. Interior decoration—Data processing. 3. AutoCAD (Computer file) 4. Computer-aided design. I. Title.
NA2728.V34 1996
720'.28'402855369—dc20   95–41307
                                                                                              CIP

Copyright © 1996 by Karen A. Vagts

All rights reserved. No part of this publication may be reproduced, stored in a retrieval system, or transmitted, in any form or by any means, electronic, mechanical, photocopying, recording, or otherwise, without the prior written permission of the publisher. Printed in the United States of America. Published simultaneously in Canada.

Sponsoring Editor: Kathleen Tibbetts
Project Manager: John Fuller
Production Coordinator: Ellen Savett
Technical Editor: David S. Cohn
Cover design: Ann Gallager
Text design: Joyce C. Weston
Set in 10-point Palatino by GEX, Inc.

1 2 3 4 5 6 7 8 9 -MA- 0099989796
First printing, January 1996

Addison-Wesley books are available for bulk purchases by corporations, institutions, and other organizations. For more information please contact the Corporate, Government, and Special Sales Department at (800) 238-9682.

Find us on the World-Wide Web at:
http://www.aw.com/devpress/

For my grandparents—
Miriam Beard Vagts and Alfred Frederick Vagts—
writers who always believed that their granddaughters
would someday join the family trade

Contents

Preface xv

What This Book Is About   xv
What This Book Is *Not* About   xvi
Assumptions   xvii
Presentation of AutoCAD software in Text and Illustrations   xviii
Production Notes   xix
Disclaimer   xix

Acknowledgments   xxi

Chapter 1 Introduction to AutoCAD and a Little History   1

A Brief History of CAD and AutoCAD   2
AutoCAD in Architecture and Interior Design   6

Chapter 2 CAD Decision Making: When and Why You Should Use AutoCAD   9

CAD Decisions As Part of the Design Firm Strategy   9
CAD Decision Making   12
   Should You Use CAD at All?   13
   If You Do Use CAD, When and How Should You Use It?   17
   When You Do Use CAD, Should You Use AutoCAD?   19

Chapter 3 Configuring AutoCAD for Design Offices   25

Which AutoCAD Platform Should You Use?   26
   AutoCAD for DOS   27
   AutoCAD for Microsoft Windows   28
   AutoCAD for the Macintosh   30

      The Multi-Platform Office   30
Selecting Hardware   31
AutoCAD on Networks   35
      Advantages and Disadvantages of Networks   35
      The Costs of Networks   39
      Configuring Networks   40
      Configuring AutoCAD for Networks   42
      Networks and Office Culture   44
Complementary Software for AutoCAD   46
      Other Autodesk Software   48
      Third-Party Add-ons for AutoCAD   49
      Microsoft Windows Applications   54
      Utilities   58
      Typefaces   61
      Peripherals and Other Hardware   61
      Training and Evaluation Tools   67
      Collecting Information about Complementary Software   67
      Tips on Selecting Complementary Software   69

Chapter 4   *Customizing AutoCAD for Maximum Design Productivity*   73

General Guidelines for Customizing AutoCAD   74
      In-House Customization   76
      Third-Party Software Add-ons   76
      External Consultants   78
      Scheduling and Budgeting for Customization   80
In-House Customization: A Guided Strategy   82
      Step 1 Configuring Your Overall Computer Operating System   84
      Step 2 Setting Up Computer Directories and File-Naming Standards   86
      Step 3 Installing AutoCAD   88
      Step 4 Customizing Your Systemwide AutoCAD Configuration   89
      Step 5 Customizing Your AutoCAD Drawing Environment   98
      Step 6 Creating Prototypical AutoCAD Drawings,
               Templates, and Libraries   107
      Step 7 Customizing Your Output Devices   115
      Step 8 Customizing Special AutoCAD Applications
               and Links to Other Software   119
Protecting Your Customization   123
Implementing Your Customization   123
Feedback and Subsequent Revisions   124

Chapter 5 Making AutoCAD a Team Player: The CAD-Confident Project Manager 125

Scheduling CAD-based Projects 127
  Recent History 128
  Availability of CAD Skills 128
  Computer/Drafter Ratio 129
  Availability of CAD Standards and a Customized CAD system 129
  Modification of Office Standards 129
  Hybrid Production 130
  Precedents for Similar CAD Projects 130
  Project Complexity 131
  Computer Speed 131
  Plotting Capabilities 132
  Plotting Quality 132
  Plotting Schedules 132
Staffing Projects 132
Refining the Project Scope 134
Delegating Tasks 136
Managing Team Communications and Coordination 139
Implementing Quality Control 140
Scheduling Output 141

Chapter 6 AutoCAD in the Design Process 145

Getting the Project: Marketing and Client Development 147
  Marketing AutoCAD-based Services 147
  Incorporating AutoCAD-based Projects into Your Marketing Materials 149
Predesign Services: Programming and Data Collection 152
  Programming and Database Creation 153
  Feasibility Studies 155
  Merging Drawing Information 156
Schematic Design 156
  Flowcharts and Diagrams 156
  Adding and Editing Information 158
  Tracking Schemes 159
  Presenting Multiple Schemes 160
Design Development 160
  Evolving from Schematics to Developed Designs 161
  Interactive Design Routines 166
  Products and Components 167

      Models and Mock-ups   168
      Geometry and Math   169
      Design Checklists   172
      Finish Color Boards   173
   Design Review and Approval   173
   Document Production   174
      Coordination and Quality Control   174
      Check Sets   176
      Redlines   178
      Links to Specifications and Cost Estimation Software   179
      Cross-Referencing   180
      Coordination with Consultants   182
   The Bid Process   184
   Construction Administration   186
      Documentation   186
      Shop Drawings and Fabrication   186
      CAD in the Field   188
      Move Coordination   189
      Punch Lists   189
   Postconstruction Activities, Follow-up, and Evaluation   190
      As-Built Documents   190
      Evaluation   191
   Promotion   192

Chapter 7   *Special AutoCAD Applications for Architects and Interior Designers*   193

   The Americans with Disabilities Act and Code Compliance   194
   Area Takeoffs   197
      Area Takeoff Methodologies   199
      A Suggested Procedure for Area Takeoffs   201
      Area Takeoffs and Databases   203
   Computer-Aided Facilities Management (CAFM)   203
      CAFM and Design Practice   205
      Selecting a CAFM System   206
   Color in AutoCAD   209
      Uses for Color Output   210
      The AutoCAD Color Palette   210
      Color On-Screen versus Color Plotting   213

Office Color Palettes   214
Commercial Symbol Libraries   216
Desktop Publishing with AutoCAD   218
   Desktop Publishing within AutoCAD   219
   AutoCAD within Desktop Publishing Software   219
   Slide Presentations   220
   File Exchange with Desktop Publishers   220
Field Conditions and Dimensioning   222
Government Work   226
Historic Preservation   229
   Documentation   230
   Analysis   235
   Construction Documents and Specifications   236
International AutoCAD   238
Layer-Naming Systems   242
   AutoCAD Layer Characteristics   242
   Typical Approaches to Layer Names   244
   Implementing Standards for CAD Layer Names   248
   Documenting Standards for CAD Layer Names   249
Modeling in AutoCAD   250
   Physical Models   250
   CAD Models   251
   AutoCAD Modeling Options   259
   General 3D Modeling Guidelines   262
   Modeling with Light   263
   Modeling with Materials   265
   Presentation Options   267
   Internal versus External 3D   268
   Some Additional 3D Tips   269
Multiple Drawings, Floors, and Project Phases   269
   Multiple Drawings   270
   Multiple Floors   273
   Multiple Phases   274
Signage   275
Working with the Metric System   276

Chapter 8 Managing AutoCAD Output   281

Production: Choosing between Internal and External Options   283
   Reasons to Purchase a Plotter   284

    Reasons *Not* to Purchase a Plotter 285
   Determining the Break-Even Point for Plotter Purchases 286
   Plotter Options 287
    Pen Plotters 288
    Pencil Plotters 289
    Inkjet Plotters 290
    Laser/LED Plotters 291
    Thermal Imaging Plotters 291
    Electrostatic Plotters 292
    Hybrid Printer/Plotters 292
   Internal Production: Equipment Selection Guidelines 293
   External Production: Vendor Selection Guidelines 297
   Options for Plot Media 300
   Configuring Plotters for AutoCAD 301
   Some Plotting Tips 303
   Plotting Reimbursables 304
   Plotting Schedules 305
   Other Output Devices 308

Chapter 9 *Administration of an AutoCAD System* *311*

   Financial Administration 313
   Budgeting for AutoCAD 313
    CAD-related Income 315
    CAD-related Expenses 316
    CAD Economies of Scale 320
    Sources of Budgeting Information 321
   Calculating Returns on AutoCAD 322
    The AutoCAD Return on Investment (ROI) Calculator 322
    Revenues and Expenses per Employee 325
    CAD Utilization Ratio 325
   Setting Fees for AutoCAD Services 326
    Scope of Fees for CAD Work 327
    Methods of Setting Fees 329
    Guidelines for Estimating Fees 331
   AutoCAD in Design Contracts 331
   CAD Deliverables 332
    Copyright Ownership 334
    Designer's Liabilities 334
    Consultants 334

Making AutoCAD More Affordable   335
Personnel Administration   340
CAD Job Descriptions and Responsibilities   340
AutoCAD Personnel: Interviewing and Hiring   341
   Existing versus New Staff   341
   Interviewing and Evaluation   342
   Salaries and Benefits   344
   Temporary Staff   345
   Service Bureaus   345
Training in AutoCAD   346
   Built-in AutoCAD Options   347
   Third-Party Software   347
   Formal Training Programs   347
   In-House Training Programs   349
CAD Performance and CAD Parity   349
CAD Managers   351
   Assigning Responsibilities   351
   Establishing the Need for a CAD Manager   352
   Hiring a CAD Manager   353
Mentoring and Apprenticing with AutoCAD   356
The CAD Standards Manual   357
The CAD Technical Committee   358
CAD Ghettos and CAD Burnout   359
   CAD Ghettos   359
   CAD Burnout   361
Routine Administration   362
Memory and Speed Management   362
   System Memory   364
   AutoCAD Memory Management   365
Controlled Obsolescence   367
   Postponing Obsolescence   367
   Unloading Old Computers   368
Security   370
   Physical Destruction   371
   Theft   371
   System Damage   372
   System Backups   372
   Insurance   373
Space Planning for the Ergonomically Sound Office   373

Chapter 10 Managing AutoCAD Data 377

Data Organization: Directories and Files 379
   Directory Organization 381
   File Access 381
   AutoCAD File Names: Typical Approaches 383
Data Input 390
   Files on Disks 391
   Other Forms of Electronic Input 395
Data Conversion 402
   Converting AutoCAD Data: Major File Formats 403
   File-Exchange Documentation 410
Data Linking and Embedding 411
Upgrading Data 414
   Planning Upgrades 416
   Scheduling and Budgeting Upgrades 418
Managing Licenses for Data 418
   Licenses and Employees 421
   Licenses and Computer Inventories 422
Backing Up Data 423
   Backup Media Options 424
   Backup Procedures 425
   Protecting Backups 426
Archiving Data 427
   How to Archive AutoCAD Drawing Files 427
   When to Archive AutoCAD Drawing Files 428
   Challenges with Archiving CAD Files 428
   Resolving Archiving Challenges 429
Protecting Data: Copyrights and Liability 431
   Copyright Protection 432
   Liability and Legal Disputes 434
Insuring Data 441

Chapter 11 The Future for AutoCAD and Design 443

Appendix A 449

Checklist for Configuring/Customizing AutoCAD in Design Offices 449
AutoCAD Drawing Scale/Scale Factor Matrix 456

Appendix B: AutoCAD Resources 457

Glossary 463

Suggested Reading 471

Index 473

Preface

What This Book Is About

Managing AutoCAD in the Design Firm is intended for anyone who works in a firm that offers architecture, interior design, or related services and who has management responsibilities for, or interaction with, AutoCAD-based projects or operations. Its purpose is to help you make better use of the resources you have and to feel more confident working with Auto-CAD users and CAD managers. It does not assume you know CAD in general or AutoCAD in particular (although you'd certainly benefit from some hands-on knowledge!).

These days, you have only to stroll past the computer books section of any bookstore or computer supplies store to see just how prolific authors and publishers of computer-related topics are. If you stop to look at the shelf containing books on AutoCAD, you'll find many books that offer help on how to draft with AutoCAD in general. Here, you can find plenty of tips on becoming a power user, developing "killer" AutoCAD routines, and other aspects of technical expertise. But if you are a designer of manager in a design firm, you may find little in these books that will help you, a non-CAD user, make AutoCAD a more accessible, productive, and profitable tool in your office. You'll find that most books on AutoCAD, for example, focus on *generic* drafting, with little attention given to the very specific standards and procedures that architects and interior designers apply in their normal practice. Moreover, these books might show you fifteen different ways to construct an arc, but you'll be hard put to find information on AutoCAD's impact on your design fees, marking program, or personnel policies—all of which are of more immediate importance to you work. While the information you need *does* exist, it is not provided with your AutoCAD software, nor is it easily retrievable from one source.

The purpose of this manual is to provide you with the guidelines for better implementing AutoCAD in your design practice. It tells you only what you really need to know about the software and won't distract you with unnecessary technical details. The world of computers in general, and AutoCAD in particular, is dynamic, and staying abreast of the latest developments in computer technology is a full-time job; therefore, this book does not presume that you should, or wish to, become an expert on the latest developments in software and hardware (that is your CAD gurus' and CAD manager's role). Nor will it bombard you with technical specifications that will be immediately obsolete. Rather, this manual provides guidance on the general, more permanent issues. It seeks to show you where to find information and what to look for. Beyond that, and perhaps more important, it seeks to give you an attitude, a conceptual structure, or what in German is called a *Weltanschauung* (a combination worldview and life philosophy), towards computers in your practice. As this book stresses in Chapters 1 and 2, success with CAD in design practice is not ultimately an issue of technology itself but rather of management and of general goals and strategies.

What This Book Is *Not* About

Managing AutoCAD in the Design Firm is neither an instructional manual nor a command reference. It won't tell you how to draw a line or a circle or even how to open up an AutoCAD file. You will see references to certain commands and screen captures of AutoCAD dialog boxes and other features, but these are provided to illustrate certain points more effectively—and to help you explain to a CAD user what you want AutoCAD to do for you.

If you are interested in particular AutoCAD features or in learning AutoCAD, a starting place is AutoCAD's own documentation; if you own AutoCAD, you already have a copy of this manual, and while it is not written specifically for architects and interior designers, it is the horse's mouth. The AutoCAD software also includes a tutorial. You can purchase any number of reference guides and tutorials from third-party developers. In addition, you can take a course at an Autodesk Training Center or at any number of schools, colleges, and training centers. See Appendix B, "AutoCAD Resources," for further information on ways to learn more about AutoCAD as well as ways to incorporate it in your ongoing professional education and development.

Assumptions

This book is directed to a specific group of practicing professionals, namely those working in architecture, interior design, or related fields. Although design firms have much in common, they vary widely, not only in the nature of their practice but also in their professional goals. Nowhere is this more the case than with regards to computers. So you are well advised to be aware of the author's attitudes about CAD-related issues.

1. CAD is an acronym for "computer-aided *design,*" not merely "computer-aided drafting."
2. Design and drafting—whether by hand or by computer—should be interrelated and integrated, rather than widely separated functions.
3. Computers can be used for design, not merely for drafting. And while computers themselves can't *create* great designs, they can help you improve upon a good design concept.
4. The ideal title for a CAD-proficient designer or drafter is just that—designer or drafter—and not "CAD operator," or "CAD drafter," or, heaven forbid, "techie."
5. Computers in your office should be configured and customized so as to be as "user-friendly," fail-safe, and straightforward as possible. Computers should be *easy* and *fun* to use, and your system should be democratic in its accessibility for everyone.
6. The ultimate goal of design practice is *not* to meet unrealistic deadlines, scrimp by on paltry fees, and burn out yourself or your employees. Rather, it is to produce *better* design, service clients more aptly, provide future designers with proper training, and, above all, to enjoy your work.
7. The practice of design rarely offers a "free lunch," but some meals are more palatable than others.

Just as there is no uniformity in the practice of architecture and interior design, so there is no uniformity in titles. Members of the architecture/design (A/D) community refer to themselves in a number of ways, depending on their training and credentials. They may be designers, drafters, architects, interior designers, interior architects, and so forth. And while architecture and interior design are professionally intertwined, they are not synonymous. Therefore, in this book, the term "designer" generally refers to both architecture and interior design professionals. When something applies to a specific subset of the design profession, more specific titles are used.

In a similar line, the term "CAD user" is applied to all those who use CAD—whether or not they are drafters, designers, or CAD managers—as distinguished from those who don't use CAD at all. The intent here is to reflect the author's goal of someday eliminating terms such as "CAD operator." The day we'll know CAD is truly incorporated into the design professions will occur when "CAD" no longer appears in job titles or job descriptions.

Presentation of AutoCAD Software in Text and Illustrations

At the time of writing, AutoCAD is available in several versions for personal computers (workstations are not common in design firms, so they are not discussed here). AutoCAD Release 12 is still available for the DOS, Microsoft Windows, and Apple Macintosh platforms. AutoCAD Release 13, introduced in 1994, is presently available for DOS and Windows.

The principles espoused in this book apply to *all* AutoCAD recent releases and current personal computer platforms. For purposes of illustration, however, the commands and illustrations shown here represent AutoCAD Release 13 for Windows. There are several reasons for this: First, presenting computer screens and commands in multiple versions of AutoCAD would be confusing to readers; since this is not a command reference, the actual appearance of specific commands and functions on particular platforms is not particularly relevant—only their capabilities are. Second, within a year or so, Release 13 for Windows (both 3.1 and 95) will probably be the most prominent version of AutoCAD in design offices. Third, the AutoCAD Windows platform is what the author recommends for design firms if they have not yet shifted from DOS to Windows (unless they are on the Macintosh, which is still more user-friendly).

Throughout this book, references are made to other Autodesk products as well as to various Windows and AutoCAD third-party add-ons. These references are merely to highlight capabilities of other software packages that you may find useful. These references do not constitute product endorsements, nor does the author suggest that these products are superior to any others; rather the brand names are among the better known examples of their product segment and are benchmarks against which to review other products. For information on AutoCAD-related products, refer to *The AutoCAD Resource Guide,* a periodical that lists software packages. Appendix B also lists sources of information on AutoCAD-related products.

This manual attempts to avoid overt use of computer jargon. In some cases, however, it can't be avoided. Please refer to the AutoCAD Glossary at the end of the book for explanations of selected AutoCAD-related terms and acronyms.

Like all software upgrades, the initial version of AutoCAD Release 13 contained "bugs." Since then, three "patches" have been issued to address flaws in the initial release, and no doubt additional patches will be released as necessary. Some of the commands that this book describes initially contained problems; most have subsequently been addressed. Since this book is not a command reference, the author does not discuss "bugs" or problems with specific commands but rather assumes that by the time you read this, the major issues will have been resolved.

Production Notes

This book was written using a Compaq Prolinea 486/66 and a Macintosh Quadra 650. Software used to produce the manuscript includes AutoCAD Release 12 for Apple Macintosh; AutoCAD Release 13 for Windows; Canvas 3.5 by Deneba; Capture 4.5 by Mainstay; Corel Gallery for Macintosh by Corel; Hijaak Pro 3.0 by Inset Systems; Word 5.1 and Excel 4.0 by Microsoft; and Windows 3.1 by Microsoft. In addition, various design firms and software manufacturers have provided AutoCAD drawing files and screen captures for use as illustrations; they are credited with each illustration.

This book was composed in QuarkXPress. The typefaces used are Palatino, Christiana, and Tekton.

Disclaimer

This book contains a number of sample forms that list names of design firms, clients, buildings, and so forth. These names are purely figments of the author's imagination and do not refer to any actual entities. Any resemblance to actual people, buildings, organizations, or places is purely coincidental.

Acknowledgments

Managing AutoCAD in the Design Firm reflects the generosity of many people who have offered technical advice, professional expertise, and enthusiasm over the past two years.

The author wishes to thank the staff at Addison-Wesley: John Fuller, Project Manager; John Webber, Type and Design Manager; Ellen Savett, Production Coordinator; and copy editor Maggie Carr. Special thanks are extended to editor Kathleen Tibbetts, who made this book possible.

Through the Registered Author/Publisher program at Autodesk, the author received extensive technical support from Lisa Senauke, Autodesk Developer Marketing; Amar Hanspol, Autodesk Product Marketing; William P. Tryon, Image Lab, Marketing Support; Art Cooney and Kevin Vandecar, Technical Support.

The manuscript had the good fortune to be reviewed by two experienced architects who are also computer savvy and skilled at grammar and syntax: David S. Cohn, AIA, is an author of several books on AutoCAD, the senior editor of *CADalyst* magazine, and president of Eclipse Software, an AutoCAD third-party developer. Roger Goldstein, FAIA, an associate at Goody, Clancy & Associates, Inc., architectural activist, and educator, offers the perspective of an experienced computer-using architect, who is at once technically adept and inherently a humanist, and enthusiastic yet skeptical about the role of CAD in architecture.

Thanks also are due to the following individuals and organizations who contributed AutoCAD-generated drawing files and other illustrations for this book:

- ARCHIBUS, Inc., especially Bruce Forbes, Carla Salizzoni, and Cynthia Zawadski;

- Earl R. Flansburgh + Associates, Inc., Kate Brannelly and Michael Bourque (and their clients);
- Goody, Clancy & Associates, Inc., especially David J. Graham and their clients;
- Griswold, Heckel & Kelly Associates, Inc., especially Jeanne Kopacz and Jay Philomena and their clients;
- Historic American Buildings Survey, The National Park Service, especially Mark Schara;
- McCarty Architects, especially Terry Graves and Michael Walters; and
- Olaf M. Vollertsen, Architectural Intern

Thanks also to the following people for providing information, advice, leads, and comments on various portions of the manuscript: Eric Benson and Steven Dill, Charrette Reprographics; Christopher Clark, Risk Management Programs, American Institute of Architects; Carolyn L. Greenberg; Julie Horvath; Hilary Lewis; Susan Lewis, Boston Architectural Center; Diane Parker, Consulting for Architects/Boston; Rachel Pike, Wentworth Institute of Technology; Eve Schlapik; and the staff at the Boston Society of Architects, the Brookline Public Library Reference Department, and the Wentworth Institute of Technology. Special thanks also are due to my former colleagues at Goody, Clancy & Associates, Inc., who provided a first-class training in AutoCAD and the opportunity to apply it to some unique projects; and Griswold, Heckel & Kelly & Associates, Inc., who furthered my education in CAD applications and management.

Finally, thanks to Dorothy, Detlev, and Lydia Vagts, who provided ongoing support, enthusiasm, manuscript review and input, and frequent gourmet meals throughout the course of this project.

1 Introduction to AutoCAD and a Little History

AutoCAD is the world's leading computer software package for Computer-Aided Design (CAD). It is used by over 1 million people in at least 115 countries and has approximately 70 percent of the total CAD software market.* Of these users, an estimated 45 percent (over 500,000) work in the architectural/engineering/construction (A/E/C) sector. AutoCAD, of course, faces competition from other CAD software, but its dominance in CAD remains strong and continues to shape the entire industry.

 The acronym "CAD" can stand for either Computer-Aided Design or Computer-Aided Drafting. In the early days of CAD, the acronym "CADD" was used whenever Computer-Aided Design and Computer-Aided Drafting were referred to simultaneously. Nowadays, architects and designers almost exclusively prefer to use "CAD," and the distinction between design and drafting seems increasingly less important.

---

*Information on the history of Autodesk and AutoCAD taken from John Walker, *The Autodesk File: Bits of History, Words of Experience*, 3d ed. (Thousand Oaks, CA. New Riders Publishing, 1989), *The Value Line Investor Survey*, and information provided by Autodesk's Public Affairs group, including Autodesk Annual Reports and the publication *Autodesk Milestones* (1993).

If you are a practicing professional in architecture, interior design, or a related field, AutoCAD affects your practice, even if you use another CAD program or avoid CAD altogether. AutoCAD has shaped the expectations, pace, and output of CAD for the architecture/design (A/D) professions throughout the 1980s and early 1990s. An estimated 50 percent of U.S. architectural firms use AutoCAD. Whether or not they themselves use the software, your professional peers, clients, consulting engineers, and even prospective employees certainly consider AutoCAD an industry standard. If your firm uses AutoCAD, your work and productivity hinge on its capabilities, even if you yourself never go near a computer, and your success as a designer, project manager, marketing manager, or firm partner or associate increasingly will be determined by your ability to interact effectively with CAD. If your firm does not use AutoCAD, you have to be able to explain that decision to clients and other parties who increasingly expect you to use it. Therefore, it behooves you to know about AutoCAD's origins and capabilities. You need not become a CAD expert, but you should acquire some basic knowledge.

A Brief History of CAD and AutoCAD

CAD first entered the mainstream of architectural practice during the 1980s, when AutoCAD and other CAD packages were first sold for use with microcomputers or personal computers. The origins of CAD as we know it today, however, were in the 1950s. In research labs such as those at the Lincoln Labs and the Whirlwind Group, both M.I.T. and the U.S. military began investigating spatial imaging technologies. Research progressed in the 1960s, with the M.I.T. Sketchpad Project and with work conducted at the General Motors Research Laboratories focusing on the use of CAD for designing automotive parts.

During the 1960s and 1970s, CAD became more prevalent in industrial design and manufacturing, where the benefits of Computer-Aided Design and Manufacturing (CAD/CAM) were recognized. For most design firms, however, CAD held little appeal. Not only did the first commercially marketed CAD systems present a graphically unappealing user interface, but their purported benefits were of limited relevance for building design. Since design schools did not teach computer courses, few architects and interior designers had any training or inclination toward using computers anyway. Perhaps more significantly, however, most CAD systems were simply beyond the pocketbooks of most design firms. The early CAD systems averaged over $70,000 per workstation, some selling at over $1 million. Only the largest design firms could contemplate such an investment. Moreover, many of these expensive workstations were dedicated solely to CAD use and could not be used for other applications.

All this changed in the early 1980s. With the introduction of small, affordable microcomputers or personal computers, small organizations and even individuals could contemplate purchasing computers. The demand for personal computers and applications was met by a number of young, feisty computer software developers, such as Bill Gates of Microsoft and Steve Jobs and Steve Wozniak of Apple Computer, who sought to make software available and useful to the average person in the office and at home. The availability of software separate from specific computer hardware created competition among computer manufacturers, thereby driving down the price of PCs.

Autodesk, the manufacturer of AutoCAD, emerged as a result of these exciting developments. Originally named Marinchip, Autodesk was founded by thirteen computer programmers whose initial goal was to produce software packages for the mass market; the software was to be able to run on various hardware platforms. The firm's original product was actually a file management system (hence the company's name, Autodesk), but AutoCAD was the product Autodesk became known for. Originally named MicroCAD, AutoCAD was developed by a programmer named Michael Riddle. Autodesk's goal in marketing this software was to sell a "word processor for drawings"* and to offer "80 percent of a mainframe CAD system's functionality for 20 percent of the price."[†]

Autodesk was formally established in 1982 with $59,000 in capital. That year, AutoCAD Version 1.0 shipped. Within the next four years, revenues grew from $15,000 to $29.5 million. By 1986 Autodesk had sold 50,000 copies of AutoCAD and employed 190 people. As founding president John Walker remarked at the 1987 Autodesk company meeting, "In the space of sixty months, we plucked the technology of CAD from the clenched fists of elitists and handed it to the tens of thousands of individuals they disdained."[‡]

As of the mid-1990s, Autodesk is one of the largest and most admired corporations in the United States. In fiscal year 1995, it had over $454 million in sales (see Figure 1.1), 1,700 employees, and 115 offices worldwide. In 1985 Autodesk was the first PC-CAD company to go public (a $1,000 investment in Autodesk's initial public offering would now be worth over

---

*Walker, *The Autodesk File*, 75.
[†]Autodesk, *1993 Annual Report*, 5.
[‡]Walker, *The Autodesk File*, 377.

Figure 1.1 A diagrammatic view of Autodesk's growth over the past decade.

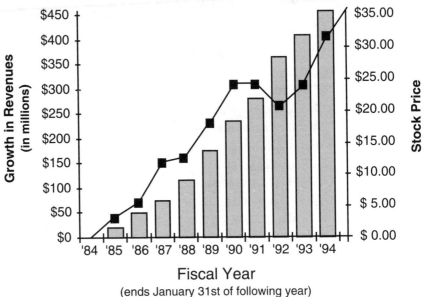

$10,000; the company currently offers shareholders an 18 percent return on their equity), and the additional support of equity financing has helped Autodesk expand both its product line and its dealer network. Ironically, Autodesk has become as large and established a public corporation as the *Fortune* 1000 computer firms it once challenged. In 1992 Autodesk appointed Carol Bartz, a seasoned corporate computer executive, as president, CEO, and chairperson; John Walker, the founding president, had long since moved on to focus once again on software development.

From the time of AutoCAD's first release, Autodesk pursued specific product and marketing strategies. First, it promoted the program's "open architecture," meaning that AutoCAD was designed to be easily adapted to whatever industry or profession it would be used in. Second, rather than tailoring the software to particular applications, Autodesk encouraged independent (third-party) developers to create programs running "on top of" AutoCAD—programs that would harness AutoCAD's features for particular "vertical" markets, whether engineering, manufacturing, or design and construction. Third, despite the rapid and continuing growth in sales, AutoCAD never became a low-cost CAD commodity. To date it has remained one of the more expensive CAD packages, sold exclusively through registered AutoCAD dealers, but it is also one of the more powerful CAD products. In short, while other CAD programs have become budget subcompact cars, AutoCAD has remained a middle- to upper-end luxury model.

AutoCAD both benefited from and fueled the growth in PC-based CAD. (AutoCAD is also available for high-powered workstations, but these are rarely found in typical design firms.) Its ability to become and remain the dominant player in this market reflects Autodesk's willingness to keep abreast of technological changes by releasing upgrades on a regular basis, the wide range of third-party applications that it encourages, and its support for training materials and programs. AutoCAD has also benefited from being recognized by all CAD users as the world's dominant CAD PC package; as discussed further in Chapter 2, design firms that use AutoCAD know they are using a product that has a wide user base and strong corporate resources behind it.

Over the past dozen years, Autodesk has supplemented AutoCAD with a number of related CAD programs, including AutoCAD LT for Windows, which is a lower-priced CAD product sold through retail outlets. In 1990 the corporation formed a multimedia division, which develops popular 3D and animation software such as 3D Studio and Animator Pro. Autodesk has also begun developing software that is targeted to specific applications and may potentially compete with products made by its own third-party developers; sample products are now marketed for mechanical engineering, manufacturing, and scientific markets. Recently Autodesk has announced an expansion into document management software; its new product, Autodesk WorkCenter, is designed to help offices control and track their electronic files and to relate such files to paper output.

In 1994 AutoCAD Release 13 was introduced. Although it was not as earthshaking a milestone as the release of the previous version (Release 12), Release 13 did contain useful new features that indicated not only that Autodesk recognized new trends but also that the company's current strategy involves starting to split AutoCAD up into more affordable and less unwieldy "modules," all of which can work with one another, with other Autodesk products, and with third-party applications. Release 13 also shipped with dual licenses for both AutoCAD for DOS and AutoCAD for Windows; providing licenses for both platforms is seen to reflect Autodesk's goal of encouraging die-hard DOS users to switch to Windows so that the DOS version of AutoCAD can eventually be eliminated.

Although AutoCAD has never lacked competition, it has recently begun to face serious contenders in the A/D community from other software packages. A primary reason behind the popularity of several new CAD products is to offer users a less pricey package. AutoCAD is simply too expensive for many design firms, especially small to medium-sized practices. Another is that many design firms use CAD solely for two-dimensional construction drafting; they don't need AutoCAD's 3D tools, open architecture, or other more sophisticated tools. Therefore they often prefer to purchase a package targeted specifically for design applications. Software programs such as MiniCad, DataCAD, and Autodesk's own retail software product, AutoCAD LT for Windows, often seem to be more cost-effective solutions. These trends may affect AutoCAD's future marketing strategy.

AutoCAD in Architecture and Interior Design

AutoCAD is widely used by members of the A/E/C professions. According to many professional surveys, it dominates in the architecture and interior design professions as well, although it has probably not saturated these markets as extensively as it has engineering, manufacturing, and industrial design. The typical design firm, which consists of an individual practitioner or small staff, may find both the initial acquisition costs of AutoCAD, as well as the subsequent training and installation, unaffordable.

Powerful CAD programs such as AutoCAD can offer many benefits to interior design and architecture firms. They can automate routine drafting tasks, provide a powerful version of overlay drafting, improve design analysis and coordination, and enable designers to test multiple design options more efficiently, in both 2D and 3D. CAD has limitations, of course. It can be more rigid than other methods, it can appear "final" prematurely, and it can under certain circumstances (especially early during the learning cycle) be slower than hand drafting. But with intelligent use of CAD, one can minimize its faults and maximize its benefits. More and more designers and related professionals are coming to recognize this.

Despite the growth in CAD usage over the past decade and ongoing improvements in CAD software, some design firms have expressed dissatisfaction with CAD. The expected benefits—both financial and professional—have not seemed to materialize for them. Some design firms now consider themselves CAD "victims."*

---

*See James R. Franklin, FAIA, "Ruminations of a CAD Victim," *Progressive Architecture*, April 1993: 59.

A decade ago—when PCs were slower, AutoCAD was DOS-based and user-*un*friendly, and output was harder to generate—dissatisfaction with CAD was understandable. By the mid-1990s, however, the CAD environment has become much friendlier and more affordable. No longer is AutoCAD the province of only highly trained CAD "operators"; rather, it can be used by anyone who is reasonably competent with computers and has access to a personal computer with a reasonable amount of RAM. Therefore, professional dissatisfaction with CAD nowadays cannot be solely attributed to obstacles created by CAD software or hardware. Rather, it has become clear that the successful implementation of CAD in a design firm is not a question of *technology* but rather of *management*. Although powerful computers, CAD wizards, and specialized training all contribute to the success of a CAD operation, ultimately successful implementation depends on how CAD is incorporated into the design process (see Chapter 6). So the effective use of CAD is something that can only be addressed by designers, design firm partners, and managers. Even when they cannot fully understand CAD themselves, design professionals ultimately are responsible for the success of CAD in their practice. The importance of management for the effective use of CAD, then, is the reason for this book.

Design firms that have used AutoCAD and computer software successfully vary considerably in the size and nature of their practice, their clientele, and their employee policies. When it comes to CAD, however, they tend to exhibit certain shared characteristics. Among them are

- a general knowledge of, and belief in, the value computers can add to their practice
- adherence to the basic computer principle of *Never Reinventing the Wheel*
- an ability to keep sight of the *long-term* benefits of CAD, even when they experience *short-term* technical difficulties
- a desire to use the benefits of CAD aggressively, for tasks such as modeling and testing different schemes as well as for automating routine design chores
- a recognition that computers can be a design tool as well as a drafting tool
- a willingness to promote and require the use of computers even if the highest level staff are personally computer-phobic or -illiterate
- a willingness to change traditional aspects of their practice to cater to the demands that CAD makes on the design process

- a willingness to invest in training staff—at all levels—in computer operation
- a willingness to provide full-fledged support—financial, emotional, and otherwise—for the efforts of CAD users, CAD managers, and others who have CAD-related responsibilities
- a commitment to making their overall officewide computer system as user-friendly and as accessible as possible
- a commitment to "mainstreaming" CAD with other computer applications and with other tasks in general and to eliminating "CAD ghettodom"
- an emphasis on *productivity*, that is, the quality as well as the quantity of output generated in a particular time frame, rather than merely on *speed*
- an ability to tie CAD in with all other aspects of their practice, from marketing and promotion to financing to human resources
- a grasp of all their available computer-related resources—equipment, software, data, and people—and ability to make the most of those resources
- a recognition that being a responsible design professional in the 1990s requires keeping abreast not only of developments in building codes, new products and materials, ergonomics, and other design issues, but also of the status of computer applications for the design professions
- a management approach that stresses strategic planning, the creation and attainment of goals, and the use of logic, even in these uncertain and sometimes unpredictable times.

As the characteristics listed above indicate, much of the successful use of CAD is not only about technology; it is also about leadership, personnel issues, office sociology, and the economical use of finite resources. Above all, it is about setting goals and pursuing a strategy for your firm.

The purpose of this book is to help design firm owners, project managers, and others in design firms manage CAD in general, and AutoCAD in particular, more effectively and profitably. This book does not assume that you, the reader, know how to use CAD or indeed have any interest in learning it. It does, however, assume that you are prepared to shift to a new paradigm of design practice and, as outlined in Chapter 2, that you and your firm have a clear professional direction.

2 CAD Decision Making: When and Why You Should Use AutoCAD

> "Would you tell me, please, which way I ought to go from here?"
> "That depends a good deal on where you want to get to," said the [Cheshire] Cat.
> "I don't much care where—" said Alice.
> "Then it doesn't matter which way you go," said the Cat.
> "—so long as I get somewhere," Alice added as an explanation.
> "Oh, you're sure to do that," said the Cat, "if only you walk long enough."
>
> —Lewis Carroll, *Alice's Adventures in Wonderland*

CAD Decisions As Part of the Design Firm Strategy

The stressful and uncertain business climate of the late 1980s and early 1990s has taken its toll on many architecture and interior design firms. Design fees have shrunk, project schedules are unrealistic, and clients are unpredictable. While they are helping their clients to plan for future needs, designers themselves often seem unable to plan or control their own futures, whether for the coming month or the coming decade.

Professional anxiety causes many design firms to deal *reactively* to their opportunities and competition. Uncertainty over cash flow, client and project stability, and staffing needs often compels designers to grab what they can get, when they can, without following a plan or strategy.

Despite the professional challenges of the past decade, a number of design firms not only have remained profitable but actually have increased revenues, profits, and staff; they have also expanded their existing client base and entered new markets. Characteristics the principals of these firms share include a clear vision of their future, a set of defined goals—ambitious yet feasible—and a willingness to reshape their firms' policies and operations in accordance with those goals.

For many, but not all, of these firms, Computer-Aided Design and Drafting (CAD) plays an essential role in their overall organizational strategy. But when CAD proves a helpful tool, it is not because of the program's innate capabilities but because its users are determined to maximize their return on investment in software, computers, and training. They are willing to make the necessary investment of funds and staff time to implement CAD within their operations, marketing programs, and personnel policies. The principals of these firms take a *proactive* and *holistic* approach to CAD. To do so does not require hands-on technical knowledge of computer drafting, but it does demand a willingness both to understand CAD's capabilities and limitations and to rethink assumptions about how work gets done in a design office.

Making effective decisions about CAD, as shown in Figure 2.1, should be done in conjunction with revisiting all other aspects of your firm's practices, including marketing and client development, staffing and personnel policies, operations, and financing. Your CAD manager and drafters implement your decisions on a daily basis, but ultimately general firm policies determine the character and success of your CAD operations.

Some of the general policy issues that shape your CAD decisions include:

- The type of work you want to do. Do you wish to design for corporations, institutions, government agencies, or individual homeowners? Do you want small-scale projects or large ones? Are you happy to accept mundane but reliable "bread and butter" renovations, or do you only want to do unique specialty projects? The kind of practice you envision may determine whether you actually use CAD at all; it often mandates exactly how you use it.

- Where you want to do your work. Running a local practice requires less technology than working regionally, nationally, or internationally.

If you work locally, you work with a smaller range of consultants, clients, and vendors; working with a broader geographical area often requires using a CAD system to communicate electronically with consultants, clients, and vendors who are scattered around the country or the world.

- What you wish to be known for. Some firms simply wish to be known for good design results, in the traditional sense. Others wish to be known for the *process* as well as the product, so they market their design *approach* as well as the end product. If you market technology, in effect, you may expand your possibilities for landing interesting design projects, but you also have to live up to your marketing approach by maintaining a constant investment in state-of-the-art software, machines, and training.

- Your preferred firm size. Do you wish to remain an individual practitioner? A small to medium-sized firm? Grow into a mega-organization of several hundred people? CAD can help you either to maintain your current size or to grow, but your target office size will shape your CAD decisions, especially if you seek a 1:1 ratio of computers to people. Size also determines how much overhead you will have to absorb, which is a key consideration when you are budgeting for CAD managers, training, and other indirect costs.

- Your preferred hiring policy. The economic climate of the past decade has forced many design firms to scale down permanent staff and to rely on consultants and other short-term staff to pick up the slack in times of heavy workloads. Many firms prefer to hire experienced CAD users on a consulting basis rather than to train their existing staff in CAD. Your decisions about hiring and training will have a big impact on how you incorporate CAD into your office.

- Your preferred financing strategy. After the boom and bust cycle of the past decades, many design firms have learned to be cautious money managers. They are more careful with contracts and more likely to avoid debt and speculative investments. Still, some firms are comfortable using debt to finance acquisitions. If you tolerate leverage, you can purchase a very expensive CAD system with the expectation of obtaining work that will pay for the financing expenses and then some. If you are financially conservative, you probably want to wait until you can purchase a system that you've already earned, so to speak, but you may be constrained in what type of system you can acquire and when you can acquire it.

CAD Decision Making

As shown in Figure 2.1, making decisions about using CAD in general and AutoCAD in particular involves traveling down a decision tree. This tree presents three levels of decisions:

1. *Should* you use CAD at all?
2. *If* you do use CAD, *when* should you use it and for which projects and applications?
3. *When* you do use CAD, *should* you be using AutoCAD?

Any firm that has already ventured into CAD has already traversed this tree and confronted these decisions to a certain degree. Even if you are

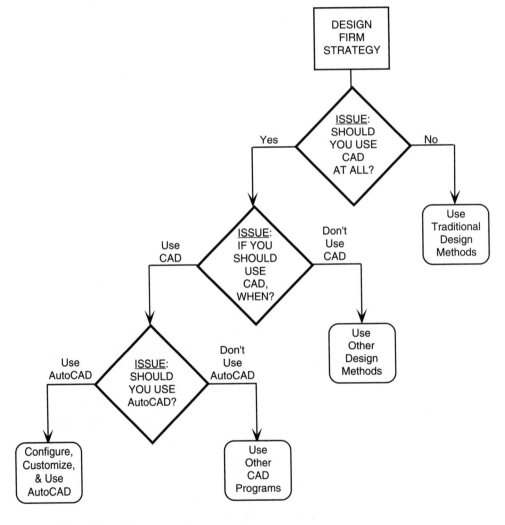

Figure 2.1 *The CAD decision-making tree*

satisfied with your current investment in CAD, however, you should always be prepared to repeat this decision-tree analysis whenever you have to make an additional investment in computer hardware or software, or whenever you are considering a new way of using computers.

In traveling down the CAD decision tree, and indeed when making any major decision regarding CAD, you should always bear in mind that successful decisions reflect a coordinated and integrated approach. As shown in Figure 2.2, successful CAD decisions seek to maximize design quality while ensuring ease of use given the available time and funds. These solutions should be cost-effective, of course, but in a *long-term* context. Many design firms have felt burned by their CAD investment because they expected immediate benefits. Like most long-term investments in technology, however, CAD requires training, testing, and time before it generates returns. With a clear strategy, the payoff time for a CAD investment can be greatly reduced.

Figure 2.2
The best CAD solutions reflect an optimal blend of the elements shown in this diagram.

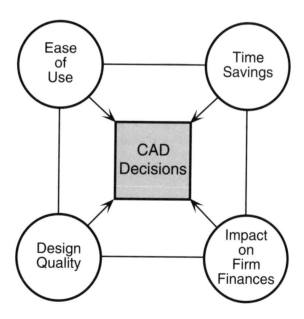

Should You Use CAD at All?

Although many design futurists predict that all design and drafting will be computer-generated in the very near future, many design firms still do not use CAD. Indeed, in a 1994 survey of its members, the American Institute of Architects found that only 58 percent of its firms used CAD, and the heaviest users were the larger firms.* Sometimes avoidance of CAD reflects financial constraints or computer phobia, but often it is a deliberate choice. Many notable, award-winning design firms choose not to use CAD because their mix of skills, clients, and work approach enables them to be successful without it.

Here are some reasons not to use CAD *at all*:

- The status quo works. If you are profitable, meet deadlines comfortably, and are keeping clients, staff, and yourself happy, you have no reason to invest in CAD unless you anticipate future developments that could change your practice, or unless you hunger for specific features of CAD.

- You work independently. Firms that collaborate with many consultants and that must provide clients with documents in electronic form usually have no choice but to switch to CAD. But if you have the sort of design practice that involves minimal collaboration with other parties, you face less pressure to switch to computer-generated design.

- You don't have the requisite CAD skills. If you can't (or won't) invest in training current employees in CAD or are unwilling to acquire CAD skilled personnel, you are better off relying on the in-house skills you already have.

- You may save money by waiting. In true inflation-adjusted dollars, the costs of CAD have dropped rapidly. A decade ago, a typical workstation cost over $10,000; now you can purchase a CAD station for $5,000 or less. And even when overall costs remain fairly constant, as with AutoCAD software, you can now acquire more value per byte in terms of speed, features, and user-friendliness as shown in Figure 2.3. Designers who waited until the early 1990s to plunge into CAD definitely spent less money not only to acquire the technology but also to learn the program because they were able to profit from others' learning curves. Declines in the cost of CAD may continue, but waiting for further price savings must be weighed against the opportunity costs of *not* using CAD.

---

*American Institute of Architects, *Architecture Factbook* (Washington, D.C.: American Institute of Architects, 1994). According to this survey of 15,000 AIA members, 58 percent used CAD; of these CAD users, 65 percent used AutoCAD. Sixty-four percent of CAD usage was for construction documents; 33 percent for conceptual design. Firms of ten employees or more, while comprising only 14 percent of the members surveyed, were on average 40 percent more likely to use CAD than were small firms or solo practitioners.

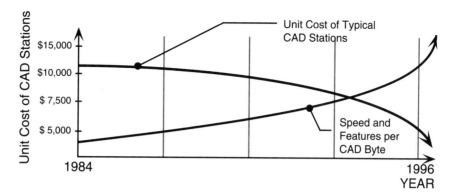

Figure 2.3
The typical CAD station of today costs less in inflation-adjusted dollars and delivers far more power and ease of use.

- You can't make the overall commitment. Serious CAD usage involves much more than merely acquiring a computer and software. It requires ongoing expenditures on training, configuring, upgrades, and maintenance. For many firms, these expenditures pay off—in part because they shift resources to CAD from other methods of design and production; for others, CAD can seem like a bottomless money pit.

- You won't change your work approach. Although the ultimate goals of design remain the same when you use CAD, the process does change. You have to rethink how you work with staff and how you plan and schedule projects. Most important, you must rework your office—both physically and sociologically—to fully integrate computers into the design process. If you don't, you'll end up with a CAD ghetto (see Chapter 9), which is worse than not using CAD at all.

- CAD conflicts with your design philosophy. Some designers have a philosophical problem with Computer-Aided Design. In fact, they consider the very concept to be an oxymoron. If you don't really believe in CAD as a design tool, don't use it. Your skepticism will hinder effective implementation.

- CAD conflicts with your design approach. Designers who successfully use computers in general and CAD in particular tend to have a certain attitude about design. They view design as a systematic, rational, and analytical process that benefits from logic, standards, and procedures. Designers who approach design as a more intuitive and unstructured process may find CAD overly restrictive.

Although many firms can successfully avoid or delay switching to CAD, the reasons to switch to it are already compelling and are increasing. Here are just a few of the reasons to use CAD or to increase your investment in it:

- Enhanced productivity. In the very short run, CAD may not seem to produce designs and drawings any faster or better than traditional methods. And if you don't configure and customize your CAD software properly, CAD can slow you down. With a properly installed and customized system, however, CAD will speed up the design process. Eventually, you will find that, in a given time frame, you can complete more projects than before while potentially improving their quality.

- Automation. The underlying principle of computer use is that you should never perform the same function more than once. You may copy, clone, modify, and so forth, but the routine, repetitive tasks should be automated. Thus CAD not only enhances productivity but saves time on the boring, routine aspects of your practice so that more time can be spent on the fun stuff—design!

- Accuracy. The most powerful CAD programs perform mathematical calculations that are extremely precise and internally consistent. You can thus generate far more accurate dimensions and calculations than you can by hand; you can also use CAD to double-check your math and geometry.

- Efficiency. Along with providing greater accuracy, CAD can help you produce designs that make more efficient use of clients' real estate, equipment, and other assets.

- Scenario testing. With CAD you can easily test different scenarios in a short amount of time. You can store, merge, and track various "what-if" options more easily. As clients increasingly expect more and better designs in shorter time frames, using CAD to analyze and compare different schemes becomes increasingly valuable (although a clear client program and design direction should limit the number of schemes that are truly necessary).

- Graphic quality and consistency. No machine can ever match the quality and grace of a pen-and-ink or watercolor rendering by a master drafter. And CAD is not a true expression of a designer's individuality. But let's be honest: much hand drafting is mediocre, with indistinct line weights and illegible letters. And drafters' individual quirks constantly thwart design firms' attempts to achieve overall graphic consistency. With proper formatting and output devices, CAD allows you to produce drawings that are of consistent quality and easy to read. Increasingly, these drawings are graphically appealing as well. And as Chapter 9 discusses, craftsmanship is as applicable to CAD as to hand drafting. In sum, for work that doesn't strictly require a personal touch, CAD can be preferable on a graphic basis.

- Coordination. Used effectively, CAD is a powerful tool for coordinating team efforts. It ties together the work of many engineers and other parties, thereby facilitating communications and updates. As designers increasingly spend more time on project coordination, they will find CAD invaluable.

- Controlled growth. Design firms work on a project basis and have constant fluctuations in their staffing needs. As any designer who survived the past decade knows, this trend has brought frequent and painful cycles of hiring and firing. Consequently, many firms are now reluctant to hire new permanent employees and so are forced to make unrealistic demands on their existing staff. CAD, however, can help sustain you during periodic bulges in workload, and the enhanced productivity it provides can enable you to keep your staff at a desirable level.

- Geographical flexibility. With CAD and other computer programs, you are less tied to working in particular geographical locations. You can use faxes, modems, and the Internet to exchange ideas, drawings, and specifications. While physical access to job construction sites remains essential, computers allow you to manage long-distance relationships more easily.

- Professional realities. While you should not follow industry trends just for the sake of doing so, you do have to confront the reality that the percentage of the A/E/C community using CAD will only increase over time. CAD-based clients increasingly demand that their designers use CAD, and engineers and other consultants also expect to communicate electronically. At best, avoiding the use of CAD will make your life more difficult in certain respects. At worst, you will be locked out of many professional opportunities.

If You Do Use CAD, When and How Should You Use It?

Firms that invest in CAD don't necessarily use it for every project or for every stage of design. They may use CAD selectively, depending on their resources and goals. Knowing when to use it is essential for planning and staffing projects effectively.

Here are some guidelines for knowing when CAD should *not* be used:

- The project is extremely simple, low budget, and fast track. Projects not requiring CAD involve minimal repetitive elements, minimal coordination, few staff, and/or a short time frame. Few drawings or presentation materials are required. Once the project is out the door, it will not be seen again. The required turnaround time seems to be less than it would take to boot up your computer. Applicable projects include a custom house, a special millwork configuration for a single office, or a custom architectural detail.

- You don't have enough computers or available CAD-literate staff. If you perceive a computer bottleneck or find that your CAD-skilled employees are committed to other projects, don't risk the possibility of being unable to produce work by computer. The resulting stress won't be worth the income you bring in, and you could impair your ability to service your clientele. (Ironically, some very CAD-oriented firms now have trouble finding "pencil-literate" staff; having staff that can draw both by hand and by CAD eases staffing pressures.)

- You can't run the project profitably on CAD. If some combination of fees, schedules, and staff makes CAD more costly to use, then financially you can't justify it. But if you continually lose money on CAD projects that otherwise meet the criteria for CAD-suitable projects—as outlined below—you need to reconsider how you set your fees and manage your CAD operations (see Chapter 9).

Here are some guidelines for knowing when CAD *should* be used:

- You have a customized, proven CAD setup. If you have invested the time in customizing your office CAD system (as described in Chapter 4) and have built up a library of workable standard drawing sheets, symbols, plotting routines, and so on, CAD will allow you to be more productive.

- The project scope is large and/or likely to expand. If the project is large, either in square footage or complexity, CAD can help you coordinate and consolidate the efforts and decision making of the team.

- The project time frame is long. If the project is likely to be hanging around the office for many months or years, CAD is the best way to "file" ongoing changes and developments. The initial time spent up front inputting data will be amortized over the project's life.

- The project involves substantial repetition and prototypical elements. If the project uses repetitive elements, such as a common building core shared by multiple floors, standard windows, or offices, CAD easily produces these elements and then allows you to revise them quickly.

- The project design shows variations on a theme. Although many designers value CAD for placing repetitive elements, such as columns or ceiling fixtures, you can also use CAD to develop variants on standard elements. For example, you can develop a prototypical layout for a research lab and then modify it to reflect individual researchers' equipment needs. AutoCAD commands such as **Ddmodify**, **Stretch**, and **Scale** enable you to modify prototypical design elements while maintaining their original essence.

- The project involves multidiscipline coordination. Increasingly, designers spend less time actually designing and more time coordinating design teams. CAD can be a powerful tool for coordinating the work of large teams and a wide range of consultants—engineers, lighting designers, urban planners, and other specialists.

- The project requires output in multiple formats. CAD permits reproduction of a computer image in an infinite range of media and scales. A single graphic file can be used throughout an entire project, from client presentations to project follow-up, and in all architectural and engineering scales, as well as in scales that have no name. For projects requiring consistent content in a wide range of formats, CAD is ideal.

- The client insists. If your clientele is primarily residential or comprised of small businesses or organizations, they may not care how you produce your drawings. But if they are medium to large corporations, institutions, or government agencies, they will care. They will demand record drawings in both print form and as CAD files. Increasingly, prospective clients expect CAD files to reflect their standards for layers, drawing formats, and other CAD features. The ability to produce CAD files nowadays determines whether you can even qualify to respond to a request for proposal (RFP) from such clients.

Note that while many design firms do rely on both CAD and hand drafting, a growing number find the hybrid approach inefficient, since it involves maintaining two production systems, merging two sets of graphic standards, and, often, managing employees differently.

When You Do Use CAD, Should You Use AutoCAD?

AutoCAD has one of the largest *international* market shares of any computer software and, according to a recent survey by the American Institute of Architects, a dominant share of the CAD market among architects. AutoCAD, however, is by no means the only CAD system used by designers, and in recent years, it has faced stiff competition from other software producers. AutoCAD is a logical choice for many design firms, but it is by no means the inevitable one.

Here are some reasons to consider acquiring or expanding your investment in AutoCAD:

- AutoCAD is a solid, proven product. AutoCAD has been on the market for over a decade and has a substantial share of the CAD business, both nationally and internationally. It is produced by Autodesk, a publicly held company with annual revenues of over $450 million, a solid financial position, vast resources for research and development, and an established distribution network. On a regular basis, AutoCAD provides upgrades that reflect user demand and developments in computer technology. Its users number over 1 million. Although the company faces serious competition from other software producers, it has the resources to respond to, or even acquire, the competition. Many other fine software products do not have backing equal to that of AutoCAD, and, in an industry shakeout, they would be less likely to survive. In short, your investment in AutoCAD is likely to have a secure future.

- AutoCAD has a substantial support network. From the beginning, Autodesk has relied on third-party developers to create software add-ons, reference manuals, and other products to enhance and facilitate AutoCAD, and to give it exposure to "vertical markets." The company supports user groups and training centers. AutoCAD is the raison d'être for many software companies, publishers, writers, and training organizations. When you buy AutoCAD, you buy into a universe of support, expertise, and related products.

- AutoCAD is highly customizable. From its inception, AutoCAD has emphasized its "open architecture," meaning that with a minimum of programming skills you can customize the program to suit the needs of many users. Although customization takes time up front, it produces a powerful system that exactly meets your needs and can be adjusted to suit projects of varying scope and content. While certain other CAD packages initially appear more user-friendly and better geared to architecture and interior design, they often resist tinkering with their base features and may ultimately prove more limited.

- AutoCAD is used by many industries. Precisely because of its open architecture, AutoCAD is used—on a worldwide basis—by engineers, industrial designers, governmental agencies, and other parties with which designers often must communicate during the course of projects. AutoCAD has become a design lingua franca and thus eases electronic communication with many other designers and related professionals. Indeed, Autodesk invented the drawing file (*.dwg*) and drawing exchange file (*.dxf*) formats (see Chapter 10), which now constitute basic file exchange standards for all CAD software of note.

- AutoCAD is highly accurate. All CAD programs offer a level of mathematical accuracy that few human drafters can match. But not all use internal calculators as powerful as AutoCAD's; this is a major reason for the product's success in industrial design and engineering. If your practice requires extensive measurements, calculations, and fine-tuned dimensioning, AutoCAD is a good choice for you.

- You can hire people who know AutoCAD. A large pool of experienced AutoCAD users exists, as AutoCAD predominates in design schools and CAD training centers. Moreover, designers and drafters often study AutoCAD on their own in order to improve their marketability. You are always likely to be able to hire drafters with some knowledge of it; if you have a more obscure CAD program, you will have to invest more in training or else draw from a smaller pool of potential employees.

- Your clients expect it. Given that AutoCAD has more than a 50 percent share of the CAD marketplace, the odds are that a good portion of the clients who insist that your work be done on CAD will use AutoCAD themselves. Even clients who don't use AutoCAD tend to feel comfortable with design firms that use it, because they recognize that AutoCAD is the CAD industry standard and that their CAD system most likely can read AutoCAD *.dxf* or *.dwg* files.

- Your consultants expect it. AutoCAD is used overwhelmingly by members of the engineering specialties. Indeed, engineers have been in some ways more aggressive about using CAD, especially for purposes beyond drafting. If you work with consultants only occasionally, you can get away with the occasional *.dxf* file conversion, but if you collaborate closely and frequently with engineers and other consultants, you will find that AutoCAD smoothes out electronic communications.

Despite the compelling reasons to use AutoCAD, some very valid reasons exist for not using it:

- You can't afford it. Bare-bones AutoCAD sells at a list price of $3,750 (your dealer may offer a discount price), and that's before you add on the direct and indirect costs of tailoring it to suit your needs. To use AutoCAD Release 13's full power, you need a state-of-the-art computer, a Pentium processor, and 24 MB of RAM, which can run another $4,000. While a typical workstation varies in price, you really need to budget $7,000 minimum per workstation. Effective use of CAD will give you a return on your investment (see Chapter 9), and purchasing additional workstations produces economies of scale, but for many design firms, such an investment is simply not financially feasible and cannot be justified in the face of other pressing expenses. As powerful as AutoCAD may be, the need to purchase it does not justify financial insolvency. (As Chapter 9 discusses, you can access many features of "plain vanilla" AutoCAD by using AutoCAD LT for Windows, which has a "street" price of under $500 and demands less memory.)

- You don't need its capabilities. AutoCAD may be a CAD powerhouse, but not every designer needs all that power. Indeed, many designers use AutoCAD solely for routine construction drafting, essentially using under 30 percent of AutoCAD's capabilities. Spending close to $4,000 for features you don't even use when you can purchase a satisfactory, less powerful $1,000 alternative is analogous to purchasing a Porsche to drive to the local grocery store when a Honda would suffice (and requires less maintenance and insurance).

- You don't need customizable software. The very trademark of AutoCAD's appeal—its open architecture—can actually irritate designers who want basic architectural drafting tools and ordinary graphic standards and who know that having too many choices can be distracting, intimidating, and ultimately counterproductive. You may find that a CAD package that offers fewer tools and acceptable defaults proves more useful than feature-laden software that expects you to establish all the parameters.

- You lack the in-house technical expertise to manage AutoCAD. Particularly if you use AutoCAD on a network or employ its more high-powered features, you must develop or acquire a level of technical computer knowledge that most designers and many design firms can't (or won't) provide. You will have to invest in training, hire a CAD manager, and/or rely on consultants to use AutoCAD effectively. You will also have to make an expenditure of time and money and have a degree of dependency on employees and consultants that you may be uncomfortable with.

- AutoCAD doesn't fit your preferred computer platform. Most design firms install CAD on personal computers rather than on workstations. With AutoCAD Release 13, you have (at present) only two platform options for personal computers: Microsoft Windows and MS-DOS. Autodesk has made no commitment to support any Macintosh platforms beyond Release 12. Yet many design firms, particularly smaller ones, prefer a Macintosh-based system, citing its user-friendliness, reliability, graphics, and desktop publishing orientation, and its easy installation and operation. As discussed in Chapter 3, you can mix Macs with PCs, but if your heart lies with the Mac, consider the several fine CAD software packages that fully support the Macintosh operating environment.

The major CAD decisions and their attendant issues as described above can never be considered just once. Both technology and the design professions are so dynamic that you should review your strategy and your CAD system every three to five years. Some decisions, such as whether to upgrade or to acquire new equipment, may justify more frequent review. The initial process of self-review and analysis is time-consuming and sometimes painful, but ultimately you will profit from the end result. If you build good reporting systems and controls into your management and financial practices, subsequent reviews of CAD policies will be easier. And you are more likely to avoid running the danger of ending up just somewhere, as Lewis Carroll's Cheshire Cat cautioned at the start of this chapter; rather, you increase the likelihood of ending up *exactly* where you want to be.

3 Configuring AutoCAD for Design Offices

The configuration of your office's overall computer system determines how useful and profitable AutoCAD will be for you. The decisions you make about computer platforms, networks, and other software packages will have a major impact on how effectively you use AutoCAD. Your configuration also shapes how you customize your system, which in turn will determine how successfully you can implement AutoCAD in your office (see Chapter 4).

If you've navigated the CAD decision-making tree, as shown in Figure 2.1, and have concluded that AutoCAD is right for you—at least for some projects—you need to travel along another decision tree in order to select the optimal configuration to suit your needs.

As shown in Figure 3.1, to arrive at the best configuration, you'll need to address three major issues:

- which AutoCAD platform or platforms to use,
- whether to use AutoCAD on a network, and if so, how, and
- which software to use in conjunction with AutoCAD.

Perhaps because of AutoCAD's strong DOS heritage, many design offices have installed earlier versions of AutoCAD without considering

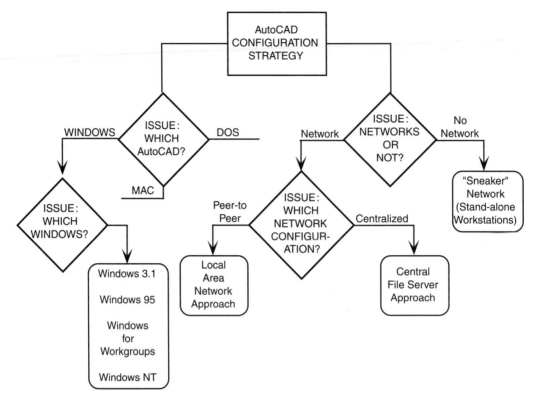

Figure 3.1 *The AutoCAD configuration decision tree*

what non-CAD users in the office do on other computers. With the development of more affordable and powerful networks and Windows and Macintosh operating systems, however, offices can easily tie together *all* their computers' users and applications. Having a standard operating system not only helps integrate CAD more into mainstream office procedures but also reduces duplicate software and file maintenance, and helps all computer users to adhere to common office standards and templates.

Which AutoCAD Platform Should You Use?

Autodesk markets AutoCAD for two major computer platform groups: personal computers and workstations. Workstations, such as DECstations and Sun-4 Stations, are generally found in the facilities management, industrial design, and engineering departments of larger organizations, especially those using powerful centralized networks. Most design firms only need—and can only afford—personal computers (PCs), so this book primarily discusses issues pertinent to using AutoCAD on PCs.

Within the realm of personal computers, designers can choose to run AutoCAD on several different operating systems. The first and still the most prevalent is MS-DOS. With Release 11, AutoCAD introduced a Microsoft Windows version, and this is now the fastest-growing version of AutoCAD. With Release 13, Autodesk gives you both DOS and Windows versions of AutoCAD under one license. Many industry observers speculate that this reflects an Autodesk strategy to encourage die-hard DOS users to switch to Windows along with much of the rest of the world; the future of DOS-based AutoCAD seems dim. A Macintosh version of AutoCAD was introduced in conjunction with the PC version of Release 11, but Autodesk has not announced plans to introduce a Macintosh or PowerPC version of Release 13.

Prior to Release 13, Autodesk's top platform consideration for AutoCAD appeared to be cross-platform compatibility, meaning that you could exchange AutoCAD drawing files among DOS, Windows, and Macintosh computers without having to convert files or lose drawing elements. Release 13 for Windows, however, relies much more on Windows-specific features, such as Object Linking and Embedding (OLE) and TrueType fonts. The more you take advantage of Windows-specific features, the more your files will lose when you transport them from Windows to DOS. Thus, AutoCAD has become more operating system–specific than previously.

Ironically, this operating system specificity has emerged just as the major operating systems are promoting file-exchange capabilities. You can now easily move files among DOS, Windows, and Macintosh operating systems, you can purchase DOS and Windows emulators to run on a Macintosh, and the Macintosh operating system now runs on UNIX. As a routine policy, all major software applications now emphasize the ability to exchange files in a range of file formats.

Following is a discussion of the most suitable versions of AutoCAD available in the personal computer platforms; their requirements are presented later in Table 3.1.

AutoCAD for DOS

AutoCAD for DOS is the oldest and most common version of the software. Since it runs in the PC environment, it can be used with a wide range of PC-based machines, so that—from an initial hardware viewpoint—it *appears* to be the cheapest configuration. Running on its own, AutoCAD for DOS provides the greatest processing speed and power and takes up the least amount of RAM and hard disk space. While the interface of Release 13 shows vast improvement, in terms of user-friendliness, the DOS version still lacks some of Windows's most helpful features, and file management can be difficult without acquiring file management software. With the DOS version, exchanging data with other software applications, such as text processors and databases, requires more effort.

AutoCAD for Microsoft Windows

This version, also PC based, operates under the popular Microsoft Windows software—a user-friendly system running on top of DOS. Version 12 of AutoCAD for Windows achieved rapid popularity in many design offices because its recognizable Windows Graphical User Interface (GUI)—based on icons, toolbars, pull-down menus, and dialog boxes that respond to mouse clicks—replicates other Windows software, so less training is required to learn to navigate the program. Release 13 for Windows also facilitates data sharing between AutoCAD and other software applications and uses Windows TrueType fonts.

The price of Windows's additional features manifests itself in a hefty demand for computer memory. As shown in Table 3.1, the Windows version requires more RAM than the DOS version does, as well more space on the hard drive, because you are required to set aside hard disk space to use as a swap file. The Windows version is also less stable than DOS.

If you don't already have Windows, you can install one of several versions:

1. "Plain Vanilla" Windows 3.1. This is the basic Windows software. It is ideal for single, independent personal computers, small networks, and networks running Windows in conjunction with network software, such as Novell.

    Windows is 16-bit software, and AutoCAD Release 13 runs in 32-bit mode, so to load AutoCAD Release 13, you first have to install Win32s Version 1.20 or later (which AutoCAD provides) to make Windows simulate a 32-bit system. You also must run AutoCAD in the enhanced mode, which requires setting up a permanent swap file. (Refer to the *AutoCAD Release 13 Installation Guide for Windows,* included with AutoCAD software, for a technical explanation of 16 versus 32 bits.)

2. Windows for Workgroups. This is a higher level of Windows designed to facilitate file exchange and intraoffice communications; Workgroups can function as a low-powered network or as an interface for working with other network software. It looks essentially the same as

"Plain Vanilla" Windows except that when you select the File Manager, you have access to other computers' hard drives. With Workgroups you can also use Microsoft Mail (a brand of electronic or "e"-mail) or other interoffice communications software.

3. Windows NT 3.5. This product is not really an add-on for DOS but rather a separate operating system with a Windows interface. Windows NT 3.5 is a 32-bit system designed to support high-powered network applications and multitasking. AutoCAD needs to be configured somewhat differently on Windows NT than on regular Windows. Windows NT has achieved popularity in larger corporations and institutions that are committed to Windows, because Windows NT is more robust and suitable for more complex tasks. Using this operating system for workstations with limited computer needs could be overkill, but with network file servers and high-powered CAD and multitasking systems, Windows NT is a good choice. A special edition called Windows NT Fileserver is available for network file servers.

4. Windows 95. The long-awaited upgrade of Windows 3.1 was formally introduced in the summer of 1995. Microsoft has promoted this version as a separate operating system rather than a DOS overlay; it may not completely eliminate DOS, but it does allow you to ignore it. Windows 95 offers a more Macintosh-like interface and greater stability; its "plug and play" feature facilitates the configuration of peripherals; and file names can now be more than eight digits long.

Windows 95 is the new version of Windows 3.1. Presumably, you will be able to run most of your existing Windows 3.1 software under Windows 95, but you should anticipate some bugs with the initial version. Moreover, your existing Windows-based applications will have to be upgraded before you can take full advantage of Windows 95's unique features. As usual, the new and improved features require more computer memory. Microsoft has a strong incentive (reducing customer support costs) to eliminate DOS, but it is probably best to ignore any pressures to upgrade to Windows 95 or future versions until the inevitable bugs are addressed and the design industry in general shifts to it; in general, hold off upgrading for a minimum of six to twelve months from the date of the software upgrade's initial release. The more extensively Windows is used in your office, the more carefully you should plan for an upgrade to Windows 95, as discussed in Chapter 10.

AutoCAD for the Macintosh

AutoCAD for the Macintosh has been the least successful version of AutoCAD, in part because Autodesk's commitment to cross-platform compatibility has limited its ability to devote resources to the popular Macintosh interface and work with features specific to the Macintosh platform. Moreover, prior to the introduction of the high-powered Quadras, the speed of AutoCAD for the Macintosh was barely tolerable. Although the use of the Quadra with AutoCAD Release 12 for the Macintosh brought vast improvements in terms of speed and features when compared to earlier machines, this version of AutoCAD has never achieved a large user base, nor has it spawned many third-party add-ons. Consequently, as of this writing, Autodesk has yet to announce plans to introduce a Macintosh or Power PC version of Release 13.

Design firms that have gravitated toward Macintosh computers have been philosophically disinclined toward AutoCAD because of its strong DOS legacy, that is, its unfriendly GUI. Those firms that have used the Macintosh version find that AutoCAD is more appealing in the Macintosh interface and easier to install and operate than are the DOS and Windows versions. You can still purchase copies of Release 12 for the Macintosh. You can also try running AutoCAD for Windows on a Macintosh PowerPC with a built-in DOS board or a Windows-simulation package, such as SoftWindows by Insignia Solutions. You cannot run Release 13 with a simulation package that emulates anything slower than a 386. At present, your options for Macintosh simulators are limited, but new products along these lines are constantly being introduced. At present, however, running Windows on a Macintosh only makes sense for mundane file management tasks, such as writing *.dxf* files; serious AutoCAD use requires a real IBM-compatible PC.

Note that many firms that are committed to the Macintosh and to more Mac-like CAD packages nonetheless may maintain one or two AutoCAD workstations in their office. That way they can service clients who insist on the use of AutoCAD, and also control the file-exchange process when exchanging *.dxf* files with engineers and other consultants.

The Multi-Platform Office

Many design firms run software on multiple computer platforms. With current file-exchange capabilities and the right network configuration, designers can share files and communicate with each other successfully, regardless of which platform each uses.

The multi-platform office often reflects financial and professional pressures as well as the legacy of earlier decisions about hardware and software. Upgrading from DOS to Windows AutoCAD, for example, requires additional RAM as well as software, and many design firms cannot afford to uniformly upgrade all their machines at once.

While often born of necessity, the multi-platform strategy must be planned for carefully; otherwise it can be an unduly expensive approach, especially for smaller design offices that may not be able to absorb the overhead involved in researching and configuring multiple platforms. For instance, more software, drivers, and external devices are required to make the various platforms work together. There would be limited savings in terms of economies of scale in training, configuration, customization, and management. If your firm has not yet made a major investment in computers or is ready to do a major overhaul, it would be best to acquire a single platform that will serve you for three to five years. If eventually making a shift to, or adding, another platform seems necessary, your system configuration in the current platform can, with careful planning, be transported to the new platform, so you won't lose your initial investment in customization and training.

Selecting Hardware

Table 3.1 shows the minimum requirements for RAM, hard drive space, and coprocessors for various AutoCAD versions. For Macintosh users, hardware selection was—at least until the recent introduction of Macintosh "clones"—a simple procedure of selecting among different models from a single manufacturer.

For PC users, however, the hardware field traditionally has been much more open. Even when focusing solely on Pentium computers (the only sensible standard for Release 13), you still face a mind-boggling range of choices. AutoCAD does not explicitly endorse any manufacturers of computer microprocessors and hard drives, although its minimum and recommended central processing unit (CPU) features narrow the field somewhat. The field for video monitors, digitizers, and plotters is implicitly narrowed because these devices must work with Autodesk Device Interface (ADI) drivers, which AutoCAD and certain third-party manufacturers provide. Compatible peripherals are listed or inferred from the appendices in the *AutoCAD Release 13 Installation Guide*.

For additional guidance, see the section on peripherals in *The AutoCAD Resource Guide*, which is included with your AutoCAD software and is updated on a quarterly basis. An experienced Autodesk dealer can also

SYSTEM COMPONENTS	AutoCAD for MS-DOS				AutoCAD for Windows				AutoCAD for Macintosh			AutoCAD LT for Windows 1		AutoCAD LT for Windows 2	
	R12		R13		R12		R13		R12		R13	Minimum	Optimal	Minimum	Optimal
	Minimum	Optimal	Minimum	Optimal	Minimum	Optimal	Minimum	Optimal	Minimum	Optimal	Minimum				
Computer microprocessor	IBM PC or Compatible; 386/486; Pentium	IBM PC or Compatible; Pentium	IBM PC or Compatible; 386/486; Pentium	IBM PC or Compatible; Pentium	IBM PC or Compatible; 386/486; Pentium	IBM PC or Compatible; Pentium	IBM PC or Compatible; 386/486; Pentium	IBM PC or Compatible; Pentium	Most Macintosh Quadras, Centris, II Models; SE/30	Macintosh Quadras or later	Not Supported on Macintosh or PowerPC to date	IBM PC or compatible; 386, 486, Pentium	IBM PC or compatible, 486 or Pentium	IBM PC or compatible; 386, 486, Pentium	IBM PC or compatible; 486; or Pentium
Coprocessor	Math Coprocessor		Math Coprocessor		Math Coprocessor		Math Coprocessor		FPU 68881 or later			Math Coprocessor		Math Coprocessor	
Operating system	MS-DOS 3.3 or later	MS-DOS 5.0 or later	MS-DOS 5.0 or later		MS-DOS 3.1 or later; MS Windows 3.1	MS-DOS 5.0 or later; MS Windows 3.2	MS-DOS 5.0 or later; MS Windows 3.1 (enhanced); 3.5	MS-DOS 5.0 or later; MS Windows 95, NT 3.5	Macintosh System 7.1 or later			MS-DOS 3.1 or later; MS Windows 3.1	MS-DOS 3.1 or later; MS Windows 3.1	MS-DOS 3.1 or later; MS Windows 3.1	MS-DOS 3.1 or later; MS Windows 3.11
RAM	8 MB	16 MB	12 MB	32 MB	8 MB	16 or 24 MB	16 MB; 20 MB for NT	32 MB	8 MB	24 MB		4 MB	20 MB	8 MB	20 MB
Hard drive space	25 MB	32 MB*	26 MB	32 MB; 80 MB free*	8 MB; 32 full installation; 37 MB w AME; swap file of 4X RAM; more space as required by drawing files size		37 MB to over 80 MB for full installation; 40 MB swap file minimum; more space as required by drawing files size*		30 MB	80 MB		8 MB; 32 full installation; space as required by drawing file size		8 MB; 32 full installation; 37 MB w AME; swap file of 4X RAM	
Video display/monitor	AutoCAD Supported Monitor		AutoCAD Supported Monitor		Windows-supported VGA Monitor		Windows-supported VGA Monitor; SVGA or graphics card for rendering and animation		Macintosh-compatible 256c min color monitor; 640 x 480 Dis min; 8 bits min per pixel			Windows-supported VGA monitor		Windows-supported VGA monitor	
Disk drives	1.2 MB 5 1/4" or 1.44 MB 3 1/2" diskettes		1.44 MB 3 1/2" diskettes or CD-ROM		1.2 MB 5 1/4" or 1.44 MB 3 1/2" diskettes		1.44 MB 3 1/2" diskettes or CD-ROM		1.44 MB 3 1/2" diskettes			1.44 MB 3 1/2"		1.44 MB 3 1/2"	
Input device	DOS compatible mouse; digitizer optional		DOS compatible mouse; digitizer optional		Windows compatible mouse; digitizer optional		Windows compatible mouse; digitizer optional		Macintosh compatible mouse; digitizer optional			Windows compatible mouse; digitizer optional		Windows compatible mouse; digitizer optional	
List price	Contact your AutoCAD Dealer		$3,995 (Disk); $3,750 (CD-Rom)		Contact your AutoCAD Dealer		$3,995 (Disk); $3,750 (CD-ROM)		Contact your AutoCAD Dealer			$495 List		$495 List	
Upgrade price															
From R12	Not applicable		$495.00		Not applicable		$495.00		Not applicable			Not applicable		Not applicable	
From R11 or earlier	Contact your AutoCAD Dealer		$695.00		Contact your AutoCAD Dealer		$695.00		Contact your AutoCAD Dealer			Not applicable		Not applicable	
Other features			License includes Windows and Windows NT versions of AutoCAD; network versions require hardware lock	Lacks features supported in Windows such as OLE			License includes DOS and Windows NT version of AutoCAD; network versions require hardware lock		Some dealers may still have copies of R12 available; no version for PowerPC yet exists			Does not support AutoLISP routines; architectural symbol libraries available; file compatibility with AutoCAD R12			
Distribution	AutoCAD Registered Dealers only		AutoCAD Registered Dealers only		AutoCAD Registered Dealers only		AutoCAD Registered Dealers only		AutoCAD Registered Dealers only			Any PC retail stores or mail-order catalogues		Any PC retail stores or mail-order catalogues	

Note: AutoCAD also supports Workstation platforms such as DEC, IBM, and Sun Workstations; these are outside the scope of this book.
*AutoCAD Release 13 Drawing files are 30 to 40% larger than Release 12 files and disk space should be allocated for that.

Table 3.1

show you compatible machines. Many dealers provide preconfigured complete systems (AutoCAD may or may not be preinstalled) at special prices. Visiting an AutoCAD dealer provides you an opportunity to view how an actual installation of AutoCAD runs on the various machines. You can also review product evaluations in AutoCAD publications such as *CADalyst* or *CADENCE;* these magazines feature the latest products and often present the results of tests conducted by product evaluation "labs" on competitive products.

Without referring to specific product manufacturers, here are some general guidelines for purchasing hardware for AutoCAD stations:

- While not every computer in your office may run AutoCAD, all AutoCAD stations should run, or have the capacity to run, other standard software in your office. This way, you can "recycle" older AutoCAD stations when you have to upgrade to faster CAD machines. Make sure that your computers can potentially run Windows 3.11 or Windows 95, for example, even if you use the DOS version of AutoCAD, and that they are compatible with your printers and other peripherals.

- Don't skimp on memory. Unless you plan to run AutoCAD only for occasional, low-level drafting, you should purchase at least as much RAM and hard drive space as recommended in Table 3.1. If you plan to become an aggressive user of 3D and animation, going up to 64 RAM and a 1GB hard drive is not extravagant. Having these features factory installed is often cheaper than adding them later.

- Buy a *reasonable* amount of expandability. Gauging what is reasonable presents a challenge, but you should assume that you will rapidly exceed the maximum capacity of your computer's random access memory and hard disk storage. Buying a Pentium with 48MB of RAM and a 500MB or 1GB hard drive to run AutoCAD Release 13 for Windows exceeds the minimum requirements but sensibly so; it gives you legroom for large drawing files and 3D applications. A PC with 2GB or more of hard disk storage, on the other hand, may be excessive; by the time you will need that amount of power, you'll want to trade in your Pentium for the next generation of PC.

- Make sure you understand the *actual* cost of expanding your computer's memory and storage capacity. You may be able to expand your RAM, for example, by adding new SIMMs (Single In-line Memory Modules) of RAM to empty SIMM sockets in your computer's coprocessor. If all your sockets are already filled, however, you will need to remove at least some of the existing modules and replace them with new ones containing more RAM, thereby effectively losing your initial investment in RAM (unless you can reuse the original SIMM on another machine). Therefore, before purchasing a computer, you should investigate the true costs of expandability and weigh them against the costs of purchasing a more powerful computer up front.

- For computers used for basic word processing and project management, a standard 13-inch or 14-inch computer monitor usually suffices. CAD stations, as well as machines used for multimedia, graphics, and desktop publishing, should have larger monitors, at least 17 or 21 inches. For stations used for on-screen presentations, a 27-inch monitor is helpful. As important as screen size is *resolution*, which is measured by the number of pixels or dots per square inch on the screen. A standard resolution for regular 14-inch monitors is 640 × 480; for larger monitors, a minimum of 1280 × 1024 or 1600 × 1200 is preferable.

- When selecting video cards for computer monitors, consider the number of screen colors you will be likely to use. For routine 2D, AutoCAD can use a maximum of 256 different colors, and many CAD users limit themselves to eight or sixteen colors; some even prefer to work with a monochromatic screen. But for 3D rendering, animation, paint, and desktop publishing, you will need monitors that can display thousands of colors at an acceptable resolution. Note that the more colors you display, the slower Windows will run; therefore, when you don't need extensive color palettes, you should use the Windows Setup Manager to switch to a more limited palette.

- Purchase a CD-ROM drive and consider getting a second one. More and more software will be available to you on CD format. CDs will save you installation time and disk storage space as well as money on software—many software vendors, including Autodesk, offer discounts to firms that accept their products on CDs. CD-recorders (CD-Rs) that enable you to "write" archival or "backup" copies of your files are now available. Although their cost is still relatively high, prices for CD-Rs and blank discs are falling rapidly.

- Consider special product bundles or bulk discounts from dealers or mail-order firms, but don't purchase a configuration that fails to meet your minimum requirements. Adding additional features to specials often proves more expensive than selecting equipment á la carte.

- Avoid purchasing equipment from too many different manufacturers. The more different makes of hardware you use, the greater the opportunity for hardware conflicts and the greater the difficulty of diagnosing sources of hardware problems. With multiple manufacturers, you will spend more time coordinating upgrades, licenses, technical support staff, shipping arrangements, and repairs than you would with a smaller number of vendors.

- Develop an office hardware standard for CAD workstations and stick with it until it becomes obsolete. Specifying uniform workstations saves money on installation, configuration, maintenance, and management as well as on learning curves for employees. You also achieve the additional benefit of employee parity—CAD drafters may complain if they have to work on a machine that runs more slowly than their peers' computers do.

- Consider purchasing at least one portable—or laptop—computer for your office. These are compact, relatively lightweight computers that can operate on batteries when electrical outlets are unavailable. Some models are now powerful enough to handle a full AutoCAD for Windows installation. Often a laptop can be used as the primary computer for mobile designers. The portablility of laptops enables them to be used in transit on planes and trains, at home, in field surveys, or on construction sites. You can take them to the offices of actual or prospective clients to conduct computer-based demonstrations.

AutoCAD on Networks

Any version of AutoCAD for personal computers can be run either as a single station or as a node in an office computer network. And if you use AutoCAD as part of a network, you can configure it in various ways, depending on your network's capabilities.

AutoCAD's basic drafting functions and interface operate the same regardless of whether or not you use a network, but your overall operating system determines how CAD users share data and peripherals. Some AutoCAD features, such as External References **(Xrefs)**, cannot be fully utilized unless you install a particular network configuration.

Advantages and Disadvantages of Networks

Advantages of Networks

With the advent of increasingly affordable PCs and cheaper software for small networks, more and more design firms find networks a cost-effective solution for running multiuser computer setups. Your firm can benefit from the following features:

- Simultaneous drafting and improved project coordination. With a network and AutoCAD features like **Xrefs**, multiple users can work on the same project, using the same information, at the same time. Moreover, because they can "see" the same drawing base as their colleagues, CAD users can coordinate their work as the project progresses.

- Systemwide maintenance. With the proper software, you can perform routine maintenance on all machines from one source. Maintenance includes automatic tape backups, scanning for viruses, removing unnecessary or "bad" files, and other tasks that are essential for ensuring effective operations. Being able to do systemwide maintenance saves enormous time and footwork and reduces the possibility that routine maintenance of one or more machines is overlooked owing to human error.

- Floating users. With networks, users are less prone to become attached to a single workstation. When it is time to regroup staff for new projects, the staff can switch desks, but the computers don't need to go with them (although they can).

- Floating licenses. Many software applications, including AutoCAD, offer *floating licenses* for network users. With floating licenses, a finite number of users can log on to the application from any networked workstation. While only a given number of users—depending on the number of licenses you own—can log on to the program simultaneously, they need not be specific individuals or work at specific workstations. For applications that are not used 100 percent of the time by 100 percent of the staff, this arrangement can save money by limiting the need to purchase additional licenses (this is analogous to having ten office telephone lines that are shared by twenty staff members). In addition, floating licenses can limit the potentially costly likelihood of software license infringement (see Chapter 10); the network can be set up to shut out unauthorized users whereas with stand-alone workstations, you have to monitor software installations more carefully.

- File transfers. With networks, you can easily move around files of any size. Relying on standard diskettes, however, means that you cannot share files larger than 1.44 MB (CAD files easily reach this size and surpass it) without using a file compression utility. With networks, you can also more easily share files created on different operating platforms; you need not worry whether your MAC-formatted diskettes can be read by a PC.

- Control of officewide standards. Particularly when they are working from a central file server, network users can share common databases and file libraries that reflect officewide standards. They can go to a single source to get templates for everything from drawing sheets to construction administration forms. Networks help limit the proliferation of multiple graphic standards throughout the office; the excuse that someone couldn't find the latest office standard on his or her computer becomes less valid.

- File control. An essential part of effective CAD management is controlling the proliferation of drawing files and ensuring that multiple versions of the same drawing file are not sitting in multiple workstations, causing everyone to panic about where the latest version is. With a network you can track drawing files more closely than you can with disks; you can structure your network so that users have to work from a given version of a file. Effective file management, however, is contingent upon a proper officewide system for file naming and structuring directories, as described in Chapter 10.

- System monitoring and evaluation. Most network software comes with utilities that help you monitor system usage, diagnose conflicts, and perform audits. These utilities allow you to easily generate reports that show—both in written form and graphically—how heavily the network is being used, and what applications and peripherals are most in demand. You can develop reports for a wide range of purposes, from tracking drawing plots and prints for client billing to planning expansions of your network. Obtaining such information with equivalent ease from stand-alone workstations would not be possible.

- Internal officewide communications. A network enables you to join the information superhighway by offering officewide electronic (e-mail) communications. E-mail can be overkill in a very small office where everyone shares a single open studio, but it can be helpful in larger or more inconveniently configured offices. E-mail can replace timely paper memos (which many people don't read anyway) and can be used for scheduling meetings and other office events as well as for submitting time sheets.

- External communications. A network doesn't have to be confined to your office; you can also communicate with engineers, clients, and reprographics services, provided they have compatible software and equipment. Your employees can work at home or at construction sites and send work into the office and vice versa. A network enables many users to share a single modem and e-mail system.

- Greater access to shared peripherals. In most offices, the ratio of computers to printers, plotters, modems, scanners, and other peripherals is high. And these machines can only be hooked up to a limited number of computers—sometimes just one—at one time. Networks, on the other hand, permit a number of computers to share printers and fax/modems, and users do not have to interrupt their peers when they need to print or plot their work.
- Plotting schedules. If you hook up a plotter to a network, you can control the sequence of CAD plots through various print monitor controls. Queuing print jobs gives precedence to users rushing to meet deadlines without forcing other users to have to resend their work to a printer at a later time.

Disadvantages of Networks

Although an increasing number of firms are turning to networks, some firms continue to use a "sneaker network" (in other works, no network). Stand-alone stations—or simple "peer-to-peer" networks—still work well in companies where the following conditions prevail:

- The number of computer users is small and stable, and there is no competition for printers, plotters, and other output devices.
- Computer users are machine-dedicated, that is, they tend to work at the same station all the time.
- Each computer user has different computer software needs. If each user has separate files and applications, he or she has less need to "speak" with other users.
- Each computer user can be trusted to maintain his or her own machine and files properly.
- You are concerned about being dependent on a network.

The one major disadvantage of networks is that as the system goes, so go the users. If the network crashes or for any reason is disabled, network users can't access anything stored on the file server. They may still use whatever files and software remain on their own hard drives, but productivity will be slowed; printing may not be possible. If you are not in a position to perform the regular maintenance necessary to protect against network downtime, you may prefer staying with a "sneaker" network.

The Costs of Networks

Many design firms, especially smaller ones, have balked at the presumed costs of installing and maintaining a computer network. If you have been reluctant to take the plunge, your anxieties are justified, but a little research will indicate that as CAD use expands in your office, a network begins to look more and more cost-effective. Moreover, the per-unit cost of networks has dropped enormously over the past ten years.

To purchase a major network such as Novell NetWare or LANtastic, you need software, network boards, and cabling. The cost of most network software and boards combined now runs as low as around $1,000 to $1,500 for a file server (not including hardware) with five to ten licenses and under $99 for additional workstation licenses. For centralized networks, you'll need to purchase a robust computer with ample hard drive space. Such machines cost a minimum of $2,500 to $3,000.

The greatest expenses networks entail are for installation, including running cables through your office and upgrading inadequate power and signal lines, and for ongoing network management. Unless you operate a simple AppleTalk network linking only a few users, you need to have a fairly advanced level of technical expertise and time to devote to running the network—and only a network or CAD manager normally can provide this service. In short, you have to absorb network *management* into your overhead budgeting.

As with other CAD-related expenses, the cost of a network at some point becomes less than the cost of not having one, so the trick is to determine at what point a network becomes indispensable. Many firms find this point occurs when the acquisition of additional computers breeds demand for additional laser printers and plotters; installing a network that enables many users to share existing peripherals may cost less in the long run than purchasing more peripherals. Other indicators are more subtle. Your firm may experience work delays owing to difficulties in moving data with disks, backlogs at printers and plotters, and computer users being disrupted by peers because two people couldn't simultaneously access the same computer file, printer, or plotter. When you start to see variations in office graphic standards and problems with file management, you may begin to be able to justify the costs of a network. An office network often serves the same function for electronic data that central file cabinets serve for paper-based information—it serves as a repository for information and permits staff to retrieve data uniformly without being dependent on specific people to store and find information.

Configuring Networks

Planning a network involves detailed discussion of cables, protocols, nodes, wide-area networks (WANS), and local-area networks (LANS). You will need to learn a new language, so it is not surprising that many design firms quite logically delegate the specification and installation of anything more complex than two or three PCs and a printer to a network expert.

Delegating the responsibility for the actual configuration and installation of a network, however, should not deter designers and managers from hands-on involvement in shaping the *topology* and *sociology* of the network. Clarifying where you want various components of your network placed and how you want them to interact is critical to your general office procedures; networks can enhance employee interaction, but they can also stymie it. Even without full technical expertise, you can state your case by diagramming your projected network. As shown in Figure 3.2, you can apply your basic programming and bubble diagramming skills to the design of your office network.

You can diagram your ideal network in an AutoCAD file and then, upon installation of the network, convert the file to a scaled working office floor plan, with graphic information about machines, electrical and data outlets, and cable layouts. You'll find such a plan enormously helpful for planning and budgeting future expansions. You can also link the information to an inventory of office equipment stored in a database.

In planning your network, you should consider the following issues and then supply the relevant information to your network consultant or manager:

- Who are the users who will be on the network, and what are their current *and* projected needs for software and peripherals?

- Will you use your software differently on a network than on a stand-alone station? Some software packages have different licensing terms for multiuser situations; some offer special enhancements for networks. Some software has built-in copy protection to minimize the use of unlicensed copies; others rely on their users' sense of ethics.

- How much memory will each software package or peripheral require on a network? A software package installed on a network file server may require a different amount of memory than is required by the same software installed on individual computer hard drives which may or may not participate in the network.

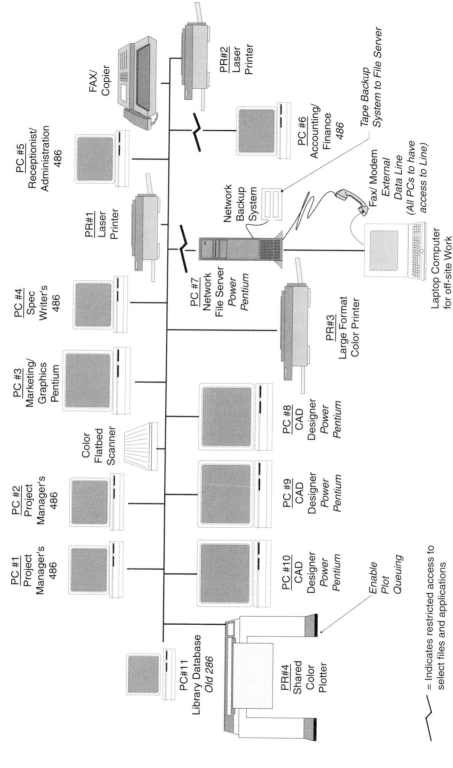

Figure 3.2 A schematic diagram of an officewide computer network system

AUTOCAD ON NETWORKS

- Will users work in designated computer teams or *work groups*, sharing files simultaneously, or will they work independently? Will different work groups need different peripherals or access to different files on the file server? Will work groups parallel departments, project teams, or "pods"? If work groups correspond to project teams, how frequently will they need to be rearranged?

- Should access to certain workstations be limited? In most offices, for example, you wouldn't want all employees to have access to accounting and personnel records. You might also want to restrict access to editable AutoCAD templates.

- What security measures should be built into your system? Do you want everyone to enter a password before he or she logs on, or should only select people need a password for select tasks?

- Do you plan to use intraoffice e-mail? If so, do you want to create zones; for example, so that only production staff, for example, will receive project memos, but marketing and accounts staff won't because they don't need to?

- Do you need to extend your network off-site, to employees' home offices, to field offices, consultants, or overseas colleagues?

- Do you want to use computers for monitoring purposes? Network software can allow you to monitor who works on what and to track hours on a project for billing purposes. You can also install utilities that assign project numbers to each file you send to print or plot; this information can then be applied to your reimbursables.

- What services are in your telephone closet and electrical box? Networks will place additional demands on your power and data capacities. You may need to upgrade your power. If you do so, buy plenty of expansion capabilities (or move to an office space that already has it); your excess capacity will be used up sooner than you might imagine.

- What are your growth projections? If you expect ongoing growth in your network, buy excess capacity up front. Many network packages price their products so that the unit cost of a ten- or twenty-license network is the same as or less than a five-license network.

Configuring AutoCAD for Networks

You can configure AutoCAD for networks in various ways, depending on your network software and the way CAD users will work together. The

AutoCAD Relaease 13 Installation Guide for Windows provides basic installation and configuration guidelines.

Peer-to-Peer Networks

Peer-to-peer networks link computers and peripherals through a chain of wiring. Users can move or copy files through the network but cannot usually work on a file that is located in another computer. No centralized computer exists, although you can designate one computer as the file server to store all working files and special applications; this file server also serves as the primary link to printers, plotters, and fax/modems.

Peer-to-peer networks are cheaper and easier to install up front. As more users join them, however, the overall system slows down substantially. In addition, this network setup may restrict your ability to share files simultaneously with peers, which limits use of AutoCAD **Xrefs** and similar file-sharing utilities.

Centralized Networks

In centralized networks, one computer is designated as the formal center or server of the network, and the other parties are "clients." All clients must go through the server to commune with other computers or to use printers, plotters, and other peripherals.

Centralized networks are generally more robust and more capable of adjusting to growth than are peer-to-peer networks. They do, however, require a dedicated file server, which should be a sturdy PC with at least 1 GB of hard drive space.

With centralized networks, you have the option of installing AutoCAD either locally on individual computers or centrally on the file server. Either way, users can share files simultaneously, even between different computers.

Note that AutoCAD creates a swap file—essentially an appropriation of hard disk space—during drawing sessions, and you have the option of assigning the hard drive of either the file server or the client computer for the swap file. The latter approach frees up memory on the file server and allows AutoCAD to operate much faster.

As previously noted, AutoCAD issues floating licenses for network installations. Instead of purchasing and installing a single or "site" license for each CAD station in the network, you purchase a network license for a specific number of network users. As more CAD users need access to AutoCAD, you can purchase additional licenses. To run a network version of AutoCAD, you must use a hardware lock (a physical device connected to a computer's parallel port), that controls access to AutoCAD. The AutoCAD hardware lock supports two major network "protocols": Internetwork Protocol Exchange (IPX) and NetBIOS. Information about installing and configuring the hardware lock is contained in the *AutoCAD Release 13 Installation Guide for Windows*. You can also obtain additional information on network installations from your Autodesk dealer or through the Autodesk FAX Information System (see Appendix B, "AutoCAD Resources").

Floating licenses can help you save money on software purchases. AutoCAD on a centralized network does not limit (unless you choose to) which workstations can access the software, only the number of users that can do so simultaneously. If your CAD-based work schedules enable users to share access to AutoCAD, you can reduce or delay the need to purchase additional licenses.

Both AutoCAD and network software packages provide a range of ways for users to log on to (access) networked applications. The options you select should be in line with officewide policies on network access and computer security. In some offices, users access a network simply by typing in the line "login"; in others, they must provide a password. You can restrict access to the entire network or portions of it, such as specific departments, applications, or personal e-mail accounts.

Networks and Office Culture

When computer networks permeate offices and more and more users rely on them for transmitting data and performing other functions, the office culture changes. People will distribute information differently, with less paper, and often there is a better sense of the overall office picture. Networks can facilitate a lot of mundane tasks as well as allow you to automate many functions, such as the collection of time sheets and billing. Implemented intelligently, networks are generally beneficial.

Networks do require changes in personnel procedures and office management. Networks make public the private quirks of individual users, and sometimes conflicts emerge. Office policies must be established on every aspect of network life from file sharing to e-mail to printing queues. Otherwise you may find distraught employees complaining about their print files being canceled or about receiving offensive electronic messages. As your network expands to include external parties, such as clients and contractors, network procedures and etiquette become even more essential. Guidelines about procedures should be written into the office employee

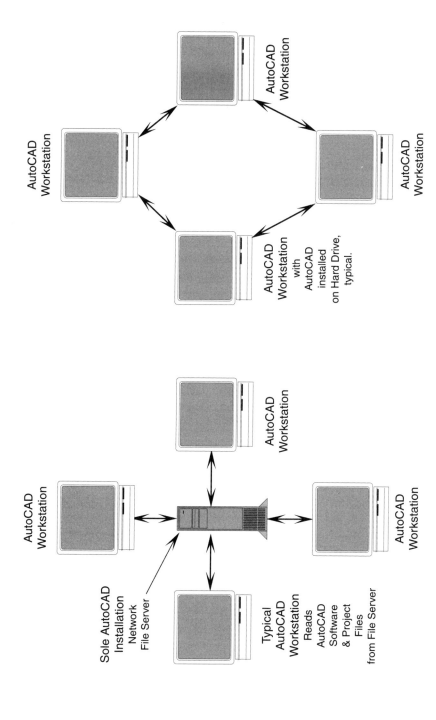

Figure 3.3 Two examples of possible "topologies" for AutoCAD on networks. The left diagram shows a centralized "hub and spoke" configuration. The right diagram shows a peer-to-peer "ring" configuration.

manuals and should become an integral part of a new employee's introduction to the office.

In addition to general procedures, another management concern should be the capacity of certain types of network installations to enable managers—and in essence the firm partners—to monitor staff electronically. With many network software packages, you can monitor who does what and for how long. You can even access other people's e-mail. In some organizations, network "big brothers" have caused employees to be demoted or fired because of their work habits as monitored by computers or because of an indiscreet message they sent over the Internet. Legal disputes about network monitoring often favor the firm, as the owner of the network, but ethically and practically this type of monitoring can be detrimental to employee relations. Until guidelines on this issue become clearer, you should protect yourself by stating up front, in your office manual, what your policies are. You might also discourage employees from treating the office e-mail system as anything other than a *public* form of communication. You may also find it helpful to acquire one of a growing number of books, such as *The Elements of E-mail Style* by David Angell and Brent Heslop (Addison-Wesley, 1994), that offer guidance on the still emerging universe of proper electronic language and behavior.

Complementary Software for AutoCAD

On its own, AutoCAD has always been a powerful drafting and design tool. Combined with other software and computer peripherals, AutoCAD's power and effectiveness increases exponentially, often outdistancing its less powerful but more user-friendly competitors.

With AutoCAD Release 13 for Windows, the possibilities for file exchange and collaboration with other software expands enormously. Copy and Paste functions, Object Linking and Embedding (OLE), and Dynamic Data Exchange (DDE) increase the value of making AutoCAD talk to other programs. At the same time, certain AutoCAD features may be redundant in Windows software packages; you can easily duplicate capabilities that you already have. So before selecting other software, carefully review their properties and compare them with your current AutoCAD version and any prospective upgrades. Considering all your software together is also helpful because all your programs compete for finite memory and hard disk space; the more resourcefully you can use your software, the better your machine will perform.

Although not every computer in your office need be a CAD station, every CAD station should have most of the basic software that other PCs have. At the minimum, these stations should have basic disk utilities, a word processor, and a spreadsheet program. Depending on CAD users' tasks and capabilities, a paint program, multimedia software, desktop publishing software, and a database program may also be necessary.

Figure 3.4 shows the major categories of software and hardware that are likely to be of use on a typical AutoCAD station. The following

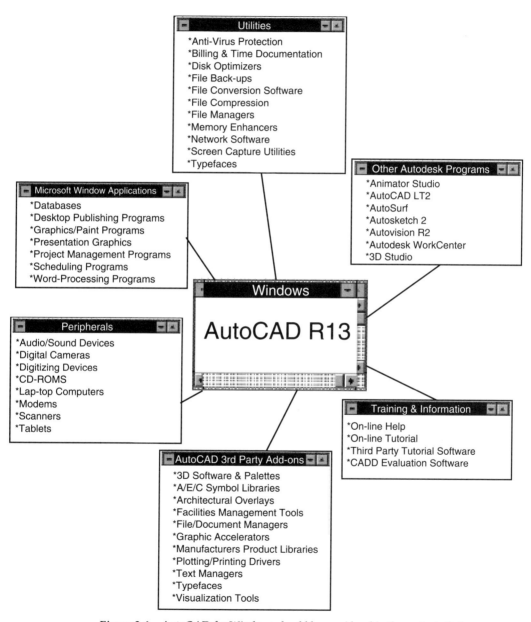

Figure 3.4 AutoCAD for Windows should be considered in the context of other Windows software on your CAD stations.

COMPLEMENTARY SOFTWARE FOR AUTOCAD

paragraphs describe particular software packages and peripherals. The products mentioned here are neither recommendations nor endorsements; instead they provide a benchmark when you are shopping around for a particular type of product.

In researching products for officewide use, consider the following product features:

- Is the software truly compliant with your preferred operating system, be it MS-DOS, Windows, or Macintosh? Just because it may run on Windows, for example, doesn't mean that a software program uses all of Windows's features. Check to make sure that the software uses the Windows Clipboard, Object Linking and Embedding (OLE), and any other Windows features of importance to you.

- What are the product's input and output options? Can the software work with files that are scanned in? Can it print files to your plotter as well as to your printers?

- What are the file conversion options? Can the software read and write AutoCAD *.dxf* files? Standard Windows file formats?

- How does the software run on networks? Does it offer special features or licenses for network users? Does it offer special features for work groups?

- What are your options for learning how to use the software? Does it come with extensive written documentation? An on-line tutorial? A demonstration video?

Other Autodesk Software

Although Autodesk's most popular software is AutoCAD, the company also produces a number of software packages designed to complement AutoCAD. Some of these products compete with offerings from AutoCAD third-party developers. Selecting AutoCAD products over others requires research and comparison of features and price. For some users, one primary value of staying in the AutoCAD family is the assurance of file-exchange compatibility, especially with upgrades.

Current products in the Autodesk family of interest to architects and interior designers include

AutoCAD LT for Windows Release 2. This is a low-level version of AutoCAD, useful for basic drafting and designing. The software reads regular AutoCAD Release 12 drawing files, but compatibility with AutoCAD Release 13 is not yet available. Version 2 of AutoCAD LT does not use AutoLISP routines or other AutoCAD customization tools, but clever users find ways to customize the software. It can be a useful basic drafting tool for designers and project managers who do not need AutoCAD's more powerful features on a daily basis. Unlike regular AutoCAD, AutoCAD LT is available through mail-order houses and retail outlets, at a street price of less than $500. Both Autodesk and several third-party developers sell add-ons, including libraries of architectural symbols, for use with AutoCAD LT.

AutoVision Release 2. AutoVision is a 3D design and rendering tool that runs *inside* AutoCAD; it actually ships with AutoCAD Release 13, but you have to purchase a license to activate it. AutoVision provides greater rendering capabilities, more materials, and more realistic shading than AutoCAD's regular 3D features do.

3D Studio Release 4. To do full-scale detailed rendering and animation, you can export files from AutoCAD to 3D Studio. You can also produce high-level presentations from scratch without AutoCAD. 3D Studio is supported by a wide range of third-party developers who sell plug-ins and add-ons for everything from libraries of architectural finish materials to animated bubbles, people, and light effects.

Animator Studio. This is a 2D audiovisual, animation, and paint package used primarily for making screen and desktop video presentations.

AutoSurf. This program is a 3D surface modeling tool for studying and producing complex 3D surfaces. It is used primarily for industrial product design but has applications for complex massing studies of buildings or for furniture design, for example.

AutoSketch. This product is the closest thing to a Windows-based paint program available from Autodesk; it allows designers to draw more loosely and intuitively than they can in regular AutoCAD. AutoSketch can read and write AutoCAD and AutoCAD LT files.

Document Management. A growing area of computer applications is document management, as discussed further in Chapter 10. Autodesk is participating in this growth by offering new products such as its Work-Center and Autodesk View software, which reads AutoCAD for Windows drawing files and manages documents, tracks drawing revisions, and monitors electronic data flow.

Third-Party Add-ons for AutoCAD

Since AutoCAD's inception, Autodesk has chosen not to produce Auto-CAD add-ons for particular professional or industrial ("vertical") applications. Instead it has supported the development of application-specific products by non-Autodesk (third-party) developers. These developers work with AutoCAD to tailor the software's broad-based capabilities to specific applications, ranging from cartography to yacht design. Although Autodesk itself does not produce these software packages, it encourages their development through the Autodesk Registered Developer Program. Registered developers have access to special software, technical support, publications, and information; this support helps ensure that the products they develop are truly compatible with AutoCAD. They also are eligible to participate with Autodesk at CAD-oriented trade shows and developer workshops and other programs.

Many Autodesk Registered Developers and Authors are architects, interior designers, and other design professionals who realized that their applications of AutoCAD could be of value to others. If you have a customized application of AutoCAD that your professional peers might find useful, or something to share in writing about the product, consider applying to Autodesk to become a Registered Developer/Author. This is an excellent potential means of increasing the returns on your investment in AutoCAD.

To keep track of the latest software and related publications, contact Autodesk for a copy of *The AutoCAD Resource Guide*, published quarterly. This handy guide describes succinctly all registered add-ons and the addresses and telephone numbers of suppliers (who must be Registered Developers). The guide is organized into over thirty major application categories, such as architectural design and layout, facilities planning, and project management. These application categories are cross-indexed by platform and suppliers. In addition to listing products, the guide also lists compatible peripherals, learning aids, training centers, and user groups.

Some of the tools that third-party add-ons introduce often appear in subsequent releases of AutoCAD as built-in features.

When considering third-party software, you must distinguish between software that runs *inside* AutoCAD and therefore requires the program to operate, and software that can *read* AutoCAD files but doesn't depend upon AutoCAD to run.

Many AutoCAD add-ons for the architectural/design professions are available and can be generally categorized as follows:

Architectural overlays. These are comprehensive add-ons or templates that customize the entire AutoCAD working environment to the needs of design professionals. These programs automate the process of establishing typical title blocks, architectural targets, and drawing sheets for different architectural scales. They also automate routine drawing tasks, such as creating layers, inserting doors, walls, and windows, and moving between 2D and 3D. These packages usually reflect industry standards for features such as layer names and architectural drawing sizes.

One of the most notable and comprehensive packages is produced by Softdesk Inc., of Henniker, New Hampshire. Softdesk provides a complete family of products tailored to the architecture, engineering, construction, and facility management professions. Its Auto-Architect add-on for AutoCAD automates drawing setup, provides wall, door, and ceiling tools, and allows you to draw in both 2D and 3D as well as to establish links to estimation databases. Softdesk also owns the popular Vertex Detailer software, an electronic library of standard architectural components for automating detailing. Auto-Architect is now prevalent enough in the architecture industry that in notices of job openings many design firms request that job candidates be familiar with it.

Softdesk products are not inexpensive, and some users find their defaults irritating, but if you are looking to purchase a full-scale customization package, these are a good reference point. Other architectural templates to investigate include Architectural Power Tools and Facade by Eclipse Software, Inc., Eaglepoint Advanced Architecture by Eagle Point, ARCHT by KETIV Technologies, CadPLUS Total AE System by CadPLUS Products, and (for landscape design) LANDCADD by LANDCADD International. You can request product information from these companies through *The AutoCAD Resource Guide.* Another good way to collect information as well as to see actual product demonstrations is to attend the annual A/E/C Systems Conference and Exhibit, which usually features an Autodesk Expo of third-party applications for AutoCAD. (See Appendix B, "AutoCAD Resources.")

Symbol libraries. These are packages of AutoCAD drawing blocks, or symbols, for typical architectural targets and other items, such as electrical, HVAC, plumbing, and structural symbols, north arrows, and other standard elements. Many libraries reflect industrywide graphic standards and may include attributes for linking to specifications databases.

Detail libraries. These are packages of typical details for wall and partition types, millwork, doors, windows, and other typical components.

Various industry associations also market add-ons that reflect their industry guidelines for products and installations.

Manufacturers' information. Manufacturers of specific building components, such as doors and windows, as well as of furniture systems, have developed symbol and detail libraries that present their product in an electronic form that is readable by AutoCAD. The drawings in these packages often come with attached attributes describing dimensions, materials, finishes, and other information that can then be imported into schedules, budgets, and orders. (See Chapter 7 for more information on these products.)

Facilities management tools. These packages, often referred to as Computer-Aided Facilities Management (CAFM) software, connect CAD-produced floor plans to databases, permitting building owners and managers to maximize space utilization in their properties. Features include area takeoffs, inventories, and the ability to generate lists for everything from equipment and furniture to telephone directories (see Chapter 7).

Document management software. These add-ons augment AutoCAD's own file tracking commands (see Figures 10.5 and 10.6), as well as the Windows File Manager, by managing and documenting files. Such add-ons not only help copy, move, and save drawing files but also track the purpose and history of each file. Many offer the option to preview a file so that you don't have to spend time opening the file to view its contents. Most now compete with Autodesk's new WorkCenter and Autodesk View products (described above; see Figure 10.7).

The most high-powered programs enable you to see the contents of an AutoCAD file, to perform minor editing of text files, and to copy, move, delete, or rename files with ease. You may also be able to annotate the files with information on who created them originally, when, and why.

File translation software. Although AutoCAD is the world's leading CAD software, it is not the only one, so odds are that a client will ask you to use or provide a drawing file that can be read by another CAD package. Although AutoCAD easily converts files to *.dxf* format, the process is not flawless, and there is software that refines the process. You can also purchase translators for importing AutoCAD files to desktop publishing software and specific file formats, such as *.cgm*, *.dxf*, *.hpgl*, and *.iges* as well as translators for reading and writing files to earlier versions of AutoCAD and for converting files between raster and vector formats (see Chapter 10).

Graphic accelerators. These are display devices that enable you to move (**Pan** and **Zoom**) around your drawing files faster and more precisely than the standard AutoCAD **Zoom** command and **Aerial View** permit. The increased regeneration speed of Releases 12 and 13 of AutoCAD has lessened the need for these add-ons, but graphical accelerators are still useful for files containing very large drawing extents and applications such as 3D and animation where drawing regeneration speed is critical. (The third version or "patch" of Release 13 includes a new WHIP driver, a high-performance driver designed specifically for Windows-based AutoCAD.)

Plotting and printer software. With the enhanced plotting features of Releases 12 and 13, AutoCAD provides sufficient plotting features for most users. But there are add-ons that make better use of plotting features, views, and line weights, and sequencing of plots on networks. You can also buy software that permits background plotting while you work on another file, as well as packages that track and log plots for billing for client reimbursables.

Text management tools. These packages augment AutoCAD's rather minimal text-processing features without your having to install a separate word-processing program. Features include spell and grammar checkers. AutoCAD Release 13 for Windows has incorporated new text commands that emulate some of the tools traditionally offered by third-party software.

Typefaces. These packages provide typefaces which augment the collection of base fonts that comes with AutoCAD. These typefaces are often designed to match popular typefaces used in desktop publishing and word-processing software; with these added choices, designers can maintain consistency between drawings, specifications, and other documents. The typefaces also help provide accuracy in the development of signage design and layout (see Chapter 7). AutoCAD Release 13 for Windows offers TrueType capabilities, so users are less likely to need to purchase type add-ons.

Visualization tools. These tools help you generate 3D and animated versions of AutoCAD drawings. They often compete directly with various Autodesk products such as 3D Studio and AutoSurf described earlier. Some of the third-party add-ons are more affordable or faster to use, although not all are as powerful as Autodesk's offerings.

Special tools. In addition to the add-ons that fall into the major categories already mentioned, there are a host of products that are unique but useful for particular applications. One such product is Squiggle by Insight Development Corporation, which literally loosens hardlined plotted drawings; this tool can be used to make presentation drawings more "schematic" in appearance.

Microsoft Windows Applications

Since its introduction in 1990, Microsoft Windows 3.0 has spawned hundreds of applications for a whole range of uses and computer skill levels. Windows has rapidly overtaken DOS as the preferred IBM-compatible operating system for computer applications in design offices.

Unlike some other professionals, architects and interior designers have not yet settled on industrywide standards regarding software. However, the A/D community definitely has gravitated toward certain packages both because of their ease of use and because certain industry-standard add-ons, such as AIA MASTERSPEC, encourage users in a particular direction. Thus, in many design offices, you'll encounter the popular Microsoft Office bundle—containing Word, Excel, Powerpoint, and Access—as well as WordPerfect, Claris Filemaker Pro, and Microsoft Project. In selecting Windows applications, it is best to evaluate the need for particular features against what your industry peers are using. Selecting fairly mainstream software facilitates file exchange with clients and collaborators and makes hiring computer-skilled personnel easier. With the strong file translations and data-exchange capabilities of Windows, however, you should not hesitate to select the software which best meets your needs, regardless of its market share.

The specific products mentioned below do not constitute product endorsements or recommendations. They are popular, mainstream products, most of which can be purchased through computer mail-order stores and retail outlets; they may not meet your needs, but they are good benchmarks for making product comparisons.

Databases

You can use database packages to generate design programming documents; schedules for doors, finishes, and other architectural components; and mailing lists. Flat-file databases, such as Filemaker Pro by Claris, can store multiple numbers of records and fields but have a limited ability to make connections between various sets of data. Others, such as Microsoft Access and FoxPro, ACI 4th Dimension, and Borland Paradox are relational, allowing you to extract information and forms based on often complex search requests (queries) and relationships among variables.

AutoCAD's SQL drivers are configured to conform to national and international standards for SQL database syntax and organization such as ANSI X3.135-1989 and ISO/IEC 9075-SQL. If you are shopping for other database software which you may wish to link with AutoCAD-compatible SQL databases, investigate how the software is organized and what, if any, standards the program adheres to.

Databases can operate independently from AutoCAD, but you can also link them to entities in AutoCAD files (see Figure 3.5) using the AutoCAD Structured Query Language (SQL) feature. You can connect, for example, door symbols in a floor plan to a schedule of door dimensions, door types, and hardware. SQL links help provide an internal consistency between plans, details, and schedules and specifications.

AutoCAD Release 13 ships with Windows-compatible SQL drivers for dBASE III Plus, Microsoft FoxPro 2.5, Borland Paradox 4.5, and Microsoft ODBC 2.0. You can download additional drivers, such as ones for Microsoft Access, from the Internet. Note that considerable skills, time, and effort are required to construct database links with AutoCAD. Consequently, many design firms fail to use AutoCAD's SQL or else purchase third-party database add-ons.

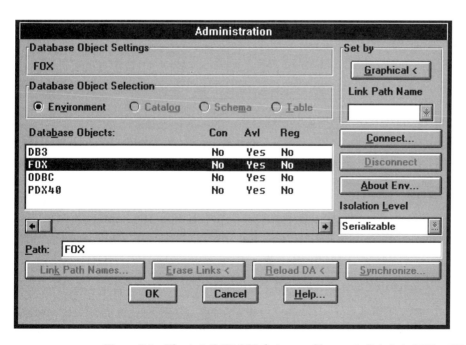

Figure 3.5 *The AutoCAD SQL feature enables you to link AutoCAD entities to external relational databases.*

With Windows pushing for data exchange through tools such as OLE, easier methods of linking AutoCAD files to databases are emerging, as described in Chapter 10.

Spreadsheets

Spreadsheet packages such as Microsoft Excel, Lotus 1-2-3, and WordPerfect's Quattro Pro enable designers to track budgets and to create schedules, shop drawing logs, and lists of various kinds. AutoCAD Release 13 permits you to literally paste a spreadsheet into your CAD drawings, thereby saving you hours of doing paste-up or replicating schedule information by hand or in CAD.

Most spreadsheet programs also enable you to perform high-powered calculations, such as return on investment (ROI), and to automate routine functions through the use of macros. Many have automatic charting capabilities; you can import the resulting charts into presentation software or even into AutoCAD to create large presentation boards.

Database and spreadsheet software often overlap in the functions they perform. One guideline for choosing one type of software over the other is whether or not you want data records presented in a range of different formats. Databases allow you to feature the same data in many different formats with less effort. So, for example, you could use a database file containing furniture specifications to simultaneously produce budget summaries, individual specification sheets, and lists of furniture and equipment sorted by their location or user. Spreadsheets are not as powerful for such a range of report formats, but they can perform more involved calculations.

Word-processing software

Design offices rely on mainstream word-processing packages such as WordPerfect or Microsoft Word to produce everything from client correspondence, meeting minutes, and marketing proposals to specifications. Although AutoCAD Release 13 for Windows now offers spell-checking and text-in capabilities, word processors handle large amounts of text better, and you can import files from a word processor into AutoCAD using the Windows Clipboard or the **Mtext** command.

Graphics and Desktop Publishing

Text, AutoCAD drawings, and other presentation elements can be merged, enhanced, and refined by using desktop publishing programs such as Adobe PageMaker and QuarkXPress. You also can import

AutoCAD-produced drawings into paint programs such as CorelDRAW, Adobe Illustrator, and Deneba Canvas. These programs enhance graphics in ways that bare-bones AutoCAD cannot; they can also convert AutoCAD drawings into camera-ready art for use in marketing brochures and design publications (see Chapter 7). A related family of products, such as Adobe Photoshop and Corel PHOTOPAINT, enables you to import and modify photographic images.

Presentation Software

Packages such as Adobe Persuasion, Microsoft PowerPoint, and Lotus Freelance enable you to present text and graphics in various formats for making presentations to clients, design review committees, and other audiences. You can import AutoCAD files, whether 2D or 3D, into these programs to produce slides, overheads, charts, and presentation boards.

Increasingly, presentation software includes multimedia programs that combine images, sounds, and action along with many types of special effects; some examples are Adobe Premiere, Autodesk Animator Studio, and Macromedia Director. These programs enable you to produce video walk-throughs and animated shadow studies, and to *morph* a client's picture with one of yours. To use all aspects of multimedia in Windows, you need lots of RAM and a sound board. (At present, it is easier to use multimedia on the Macintosh platform.)

Project Management and Scheduling Tools

Project management software enables you to plan, schedule, and budget for design projects. You can use such packages to track projects, anticipate staffing crunches, and monitor fees. Some project management packages just let you monitor one project at a time; others allow you to combine projects officewide. You can generate critical path method schedules, resource loading diagrams, and other graphics that help not only in internal planning but also for reviewing project plans with clients, consultants, and contractors. You can plot your diagrams on a color plotter to produce eye-catching large graphic boards.

Mainstream project management packages such as Microsoft Project and Claris MacProject Pro do not permit direct links to AutoCAD, but you can track CAD usage by specifying your CAD drafters or your CAD stations as project resources. In addition to mainstream project management packages, you can purchase third-party software packages, such as AutoProject by Research Engineers, Inc., and Suretrak Project Manager by

Primavera Systems, that link scheduling data to AutoCAD-based graphics and that address specific issues of concern to the A/E/C community.

Complementing project management software is a growing range of software packages known collectively as Personal Information Managers (PIMs). These packages combine electronic calendars and address books, "to do" lists, and other planning tools. Although these packages are often targeted primarily for individual users, PIMs can also be used to organize events, schedule employee vacations, and track mailings on an officewide or work group basis.

Windows Libraries

In addition to specific applications, you can easily accumulate an unlimited number of libraries of all sorts of graphics, typefaces, sounds, and other accessories that can be inserted into many different packages. Some of these are fun but nonessential, such as endless patterns of "wallpaper" for your Windows desktop screen or exotic start-up sounds. Others consist of stock photographic images of landscapes, materials, and patterns; some are junk, but others are useful. Before using these items for commercial purposes, make sure that the vendors of these packages specify that their images are royalty free.

Utilities

Utilities are programs that are accessible to your entire computer system, whether a single user station or an entire network. These programs can help protect your work, maximize your memory, and make your machine easier to use. If properly used, they will maximize your computer's performance and protect your files against damage and loss; improperly installed, they can wreak havoc throughout your system.

Antivirus Software

These programs detect computer viruses—computer programs that can destroy individual computer files or entire hard drives. A good antivirus program can save you from losing hours, if not months, of work. It can also help protect you against a cousin of viruses: Trojan horses—small programs that do something useful or interesting on your screen as a distraction while they invisibly corrupt your operating system. Antivirus software is essential for any computer that uses files from other parties; it is critical for networks, especially for network file servers.

Backup Software

These programs send copies of your application and project files to diskettes, writable CDs or tapes, or other media, so that you have an extra copy of these files in the event of a computer crash or in case you change or delete the original file on your hard drive. Backup software acts as a sort of insurance, and in some cases it may be required as a condition of winning a design project (see Chapter 10).

The importance of antivirus and backup software cannot be stressed enough. Such utilities are absolutely essential for any design firm that relies on computers to generate its work. In such offices, upper management must insist on daily use of these programs and assign someone, such as the CAD manager, to monitor their installation and use.

File/Document Management Software

As previously described, these programs enable you to find, name, and organize files and to establish file directories that fit in with your firm's overall filing organization (see Chapter 10). Unlike the aforementioned AutoCAD add-ons, these are systemwide utilities intended to manage *all* your files. Such programs are particularly helpful if you use DOS-based AutoCAD, but they are also valuable with Windows- and Macintosh-based versions, especially as computer files proliferate throughout your office.

Optimization Software

These utilities keep your hard drives clean and properly organized so that computer applications operate more effectively and swiftly. Such programs usually also detect "bad" files and restore deleted files.

File Conversion and Translation Software

File converters enable you to translate files between one software package and another. As previously mentioned, many file conversion and translation programs are designed specifically for use with AutoCAD. Although most Windows-based software files can fairly easily be retrieved and manipulated by other Windows programs, file conversion and translation utilities are helpful for routine, highly specific file conversion needs. They also help you exchange files between Macintosh, DOS, and Windows applications.

File Compression Software

File compressors enable you to shrink large computer files so that they can fit on a diskette or take up less storage space when archived. Given the expanding size of AutoCAD drawing files, these utilities are essential. File compressors also speed up the transmission of files over modems and networks. A well-known DOS-based file compression program is PKZIP.

Memory Enhancers

Memory enhancers, such as RAM Doubler by Connectix, reorganize your existing memory so your computer operates more efficiently and can work with more files and applications running simultaneously. While these programs may initially be cheaper than acquiring more RAM, they are not a long-term substitute for more RAM, and you can encounter serious conflicts if your memory enhancer is not carefully installed. These applications cannot be used to create additional disk space for the swap files that AutoCAD and other Windows applications require (swap files are portions of the computer's hard drive set aside for Windows to use to store information temporarily).

Network Management Tools

Although all network packages come with operating software, some offer features specifically for network managers. You can purchase additional programs to make the network configuration more user-friendly and to facilitate officewide functions like e-mail and tape backups. Such software contains or augments many of the network utilities described here and provides an overlay for the enhanced capabilities that a network provides.

Photographic Image Managers

You can store photographic and slide images on your hard drive or on CD-ROMs. Kodak, for example, can develop photographs and slides directly onto CD-ROM rather than onto regular slides. Photo-CD managers can be a useful tool particularly for marketing personnel, who need to store and retrieve images easily for assembling client presentations.

Screen-Capture Utilities

Screen captures enable you to literally "grab" a piece of or an entire computer screen, convert it to a graphics file format, and then embed it in a word-processing or desktop publishing program, or other software. Most of these utilities enable you to take the "snapshot" in black-and-white or color and then to resize it.

Screen captures are useful for acquiring small images that can't easily be obtained through other file exchanges. They also help generate illustrations for written documentation for computer-based procedures, especially when you are describing the contents of Window dialog boxes. Many firms use screen captures in CAD documentation and training.

Time and Billing Documentation

These utilities monitor who uses which software and for how long. They can generate reports for monitoring, scheduling, and billing work. Note that for the reports to be accurate and effective, users must provide a project number when opening a file and must be careful about closing out a project file before taking a work break.

Typefaces

Under Windows, you can access an enormous range of typefaces or fonts. When Windows is installed on your computer, it automatically loads a set of fonts. You can purchase additional sets both from Microsoft and from a number of third-party font "foundries."

When using fonts, it is important to distinguish between system-based fonts, such as TrueType fonts, which print out to any device that supports Windows, and printer fonts, such as PostScript typefaces, which require that you use PostScript-compatible printers and font files. You must also distinguish between systemwide fonts (such as TrueType, which when installed on a computer, can work with almost any Windows application) and fonts that can only be used with specific applications. Although AutoCAD for Windows can now use PostScript and TrueType fonts, it relies primarily on its own proprietary fonts, which can't be accessed by most Windows applications.

Literally thousands of third-party typefaces for TrueType, PostScript, and other font types are available. When installed, however, typefaces consume hard drive space and slow down operating speed. Moreover, for every typeface that is distinctive and attractive, there are twice as many that are graphically unappealing and redundant. So Windows users can benefit from installing only a finite library of fonts containing only those typefaces that meet office needs and design standards.

Peripherals and Other Hardware

Peripherals are hardware devices that are plugged into your computer to enable you either to generate computer output or to input data.

For Macintosh users, peripherals have been easy to deal with since compatibility is almost assured for any peripheral designed for this platform. For Windows users, peripherals have often been troublesome; inevitably, your computer needs a cable, driver, or board other than the one you currently have. Windows 95 introduces a "plug and play" technology designed to improve Windows's compatibility with peripherals and to reduce the effort needed to configure each device for your software packages.

Over time, popular peripheral devices often become standard built-in computer components. Just a few years ago, CD-ROMs were special and highly expensive external devices; now new PCs include at least one CD-ROM drive, and many models offer ports for two. Similarly, modems are now built in to many computers, so users don't need to purchase separate machines. Since peripherals consume both money and desk space, be cautious about what new technology you buy. You can spend a lot of money on new but quickly obsolete devices, when, if you had waited a year or so, you could have acquired the same technology more affordably and with substantial refinements.

Audio/Sound Devices

A growing and exciting development is the use of sound in computers. If your computer contains a sound board, you can use sounds to enhance graphic and animation presentations and to emphasize commands and events. Speakers attached to your computer can enhance presentations (or allow you to listen to CDs). You can also use audio devices for conducting conferences and attach musical instruments for composing themes to accompany your design presentations.

CD-ROM Drives

When AutoCAD first emerged, drawing files and other data were transmitted on 5 1/4-inch floppy diskettes; later 3 1/2-inch diskettes became the norm. In the 1990s, a growing number of manufacturers are providing software on CDs, and often they provide financial incentives for you to accept their software on CD-ROMs. You need to invest in a CD-ROM presser (or recorder) to create or "burn" your CDs in order to copy files to these disks (which you can't erase or change), but even without one, you can retrieve software, clip art, and other files from them. CD drives can either be installed in your computer or be purchased as separate external devices.

Digital Cameras and Videos

A fairly new technology, digital cameras operate much like regular film cameras but instead of sending the film out to be developed, you plug the camera directly into a computer port and copy your camera images onto your hard drive to use in various applications. The technology is still being refined and does not yet give you as many images or as high a quality as you can expect to see in a few years. But these cameras can serve a range of useful purposes in your office; for example, you can paste photographs of job sites into client presentations or produce an illustrated directory of your employees. You can also use still images from home-quality video cameras.

Input Devices

Designers use input or pointing devices to give commands to computer software. Strictly speaking, the only input devices you usually need are a standard PC mouse and a keyboard. However, you can use more high-powered mice or pucks, which come with as many as forty buttons and the ability to program macros; these devices are designed to reduce the time spent using pull-down menus and toolbars and typing out long commands, although most users find keyboard shortcuts and simple mouse clicks faster. You can also use electronic pens for tracing, freehand sketching, and writing signatures. Many input devices can be used with digitizing pads. A growing number of input devices are also available for users with physical handicaps.

Digitizer Pads or Tablets

Digitizers are "absolute" pointing devices that, with a pointing device, relate points on a digitizer pad to points in a computer file. When digitizers are calibrated, they can establish a reasonably accurate correlation between coordinate points on a digitizer tablet and those in an AutoCAD file. Digitizers vary in size—from 6" × 9" to 60" × 120"—and can include a range of special features, including backlit surfaces and voice recognition. Although most pads are rigid, others can be rolled up and stored in tubes. Although digitizers can be used for freehand drawing and loose tracing, they can also be used for applications such as photogrammetry where precision is key. For such applications, you can purchase tablets that offer a high resolution, in terms of lines per inch (LPI), and an accuracy of 0.01 to .003 inches. Digitizers can also contain icons for inputting drawing commands.

Along with its software, AutoCAD includes drivers for digitizer tablets produced by CalComp, Hitachi, Kurta, Summagraphics, as well as a generic driver for tablets that are Wintab compatible, that is, tablets that allow you to select items both on the tablet and on the Windows menu screen. AutoCAD also includes a rigid plastic template for 12" × 12" tablets, which is what most design firms use. Additional digitizer products are referenced in *The AutoCAD Resource Guide*.

Modems

Modems establish connections with other computers and allow you to send files over telephone lines to consultants, clients, plotting bureaus, and other parties on a worldwide basis. The price of modems has dropped substantially in the past five years, and now you can also readily purchase computers that contain built-in modems.

Modems require a telephone line. If you use your modem only occasionally, you can borrow a line from an office phone or install a telephone line sharing device; frequent modem use, however, justifies establishing a separate line and modem number. If your modem is on-line only at specific times, you should let outside parties know.

In selecting a modem, you often are confronted with a bewildering array of technical specifications, but this should not distract you from focusing on two criteria. The first is the modem baud rate—how much information the modem transmits per minute. You should opt for the fastest rate you can afford because the industry standard constantly accelerates. The second specification to look for is the required transmission *protocol* you need to communicate with various parties; your modem software determines the protocols that will be available to you. If you work with certain parties on a regular basis, you should ask them what baud rate and protocol they require to receive files from you. To access services on the Internet you need to meet specific protocol requirements. The best modems work with a range of protocols and some will automatically adjust your modem's baud transmission rate to comply with that of the other party.

Modems generally ship with software, but some software is more advanced than others. Desirable modem software features include the ability to let multiple users access the modem line to send and receive faxes; to "batch" e-mail to multiple users; to keep address books of frequently accessed modem numbers; and to keep transaction logs. The most sophisticated software automatically adjusts your modem to respond to different requirements for baud rates and protocols and may include Optical Character Recognition (OCR) technology, which enables faxes you receive by modem to be edited by word-processing programs. Some software includes preformatted transmittal forms; if possible, substitute one which replicates your office standards for correspondence.

Printers, Plotters, and Other Output Devices

As discussed in Chapter 8, output devices such as printers and plotters are essential if your work is to be distributed to the public (although the increased reliance on modems, networks, and audiovisual software does limit the need for certain "hard" output). The prices of inkjet and laser printers and plotters have come down considerably over the years, and printers have begun offering enhanced capabilities such as color, more dots per inch (DPI), and trays for larger sheets of paper and envelopes.

When selecting output devices, a key issue is the type of printer drivers they can read. For many Windows applications, using TrueType fonts is adequate, but compatibility with PostScript is also often essential. AutoCAD has its own plotting requirements, and it makes sense to find out for which plotters and printers AutoCAD provides drivers, so you can save time on configuration and reduce compatibility problems.

Although plotters and printers are the most typical output devices in a design office, you can also purchase machines that, in conjunction with AutoCAD and similar software, scribe and cut materials such as metal, wood, and plastic for models, signs, and other applications.

Scanners

Scanners enable you to copy graphics, text, and other information from paper or another nonelectronic format into your computer. Scanning saves hours of input time and, for complicated graphics, provides a greater degree of accuracy and information than tracing with a stylus. Scanners can be hand-held, small devices that can scan only a small area at one time, or flatbed, larger machines that work similarly to a photocopying machine and copy from legal-, letter-, or ledger-size originals. Large-format scanners resemble plotters and are designed to scan over-size graphics and architectural and engineering drawings. The prices of flatbed and hand-held scanners have dropped radically over the past few years, and a good-quality color flatbed scanner can now be had for under $1,000; using a service bureau to scan files has therefore become much less economically sensible. The large-format scanners are still expensive (usually upwards of $7,000) and thus are still more likely to be found in service bureaus than in design offices.

Scanners work in either black/white/gray (monochrome) or color. The best (and most expensive) machines give you a range of resolutions, measured in DPI. Many scanners come bundled with photo-processing software, like Adobe Photoshop, which enables you to size and manipulate the scanned images in various ways and then save the resulting file in a standard graphics file format. Note that you cannot scan images directly into AutoCAD files; you must convert the image to a *.bmp*, *.tiff*, or Windows Clipboard graphic, or else "vectorize" the image so it can be converted into an AutoCAD-usable *.dxf* file; the latter requires special conversion software.

Power Protectors

All electrical and data equipment is vulnerable to power spikes, surges, brownouts, and other problems related to power lines. Without protection against power irregularities, you might at best lose your current work, and at worst find your equipment destroyed. Every major device in your office, including CAD stations, printers, network fileservers, and modems, should be attached to surge protectors or other power protection devices. Many manufacturers of these devices not only offer lifetime warranties but also offer policies that cover the replacement costs for equipment destroyed by a power surge.

While surge protectors are essential, your machines are less likely to be vulnerable to power-related faults if they conform to industry standards, such as UL and CSA, for electrical equipment. Devices using data lines generally should conform to requirements set by the Federal Communications Commission (FCC) or the Canadian Telephone Network.

Ergonomic Accessories

As designers spend increasing amounts of time in front of computers, they—like other computer users—have become concerned about the physical side effects of working so intensively in front of a machine. A range of ergonomic products now addresses these concerns; they include special mice, curved keyboards, and other input devices, wrist support pads, adjustable keyboard trays, and monitor supports. Voice recognition software is also becoming available for various applications, including drafting and word processing.

For all that computers can produce physical ailments and antisocial behavior, they also provide enormous opportunities for people with physical and mental limitations to participate more fully in the world around them. Computers permit paraplegics to draw, paint, and compose music; they enable autistic children to socialize in a controlled environment; and they help wheelchair users travel throughout the world intellectually. Given the mandate to create buildingwide accessibility and to comply with the Americans with Disabilities Act (ADA), designers should always be aware of ways to encourage access to computers, whether for their employees or their clients.

Training and Evaluation Tools

One of the most important components of an effective office computer configuration is building in ways to help users learn and refine their knowledge of software. Even computer gurus need access to information when they are pushing the envelope of their knowledge. Some helpful tools come imbedded in software; others can be obtained for free from the right sources or purchased separately.

Most software and hardware products include a manual for both installation and reference. Publishing manuals is an expensive proposition, and manuals quickly become obsolete, so a growing number of manufacturers are encouraging alternative forms of reference, such as built-in on-line help and tutorials, or manuals on CD-ROM. Many manufacturers have become parsimonious about mailing updates and offering telephone-based technical support; instead, they encourage you to download information from their electronic bulletin boards via the Internet or to call their automated FAX information systems.

Manufacturers' literature often summarizes essential information, but it rarely enables users to master the software; such literature tends to be extremely generic, so it is of little value for focused applications such as architecture and interior design. So offices should make available software tutorials and books that augment what manufacturers provide. For Auto-CAD in particular, you can find teaching aids listed in *The AutoCAD Resource Guide*.

In addition to training, design firms need to be able to evaluate their employees' performance on CAD (see Chapter 9). The ultimate evaluation criterion is the ability to meet deadlines while producing quality work, yet you may find that even computer wizards have knowledge gaps. Both Autodesk and other manufacturers provide testing aids and tutorials; dedicated AutoCAD and Windows users can also study and apply for certification from both Autodesk and Microsoft (see Chapter 9).

Collecting Information about Complementary Software

Obtaining general information on available products for Windows or AutoCAD is *not* difficult. Once you've registered your purchase of computer hardware or software, you will be inundated with catalogs and product literature. Finding information and clarifying what products are best for you is somewhat more mysterious. Aggressive use of the following resources, however, will help you narrow your scope.

- *The AutoCAD Resource Guide.* As previously mentioned, this regular Autodesk periodical lists a wide range of products that work in conjunction with AutoCAD. This highly useful directory provides the range of available products; you can call or fax the listed manufacturers for additional information or mail in the Reader Information Card. The guide is included with AutoCAD software or can be ordered directly from Autodesk.

- Magazines about AutoCAD. AutoCAD has spawned several publications, the most notable of which (in the United States) are *CADalyst* and *CADENCE*. Apart from being filled with advertisements and new product descriptions of all sorts of AutoCAD-compatible items, these magazines provide critical reviews of new releases of AutoCAD and of various groups of related products. They also conduct lab tests for speed and performance comparisons and describe real-life applications for CAD products.

- General magazines on computers. For general information on the computer world in general and Windows in particular, you can subscribe to a host of magazines. Some, such as *Byte* and *InfoWorld*, are fairly general; others, such as *PC Magazine*, *Windows*, and *MacWorld*, are platform specific. Some are weekly, some monthly. All offer product reviews and comparisons, manufacturers' information, and previews of forthcoming products.

- Design publications. As a matter of course, most design magazines such as *Architecture, Architectural Record, Interiors,* and *Progressive Architecture* now address computer topics. Their New Product columns regularly include computer products, for which you can obtain additional literature by submitting a Reader Service mailer. These publications also conduct professional surveys covering computer usage, CAD salaries, and other industry indicators. They frequently feature articles outlining working examples of designers using CAD, and thus are highly useful ways to keep in touch with industry trends.

- Computer trade shows. In the major U.S. cities, annual computer trade shows are conducted by both general organizations and specific manufacturers. DecWorld, MacWorld, Microsoft Windows—these are just a few of the trade shows you may attend. For design professionals, the most important trade show is the annual A/E/C Systems Conference, which not only presents products but features seminars on all aspects of computer use in the design and construction fields. Trade shows are a great way to see in a single place a wide range of products that are not usually found together under other circumstances; viewing system configurations at a trade show saves enormous research time, and you may also have the opportunity for hands-on testing of new products.

- Local user groups. Many major U.S. cities are home to user groups that have reason to keep abreast of CAD developments. Some are generic AutoCAD groups, such as the North American AutoCAD Users Group (NAAUG); others are subgroups of local branches of the American Institute of Architects (AIA), the International Interior Design Association (IIDA), and other professional design organizations. Well-informed user groups provide a great source of information on actual products; many often invite product manufacturers to do product presentations. A number of these groups also maintain computer on-line bulletin boards that distribute software demonstrations and share user feedback.

- Authorized AutoCAD dealers. Many AutoCAD dealers offer a range of AutoCAD-compatible products, often including those produced by AutoCAD third-party developers. A growing number of AutoCAD dealers are specializing in vertical markets; a dealer that focuses on architecture and interior applications can provide information and demonstrations on products tailored to your needs. AutoCAD dealers may also refer you to peers who use products that you may find suitable.

- Software stores and mail-order houses. Retail outlets that specialize in computer products are most useful for supplying accurate prices and product availability. In general, they vary in their knowledge of actual product use, and often they have very little to say about applications for the A/E/C profession. AutoCAD and many of its third-party add-ons are not sold through retail outlets (although AutoCAD LT is); retail outlets are more useful for general Windows applications.

Tips on Selecting Complementary Software

- Review all product features carefully. As individual software packages expand their capabilities, they tend to replicate features available on other programs. Powerful word processors, for example, now offer many features similar to those found in both spreadsheet and desktop publishing programs. Often project management programs can perform scheduling tasks often as well as electronic personal calendars. Database software such as Claris's Filemaker Pro and Microsoft Access can replicate programs that generate everything from client rosters to invoices. Many AutoCAD add-ons now mirror features of mainstream Windows programs. So before acquiring software in every major category, inventory the capabilities you already have in existing software and be on the lookout for software redundancy.

- Don't buy a product merely for the number of features or star ratings it earns in a magazine article. Improvements and upgrades are ongoing, and within a short period of time competitors will adopt popular features. What can be equally important is the manufacturer's stability and commitment to ongoing development of its products as well as the availability of technical support, length of warranties, and quality of documentation. The costlier a product or the more your office comes to depend on it, the more attention you must give to factors beyond the capabilities of the software itself.

- If your office has not already made a heavy investment in a particular word processor, spreadsheet, database, or other general Windows applications, consider purchasing a bundled collection—or "suite"—of programs from a single, substantial manufacturer; suites often sell at a substantial discount compared to the total cost of purchasing each program separately. Popular bundles include Microsoft Office, Lotus SmartSuite, and Novell PerfectOffice. While the individual applications may not be the most powerful in their category, the bundles do offer cross-application compatibility and similar graphic interfaces, which can save enormous time when you are training users or combining information created in different programs. Some suite packages come with an overlay, such as Microsoft Office Manager (MOM), which offers additional multi-application features.

    A related option is a "works" program, such as Microsoft Works and ClarisWorks, which combines word processing, database, spreadsheet, charting, drawing, and painting tools in a single application. Works programs aren't as powerful as applications that focus on just one set of tools, but they can be easy to work with; works files often import easily for use in other software made by the same manufacturer.

- For your most heavily used applications like word processors, databases, and spreadsheets, invest in programs that have a strong market share. You then increase your chances of being able to hire employees who are trained in the use of that particular software and of finding courses, reference books, and training tools.
- If your office uses some combination of DOS/Windows/Macintosh operating systems, select products that run on all the platforms used in your office. This will help you transfer files among different computers and speed up training time.
- Major upgrades of popular products often incorporate new features that duplicate the functions in software you already own. Periodically review your software inventory to see if you should eliminate software that has become redundant.

4 Customizing AutoCAD for Maximum Design Productivity

As previously stated, AutoCAD is a *generic* drafting package, known for its open software architecture. It is designed to address the drafting and design needs of users performing many functions, in many industries and professions. Part of its power, and the basis for its popularity, is the overwhelming number of options AutoCAD offers for doing everything from creating text and dimensioning styles to producing 3D views. All elements of drawing can be customized. Indeed, the options are so extensive that every copy of AutoCAD Release 13 ships with a 683-page *AutoCAD Customization Guide*.

All these customization options can impede productivity, however, if you must start every new project from scratch. Just as hand drafters speed up their work by using vellum and film with preprinted sheet borders, titles, and libraries of typical symbols and details, so CAD drafters can speed up their work by customizing the software beforehand to suit their own (or their office's) graphic standards and anticipated needs. The nonbillable staff time—whether it takes weeks or months—that is spent on initial customization will more than pay for itself with subsequent projects.

Staff at most design firms are well aware of the importance of creating standard CAD-based drawing sheets and symbol libraries. Only experienced AutoCAD users, however, are familiar with *all* the features that can and should be customized. Experience is the key here because AutoCAD offers no single customization path for everyone to follow. Some customizable features are straightforward; others lurk in obscure commands and ASCII files that many drafters never come across. Therefore, to address the most important features, you must take a systematic approach to customization, and your approach should embody the following principles:

- Ideally, no entity or text string should ever be entered in a CAD file more than once. Don't reinvent the wheel.
- Default first to the CAD resources you already have (parsimony is implicit here). Resort to additional software only after you've fully exhausted the capabilities of your existing tools.
- CAD should be configured to be as easy, fail-safe, and fun to use as possible.

Never lose sight of the fact that customizing AutoCAD does not set your graphic standards and drafting processes in stone but rather speeds up the routine tasks so that more of your time can be spent on creative and exciting design work. Customization also creates a distinct graphic image that should reflect your firm's overall design philosophy and marketing program.

General Guidelines for Customizing AutoCAD

The following section outlines the three basic approaches to customizing your AutoCAD software. Before you consider any option, however, note that *all* customization paths entail the following preconditions:

- that *every* firm member—partner, project manager, job captain, drafter—who has a say in the design and production process and the establishment of graphic standards be involved in the customization process. Spending massive amounts of unbillable staff time to produce templates and standards is pointless if the standards are not accepted within the firm.
- that strong communication channels exist between the staff establishing the standards and the person(s) responsible for customizing the CAD system. Procedures must be established to provide ongoing feedback and communication—whether through meetings, memos, or other procedures.

- that the customization be documented on paper. The standards that are set, whether for dimension variables, pen menus, sheet layouts, and so forth, should be documented and accessible to the project team, in usable formats such as an office CAD Standards Manual (see Chapter 9). There are several reasons for this: First, although Release 13 is the most user-friendly and least obscure version of AutoCAD to date, the rationale behind your work may not be immediately apparent; having hard-copy standards saves you from having to explain the process. Second, if changes are made to the defaults, the computer system fails, or you decide to switch to another CAD program, written standards makes it easier to replicate the customized defaults.
- that time be allocated to "beta test" the new standards. The real test of customized CAD drawing standards is how they work and look when they're put to actual use. The standards should also be easy to use and stable, that is, not prone to cause computer crashes. Ideally, your new standards should be tested by someone who did *not* produce them.
- that the standards be tested under reasonable circumstances. Although the ideal test is a real project, do not use a project with extremely tight schedules and deadlines or unusual requirements. Use projects that have enough slack time in the schedule to permit experimentation and tinkering. And don't erase the previous standards from your CAD system until you have ascertained that the new ones are reliable and acceptable.
- that all computer users in the office be required and committed to using the new standards. A high-level person, such as the CAD manager, should be given the authority to enforce adherence to these standards.

There are three basic paths to customizing your AutoCAD system:

- doing in-house customization
- using third-party software add-ons
- hiring AutoCAD consultants

The approach you choose depends on your firm size, budget, staffing, project needs, and the stage you are at in terms of CAD use as well as your long-term goals. These three paths, moreover, are not mutually exclusive. You may decide to use in-house staff to establish 2D basics, for example, and hire a consultant to program a 3D project. You might augment your in-house efforts with an add-on software package that offers features that your own staff can't produce as quickly or as easily.

In-House Customization

In-house customization relies on existing staff and resources to customize AutoCAD. Regardless of which path you select, this approach is always required to a certain extent; all firms must spend time tailoring AutoCAD to reflect their preferences for title sheet layout, typefaces, and so on.

Using in-house skills is a good way to ensure that your system defaults reflect the way your firm works; only members of your staff fully understand your approach to design, production, and project management, so they are more likely to develop acceptable office standards. Your CAD drafters, however, must be well versed in computer topics such as AutoLISP routines, SQL databases, and DOS and Windows programming if you want to take advantage of AutoCAD's more powerful features. If your staff members don't have these homegrown skills already, be prepared to take some of the money you'd save on the other customization options and send your CAD drafters and CAD manager(s) to a good training program (see Chapter 9).

For firms that are aggressively expanding and that want to make a long-term commitment to AutoCAD, in-house customization may be the most cost-effective strategy. By doing your own customiztion you produce a unique AutoCAD configuration that addresses your needs and standards; you don't have to buy additional licenses (as you would with third-party software) every time you add a CAD station to your office system; and you can spread the initial cost of customization over a large number of stations. You also are less dependent on external consultants, whose availability is often unpredictable. In-house customization is also often the best way for staff members to learn about how AutoCAD actually works and to incorporate what they have learned from previous CAD projects.

If your in-house programming skills produce AutoCAD add-ons that seem unique and of general interest, you may be able to market them to others and thus earn a more explicit return on your investment. You may even qualify to participate in the AutoCAD Registered Developer Program, which supports the development of products that augment AutoCAD.

Third-Party Software Add-ons

Since its introduction, AutoCAD has spawned hundreds of software "add-ons" by third-party (non-AutoCAD) developers. As mentioned in Chapter 3, there are a great number of add-ons for the architectural/engineering/design professions. These range from comprehensive A/E/C packages that install drawing sheets, layering systems, and architectural targets to libraries of symbols for the electrical, plumbing, and structural trades. Many manufacturers of windows, doors, office furniture, and other specialties also distribute software to facilitate (and encourage) specifying and ordering their products. The list price of these packages can range from zero (free) to under $100 to well over $5,000—more than the cost of setting up one CAD station—and the extent and quality of technical support and documentation also varies from package to package.

A primary advantage of using add-ons is that you have the opportunity to acquire the benefits of high-powered computer programming skills more cheaply than if you paid to train existing staff or hired a consultant; the developers' costs are spread over many users. Many of these programs, designed by your peers, produce blocks, AutoLISP routines, databases, and other features that anticipate your needs and speed up your most routine drafting tasks, ranging from drawing sheet setup to door and hardware schedules. And these packages often reflect design industry standards. Such packages also provide direct links between drawing elements and related specifications and schedules; you would need sophisticated programming skills to create such in-house customization features yourself.

While add-ons are often extremely cost-effective solutions for firms with limited in-house programming skills, please keep the following in mind:

- While the primary goal of AutoCAD add-ons for the A/E/C professions is to make AutoCAD more accessible to designers, you do need training and hands-on experience in order to use them effectively. You should set aside part of your budget for training, along with the purchase costs of the add-ons.

- Everyone who uses these packages should review the entire package and understand how the custom features *really* work. Otherwise, you may waste or misuse the program's most powerful features. Many add-ons come with computer-based tutorials or demonstration videos that guide you through the programs.

- You shouldn't *have* to accept the software's defaults. If its layering system or notion of a north arrow seems inappropriate, for example, you should be able to change them. Most add-ons permit additional customization of *their* standards to suit your specific needs; the ease with which you can do so may vary.

- You should be able to use the drawings created with these add-ons in "plain vanilla" AutoCAD. Some add-ons use their own text styles, External References **(Xrefs)**, and other features that cannot be accessed without the support files provided by that particular add-on. These features can produce irritating results when the drawing files are opened in CAD stations that lack the add-on, and they may also create problems during file conversions. You should keep such features in mind when distributing CAD files to consultants and clients who don't use your particular add-on.

- Especially with regards to the most expensive add-ons, make sure that the software producer has built an established client base and track record. Look for companies that are going concerns, can provide adequate customer support, and are committed to providing upgrades that parallel AutoCAD's own upgrade schedule. Try to get names of users in your local area to contact.

- Be certain you understand the software's licensing policy. Many third-party developers follow Autodesk's one license–one user policy; others may give discounts to network versions or multiuser environments. What may seem a steal up front when you have only one or two workstations can become a very expensive proposition as CAD stations proliferate throughout your office.

External Consultants

Along with software add-ons, consultants have become a major part of industrywide efforts to make AutoCAD productive in the A/D community. Consultants generally will be brought onboard for a designated period of time and for specific purposes.

Consultants vary enormously both in their professional background and in the type of services they provide. Some are computer hardware dealers who have developed knowledge of particular clients' needs; others are trained designers who have gravitated toward computers. Some market their own software add-ons, which they install and customize to suit your needs; others will review your operations and build your firm a new system almost from scratch.

Working with consultants for a few months can be more cost-effective than hiring a permanent staff person, especially if you won't require their expertise on a full-time basis once customization is complete. In addition, experienced consultants with a design background bring valuable insights on how other design firms work and where the design industry is going with CAD. They bring a certain objectivity and a fresh eye to the customization process, a perspective that someone who is caught up in the daily operations of your firm may not have.

External consultants must be hired with the same selectivity as you would use in choosing a potential full-time employee. Take the following steps to find a suitable consultant:

- Clarify the exact nature of the services the consultant is to provide and the time frame for delivering them. Is the consultant to come onboard for a finite period and then hand over a completed package for you to manage, or is he or she to be "on call" indefinitely?

- Determine which is more important to you, *design* experience or *computer* expertise. Relatively few consultants will offer complete expertise in both.

- Ask for references and call those references. Find out if the consultant's services really produced the benefits that the consultant is marketing—and whether those are the benefits you need. Did the other clients find the consultant sympathetic, articulate, and reliable?

- Make sure the consultant, regardless of his or her professional background, speaks English and not *computerese*. Anyone who speaks solely in terms of bits and bytes, DOS, and other high-tech dialects will impair the customization process with your colleagues—especially with CAD skeptics. In addition to communication skills, the ideal consultant should exhibit patience with those who are computer illiterate.

- Write up a job description, if not a contract, that clearly spells out the scope of services, goals, time frame, and method of payment, such as hourly, flat-fee, and so on.

- Request written documentation of the configuration that the consultant develops for you. And ensure that at least one member of your permanent staff understands the consultants' work. Even if the consultant promises to be available indefinitely, twenty-four hours per day, don't become dependent on him or her to keep your CAD system functioning.

Even if you choose not to employ consultants to help with the customization process, you may find that they are helpful for discrete tasks for which no in-house talent is available. Creating a complicated 3D model, an animated video walk-through, or a special database are typical sorts of projects with which consultants prove helpful.

You can find consultants through many ways (see Appendix B, "AutoCAD Resources"), including

- the aforementioned *AutoCAD Resource Guide*. Some of the professionals listed here may not be formally listed as consultants, but often they can provide the support you need.

- your AutoCAD dealers, who may know of consultants serving the area (or provide such services themselves)
- AutoCAD Training Centers (ATCs) and local CAD institutes (see Chapter 9)
- local chapters of the North American AutoCAD Users Groups (NAAUG)
- local chapters of the American Institute of Architects (AIA), the International Interior Design Association (IIDA), and other professional groups, especially those that have committees dealing with technical issues
- the Internet, including the AutoCAD ACAD Forum on CompuServe and AIA*Online* (see Chapter 10)
- placement firms or "headhunters" that cater to the architectural and interiors professions (see Chapter 9)
- your peers in other A/D firms.

Often fellow designers provide the best leads and recommendations for hiring consultants. Whenever you have a choice, try to find a consultant who knows not just AutoCAD and Windows but also is familiar with the A/D professions. A consultant who only knows CAD or only knows it in the context of other specialties will not be able to offer as much information as someone who understands the current professional issues facing architecture and interior design.

The Internal Revenue Service sets stiff criteria for defining what constitutes a consultant. The criteria include the ability of consultants to set their own hours, use their own equipment, and bill you. If your expectations for consultants conflict with this definition, you are better off hiring someone and putting him or her directly on your payroll, whether or not as a permanent employee, or contracting with a consultant through one of the growing number of placement agencies that specialize in providing people with both architectural and computer skills.

Scheduling and Budgeting for Customization

Scheduling and budgeting for customization is difficult because of the many variables involved, including the nature of the firm's practice, the extent of the standards to be developed, and the availability of CAD stations and staff to do the work. The process always takes longer than firms expect, which is often less owing to the time required to input data into the computer and has more to do with resolving the issues that customization forces firms to address.

Factors you should take into consideration in choosing a customization strategy and scheduling and budgeting for it include the following:

- whether or not your existing graphic standards can be replicated on CAD or whether you plan to use customization as an opportunity to overhaul your current drafting standards
- the number of staff involved in reviewing and approving the results. A single staff member may review the standards more quickly, but feedback from a committee may provide a better sense of what the entire office needs.
- whether staff members assigned to the task are dedicated to the projected or whether they are constantly being pulled away to meet project deadlines and fulfill other responsibilities
- how CAD is to be used in your firm. If you only intend to use CAD for drawing plans, you will need less time than if you want to use it for elevations, details, and 3D.
- whether you plan to build a customized CAD system all at once or in phases

No matter how effectively you customize your system, customization is useless unless your staff is trained to implement it. Your customization budget should include training time for all CAD users and, *ideally*, for designers, project managers, and others who work on CAD-based projects.

Should CAD Force a Revision of Graphic Standards?

The conversion to a customized CAD system leads many design firms to question graphic standards that they have long held dear and that have been developed to accommodate many styles of hand drafting.

Several aspects of CAD validate revising your standards. First, you do not have to allow for *individual* approaches to pen weights and lettering. Provided you select a legible typeface, you can use smaller type sizes, and with smaller type you can scale down architectural graphics and symbols. Second, you can consistently apply an unlimited array of hatch patterns, tints, and tones. Third, the ease with which you can plot in an infinite number of scales may change your standards for the paper sizes and drawing scales for various types of drawings.

While CAD should present an opportunity to reconsider and improve your standards, don't take the changes so far from whatever constitutes industry norms that communication—which is the ultimate purpose of design documents—is impaired. You always want to know that in a pinch you can replicate your CAD-generated work the "old way"—by hand!

A well-planned training presentation, perhaps augmented by informational AutoCAD slide shows and sample files, will reduce training time. Of great importance is showing users where template files, Scripts, and other prototypes can be found on the office computer system. The CAD manager's responsibilities should include training as a matter of course.

The presentation that you develop for your current users can be used for future new employees and consultants. It can even be a useful marketing tool to show prospective clients how you have made your AutoCAD system unique and how you use it to assure quality control.

In-House Customization: A Guided Strategy

Even when you hire a consultant or purchase third-party add-ons, you must nonetheless perform some in-house customization yourself. Even if you rely on others to do the actual work of customizing AutoCAD, you are still responsible for specifying graphic standards and computer configurations that will affect general office operations. Understanding the customization process also builds confidence in the final results and provides a foundation for directing customization of other Windows software that your firm uses.

Much in-house customization of AutoCAD is not *technically* challenging. Most competent AutoCAD users can do it for routine tasks. Where necessary, step-by-step "how to" guidance is provided in AutoCAD's own reference manuals and other instruction guides. Only with network installations and more high-end applications is greater expertise required (you can always hire a CAD manager or CAD consultant for that).

What is the challenge for most firms is allocating the staff time and resources to plan the process, make the choices, and generate documentation. Customization requires not only a CAD drafter's time but also input from anyone who has a say in the firm's graphic standards or design and production procedures. It often involves soul searching, discussion, and meetings, as well as conflict resolution. The process always takes a minimum of several weeks, sometimes months. Once the major customization tasks are completed, however, the benefits can last for years.

Customization *does* involve nonbillable hours. Many CAD basics, however, can be established while setting up CAD drawing files for an actual project. Other elements may be taken from completed CAD-based projects. Thus, you should be able to bill portions of your customization to design projects along the way.

Ideally, AutoCAD should be customized concurrently with any other Windows applications that you use extensively. Your entire collection of office-standard presentation and construction documents—drawings, schedules, databases, project management forms, and so on—will benefit if you also review the applicable software simultaneously. In addition, joint customization will provide insight into which functions are best done in AutoCAD, which are best in other software packages, and even which should be done by hand. And you will benefit further if you plan to link AutoCAD to other Windows programs.

To customize AutoCAD, you need to complete eight basic steps:

1. Configure your overall computer operating system (Windows and DOS).
2. Set up your computer directories and file-naming standards.
3. Install AutoCAD.
4. Customize your systemwide AutoCAD configuration.
5. Customize your AutoCAD drawing environment.
6. Create prototypical AutoCAD drawings, templates, and libraries.
7. Customize your output devices.
8. Customize special AutoCAD applications and links to other software.

A detailed checklist of these eight steps is provided in Appendix A. You can clone it to develop your own checklist.

Customization is not as linear a process as these steps might suggest; some circularity or iteration occurs because the issues you address at each step may necessitate revisiting previous steps. Feedback from experience on actual projects also affects the process. But following a systematic approach ensures that the process remains comprehensive and focused. Some steps are sequential, however; you cannot create prototypical drawings, for example, until you define your typefaces, layers, and so forth.

Although you may not be responsible for the hands-on process of customizing AutoCAD, you will nonetheless need to understand, and make contributions to, all the major decisions that arise during customization. Many of these choices affect not only productivity but also marketing, staffing, and interaction with other computer applications. The following paragraphs describe typical issues to consider for each step of the process prior to embarking on customization.

✔ *Step 1.*   ***Configuring Your Overall Computer Operating System***

Before you can even install AutoCAD, you must install and configure your operating software. For AutoCAD Release 13 for Windows, you must install MS-DOS 5.0 or later, and Microsoft Windows 3.1 or Windows for Workgroups 3.1 or later. You can also install AutoCAD Release 13 on Windows NT. Installation instructions are provided in the *AutoCAD Release 13 Installation Guide for Windows* and *Installation Instructions for Windows NT*, both of which are included with your AutoCAD software.

How you configure your Windows operating system depends on what kind of hardware you have—CPU, monitor, printers, and other components of your workstations—and on your overall computer application needs, including all the other software that you want to be able to use. If you run Windows on a network, your configuration options increase. You can access configuration options through the Windows Control Panel and Setup dialog boxes (see Figure 4.1).

 You can install the first three versions or patches of AutoCAD Release 13 (but not Release 12) on Windows 95. However, Autodesk does not provide technical support for such a configuration and does not recommend it for serious document production. Autodesk is releasing a new version of Release 13 (patch 4) which is designed to work specifically with Windows 95 and for which technical support will be provided. It is expected that, with Windows 95, AutoCAD's performance will actually improve because both are 32-bit applications, whereas Windows 3.1 is 16-bit.

In configuring Windows, however, you must anticipate AutoCAD Release 13's system requirements. Not only must you have the requisite

Figure 4.1
The Windows (3.1) Desktop Control Panel, showing the different components of your operating system, all of which can be modified

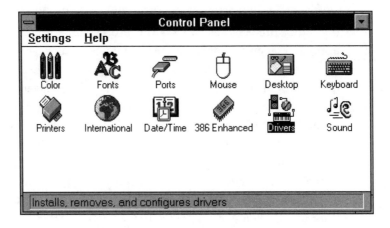

Figure 4.2
The Windows Virtual Memory dialog box

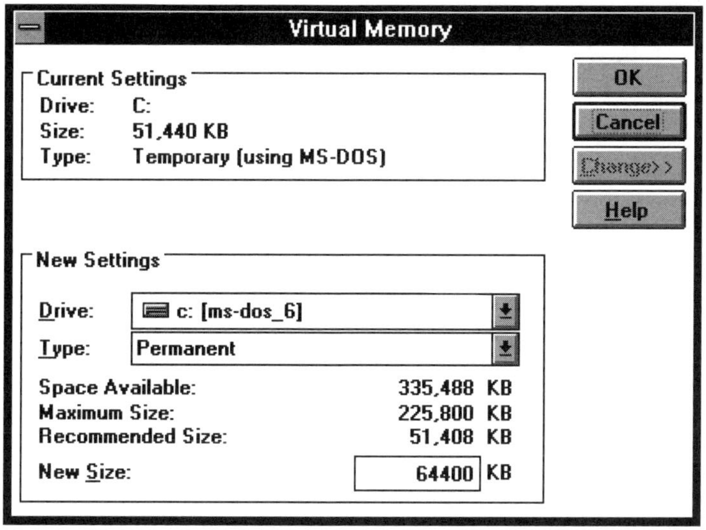

hard drive capacity and RAM (see Chapter 3), but you must also give AutoCAD at least 40 MB (64 MB recommended) of hard disk space for swap files (temporary files created on the hard drive). You will need to adjust Windows Virtual Memory(see Figure 4.2) to give AutoCAD as much swap file space as possible; the Windows default is four times the physical memory, but eight times is recommended for larger files and 3D applications.

You also must run Windows 3.1 in enhanced mode and with Win32s Version 1.20 or later (which is provided with your AutoCAD installation software). If your disk space on a particular computer is limited, you may want to install AutoCAD on another machine; alternatively, you may choose not to use that machine for other applications. You can also switch to the DOS Version of AutoCAD, which comes with AutoCAD Release 13 for Windows, but then you'll lose the benefits of using the Windows version.

Microsoft Windows 95 promises to run most Windows 3.1 applications without a hitch. But it will substantially change the Graphical User Interface, and new features will affect—mostly positively—file names, the installation of peripherals, and other Windows utilities. Other features, such as the Card File, are deleted. So if Windows 95 is in your near future, consider its features when planning your customization so as to facilitate conversion of your AutoCAD system to the new version of Windows.

In configuring your computer, you also need to decide whether to partition your hard drive (see Step 2).

> **Personalizing Computer Workstations**
>
> The user-friendly Graphical User Interface (GUI) of Windows and Macintosh computers permits users to develop highly personalized workstations with individual color schemes for screens, dialog boxes, and toolbars, as well as customized filing systems. These customization options encourage users to "bond" with their computers and, as a consequence, to take responsibility for maintaining their own computers, exploring their capabilities, and becoming less dependent on others for technical assistance. Computer users can, however, become possessive about their machines (as they do about their desk drawers) and can object to others specifying how their computer should be configured. In offices with floating users or with networks, a proper balance must be maintained between individual preferences and officewide standards for computer configurations.

Even when you have ample RAM and hard disk space, you'll do well to customize your Windows installation so that your hard drive is a perfect example of Miesian minimalism. Don't install applications and accessories that you won't need or use. Get rid of superfluous files, including printer and plotter drivers for machines you don't have, screen background patterns (called wallpaper), games, tutorials, sample files, and typefaces. The minimalist approach has several advantages:

- You free up valuable hard drive space and RAM, which AutoCAD will happily appropriate.
- You minimize possible conflicts with other hardware and software.
- You discourage the use of typefaces that don't reflect your office standards.
- You reduce the time and disk space required to back up your system.
- You reduce the temptation to play with the infinite choices for screen savers, screen background patterns (wallpaper), games, and other fun but nonessential features.

✔ **Step 2.**   *Setting Up Computer Directories and File-Naming Standards*

Before you install AutoCAD—or any other software application—you must establish a filing system for your hard drive directories and files. Organizing your application and project files properly not only speeds up your work but protects important AutoCAD application and project files from being damaged or inadvertently erased. The importance of computer housekeeping magnifies if you share AutoCAD over a network. (See Chapter 10 for additional guidance on file naming.)

If you use AutoCAD Release 13, you work with multiple files and subdirectories. For optimal performance these files must be kept together in particular directories (or Windows folders); AutoCAD defaults to the directories R13/WIN for files applicable to Windows and R13/COM for files applicable to both Windows and DOS. Figure 4.3 shows the default order of the subdirectories and files as they appear in Windows.

Project drawing files should *not* be kept in the Release 13 directory files. Otherwise, you run the risk of damaging or overwriting AutoCAD application files. Separating application files for DOS, Windows, and other software is also advised. You can create separate directories for each major application or sets of applications. Some offices partition their computer hard drives so that AutoCAD folders stay on a separately designated drive. For example, you might partition your hard drives as follows:

C: DOS

D: Windows applications (other than AutoCAD)

E: AutoCAD application files (R13 Directories)

F: master libraries and templates

G: project files

In addition to setting up directories, you should establish a policy for naming your computers and peripherals. Naming your computers becomes especially important when you run a network, since the network

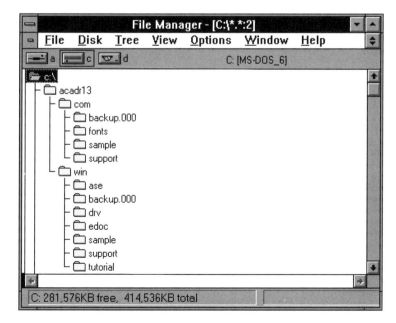

Figure 4.3
The AutoCAD directory file structure in the context of Windows 3.1 File Manager

IN-HOUSE CUSTOMIZATION: A GUIDED STRATEGY

> **Partitioning Hard Drives**
>
> Ask around and you'll find that the issue of partitioning hard drives is controversial. Many computer managers partition their hard drives in order to optimize memory and to protect important application files from "sticky fingers." Others feel that partitioning causes more maintenance, especially with backups and other procedures, and complicates workstation configurations. No industry consensus exists on this score, but the odds are that if you have many workstations and/or a network, you will be partitioning your drives.

software will be programmed to look for certain machines to perform various operations. Many firms have simple names like PC01, PC02, Printer 1, and so forth; others have fun and name their machines after their favorite designers. Whatever nomenclature you use, apply it consistently, not only when you are configuring computers and networks, but also in interoffice dialogue, inventories, and floor plans. Labels with the designated name should be affixed to each machine.

✔ **Step 3.** *Installing AutoCAD*

Installing AutoCAD on Windows-based Computers

Once you install and configure your overall operating system, you are ready to install AutoCAD. Release 13 provides you with three options for installing AutoCAD for Windows (see the *AutoCAD Release 13 Installation Guide for Windows*).

Typical installation. This option installs most AutoCAD files and is recommended for those just learning AutoCAD and for those who are unfamiliar with its capabilities (it includes a tutorial).

Custom installation. This option allows you to pick and choose selected features; it is recommended for firms that already have Release 13 installed on other workstations, are familiar with AutoCAD, and/or know exactly what files they need to use.

Minimal installation. This option is recommended for machines that barely meet AutoCAD's minimum memory requirements, for less powerful laptop computers, for offices where AutoCAD will be used rarely, and for specific, limited purposes, such as reading and writing *.dxf* files.

Installing AutoCAD on Networks

If you install AutoCAD on a network (see Chapter 3), the installation requirements become more complex, and network expertise is required, whether from a CAD manager, a network consultant, or your AutoCAD dealer. There are some key things you should know, about installing AutoCAD on networks:

- You must decide in advance whether to install AutoCAD on a file server or on individual CAD workstations. The configuration options you select determine how users work with custom libraries; External References (**Xrefs**); OLE and SQL capabilities; plotters, printers, modems, and tape backup devices. Your configuration also affects how you manage your AutoCAD licenses, as AutoCAD on centralized networks uses "floating" rather than site-specific licenses. Ideally, you should explore the sociology of your AutoCAD network and map out the configuration conceptually before beginning the installation process (see Chapter 3); switching from individual to networked installations involves computer downtime and system reconfiguration; it is not a casual process.

- During drawing sessions, AutoCAD grabs a portion of the hard drive space as swap space. If AutoCAD is installed on a central file server, you can specify the hard drive of either the file server or the local client computer as the target swap drive; the latter approach frees up memory on the file server and therefore speeds up network operations.

- AutoCAD is set up to prevent people from using the program on unlicensed workstations. You have to have an authorization code to set up additional workstations on a network.

- Networked AutoCAD stations require a hardware lock (as do work stations using international versions of AutoCAD). This is an actual piece of hardware that attaches to your network file server computer.

- You should enable AutoCAD file locking (see Step 4) so that multiple users cannot work on the same file simultaneously (although they may be able to view it at the same time).

- Plan on shutting down the network each time you add a new CAD station to the network.

✔ **Step 4.** *Customizing Your Systemwide AutoCAD Configuration*

Once AutoCAD is installed on a computer, you need to configure the overall AutoCAD working environment. This step shapes your general working environment, regardless of the particular AutoCAD file you are using. Although some aspects of the working environment may be adjusted while you are using a particular drawing file, establishing a standard configuration is beneficial, especially since it will shape network configurations and officewide procedures.

Many variables shape the interaction between AutoCAD and its users. The more you customize AutoCAD's features to address your needs, the more productive you'll be. Developing these defaults is particularly important in an office environment where users share workstations and drawing files; an officewide standard eliminates the learning curve involved when someone starts working at a different workstation. Many AutoCAD defaults are appropriate for a design firm, yet others—such as the default use of engineering units rather than architectural units—are not.

The actual commands required to configure the AutoCAD working environment are not difficult to use; what is difficult is carrying out the configuration process *systematically* and *comprehensively*. No singular dialog box or editable file contains all the variables that you can customize; indeed, many experienced AutoCAD users are not aware of all the features that can be customized. The **Environment** Page of the **Preferences** dialog box in AutoCAD for Windows Release 13 (Figure 4.4) guides you to many key choices, but you can also use the various commands in the **Option** menu and the **SetVar** command. In addition, you can configure additional features by using a text processor to edit various ASCII-based support files such as *Acad.pgp*, *Acad.mnu*, and *Acad.ini*.

In configuring the AutoCAD working environment, you'll need to address two major issues: the underlying system operating variables and the defaults for AutoCAD commands.

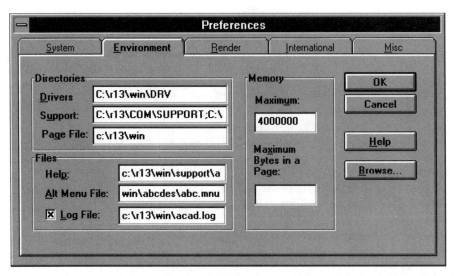

Figure 4.4 The AutoCAD **Environment** Page of the **Preferences** dialog box.

The Underlying System Operating Variables

You can adjust the system operating parameters through either the **Configure Operating Parameters** option in the **Config** menu or other commands (refer to the *AutoCAD Release 13 Command Reference*). Here are some of the key parameters that affect general operations.

- **Default Drawing Name** (Initial Drawing Setup). Ideally, this should be your office's standard drawing title sheet (see Step 6) or a blank drawing file containing office standard units, typefaces, drawing limits, and other customized defaults.

- **Default Plot File Name.** If you write plot files (rather than send the file directly to print), you must select default names for your plot files and the directory in which you will store them. Default names become especially important with networks; if plot files are automatically written to the same plot file directory, you will overwrite other plot files if they share identical names. Depending on your plotter's capabilities, you may also send the plot files to a spool folder, which will plot the files as soon as the plotter is free.

- **Automatic Saves.** You can use this feature to specify whether AutoCAD automatically saves the active drawing file, and if so, how frequently (from every 1 minute to every 10 hours); a minimum frequency of 15 or 30 minutes is suggested. Automatic Saves (also called AutoSaves) do not save to the working drawing file but rather to a special backup file; as such, it is really an emergency backup rather than a substitute for regular file save using the **Qsave** command. The default name for the AutoSaves file is **Auto.sv$**; you can rename it. Because AutoSaves can interrupt drawing commands, they tend to irritate CAD users, who often disable the **Autosave** option as a consequence. The AutoSave feature should be disabled *only* if CAD users are *extremely* disciplined about saving their work frequently (especially after completing complex tasks that cannot be easily replicated). Otherwise, users will face the risk of having to replicate work that was lost during an unexpected computer crash or because of careless file-saving habits.

- **Full-time CRC Validation.** This option generates a Cyclic Redundancy Check (CRC) to monitor possible errors and corrupted files. This option is useful if you frequently share files with other parties or are constantly reconfiguring your hardware or network.

- **Automatic Audit.** When you use the **Dxfin** or **Dxbin** commands (see Chapter 10) in conjunction with the **Automatic Audit** option, AutoCAD will check the drawing files created from *.dxf* and *.dxb* files for errors and conversion problems. This option is useful if you are constantly exchanging *.dxf* files or have experienced problems in doing so.

 Antivirus software packages available for Windows or DOS do not perform AutoCAD CRC or Automatic Audits, nor do CRCs and Automatic Audits in any way serve as substitutes for installing antivirus software for your computer system in general.

- **Log-in Name.** This option sets the default log-in name, which provides access to locked files. This name can be either a secret password or else a shared name that all CAD users know. The selection and publication of log-in names should reflect office policies on file access.

- **File locking.** When file locking is specified, an opened drawing file cannot be accessed by another user. File locking is essential in a network environment where users share files. Files can be unlocked using **Unlock Files**, but this command should never be used cavalierly.

- **Spelling dictionaries.** AutoCAD Release 13 includes a spell-checking capability and thus a spelling dictionary. With the *Dctcust* and *Dctmain* **SetVars**, you can specify a customized dictionary that reflects office-wide standards for abbreviations, product names, staff and client names, and other important defaults. The spelling exceptions contained in the AutoCAD dictionaries should correspond to customized dictionaries that you develop for word processing, spreadsheets, and other Window applications that offer spell-checking capabilities.

 Custom user dictionaries are often developed **passively** and on an ad hoc basis, as users come across words that need to be added, deleted, or modified in standard spell checkers. To help ensure proper and consistent spelling, take an **aggressive** approach and create up front a dictionary which contains words and phrases as well as names of staff, clients, consultants, and building products which your office uses routinely but which aren't likely to be found in standard dictionaries. AutoCAD dictionaries are ASCII-based text files, identified by their .cus file extension, and can be edited in most processing software. Since spelling dictionaries files developed by other programs are usually also editable text files, you can "copy" and "paste" words and phrases between various applications, thereby increasing consistency across all your documents. Adding frequently used terms to custom dictionaries also saves time, as spell checkers normally will not question any word which is already stored in the dictionary.

- **Default typeface.** The *Fontalt* and *Fontmap* **SetVars** specify default fonts to be used when AutoCAD cannot find a specified font file. Ideally, this default font should reflect an office-standard font; it can be a Windows TrueType font. Specifying a default typeface encourages consistency. It also helps you plan for text formatting if you import text from Microsoft Word or other text processors on a regular basis.

All typefaces are saved in separate font files. You can both save valuable hard disk space and ensure consistency in typefaces if you remove unused font files for AutoCAD, PostScript, and Windows TrueType fonts. While certain basic fonts cannot be removed, some of the more esoteric or visually unacceptable ones should be.

Defaults for AutoCAD Commands

AutoCAD has always offered several options for inputting commands; with Release 12 for Windows, additional options in the form of toolbars, icons, and well-featured dialog boxes were introduced. CAD wizards rely on combinations of these methods to enter commands speedily and accurately; they modify options throughout a drawing session, depending on the functions they use and their requirements for speed and memory.

AutoCAD Release 13 offers the following options for command input. Some commands can be accessed by almost all methods; others can be accessed by only one or two methods. With AutoCAD for Windows's emphasis on toolbars and dialog boxes, computer mice are assuming more importance relative to other input options (see Figure 4.5).

Keyboard commands and the command line. Keyboard commands are commands that you initiate by typing the command word; these are observed through the command line window (Figure 4.6). For example, you type "Line" to draw a line and "Plot" to specify plotting requirements. An important subset of keyboard commands are *command aliases*, which are abbreviated keyboard commands, such as "L" for **Line** and "U" for **Undo**. In addition to AutoCAD's regular keyboard commands and aliases, you can program additional aliases by editing the *Acad.pgp* file.

Pull-down and side menus. You can issue commands by accessing pull-down menus and (in DOS) the side menu with your mouse. The Windows menu is more compact than the DOS version because of Windows's greater reliance on use of icons and toolbars and because many commands can be replicated by keyboard shortcuts. Pull-down menus have assumed increasingly less importance in AutoCAD, but you can customize them to reflect

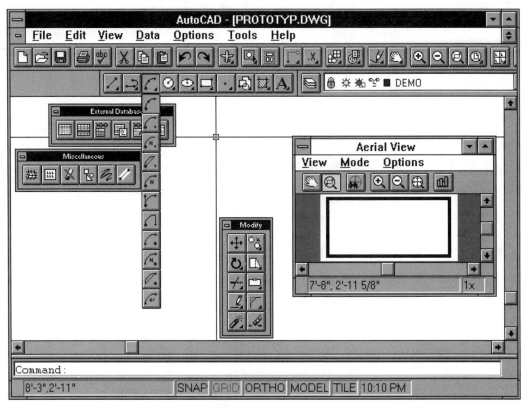

Figure 4.5 The AutoCAD Release 13 for Windows on-screen working environment. With such an array of floating toolbars and other elements, your electronic drafting surface can be completely hidden.

Figure 4.6 The AutoCAD Command Line window showing the command "line" which has been typed on the keyboard. The Command window shown here is "docked," that is, fixed in a size and location relative to the working screen, but you can also have the command line window "float" around the screen.

your most frequently used commands through the **Menu Customization** dialog box (Figure 4.7). Save modified menu defaults in a new file with a *.mnu* extension.

You can create different menus for different types of work. For example, you could have one menu for routine construction drafting, another for 3D modeling, and a third for schematic design.

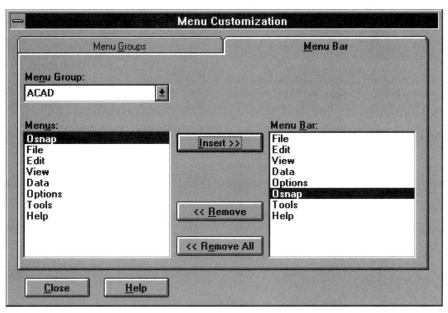

Figure 4.7 The AutoCAD **Menu Customization** *dialog box enables you to modify pull-down menus according to your needs.*

Pointing devices and digitizing tablets. You can issue commands by selecting an icon on a digitizing tablet with a mouse or puck. AutoCAD provides a plastic tablet overlay to place on your tablet (see Figure 4.8), and you can program commands into it by editing the *Acad.mnu* file. Many AutoCAD users find that using digitizing tablets for routine commands slows them down, so they use the tablet primarily for symbol libraries and special automation routines. Although tablets have their uses, they *are* optional (especially with AutoCAD for Windows), and many design firms don't use them (the larger tablets also take up considerable desk space).

Although a standard Microsoft mouse suffices as an input device for most Windows users, you can purchase special mice with up to forty separate buttons, each of which can be programmed with a speed-saving macro. Most drafters, however, find one-handed keyboard shortcuts faster than mouse picks and rarely work with more than two or three mouse buttons.

If you do use a digitizer pad, you probably also have a multibutton puck and can program shortcuts (in addition to those that AutoCAD provides).

AutoCAD toolbars. Like other Windows programs, AutoCAD increasingly relies on toolbars and menu icons. The toolbar palettes that Release 13 supplies can be modified, reorganized, and expanded to support your

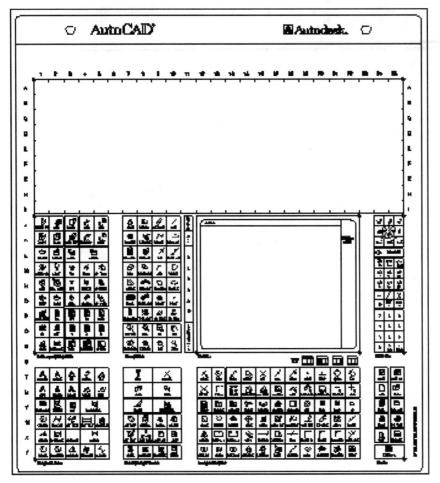

Figure 4.8 The standard AutoCAD tablet format. All cells, including those in the blank upper rectangle, can be customized for additional commands, symbol libraries, or AutoCAD add-ons.

needs. The icons in these toolbars represent both simple command aliases and complex AutoLISP routines. You can use the icons provided by AutoCAD or design your own. In addition, you can create flyouts: icons that when touched display a set of related icons (see Figure 4.9). For example, a customized flyout for door symbol libraries could contain a nest of icons representing blocks for various door types.

Figure 4.9 *Increasingly, AutoCAD commands are accessed by toolbars. Shown here is the **Dimensioning toolbar** including a flyout menu for different radial dimensions tools.*

 Toolbars and flyout menus are fun and easy to use, but don't go overboard with them. They take time to load into the program, and they cover up the drafting screen. They should be hidden when not needed.

Dialog boxes. Like other Windows programs, AutoCAD increasingly relies on dialog boxes (see Figure 4.10). These are initiated with keyboard commands or menu selections beginning with "Dd," such as **Ddmodify** and **Ddlmodes**. Dialog boxes allow you to view and specify many options simultaneously, and they increasingly replace the tedious command line routines that force you to start an entire command routine again if you enter an incorrect specification. If you are skilled in ADS applications and AutoLISP programming (see Step 8), you can design your own dialog boxes by writing ASCII files in Dialog Control Language (DCL); the *AutoCAD Customization Guide* provides useful pointers for effective dialog box design. You can also use the **Dlgcolor** command to change the colors of existing dialog boxes.

CAD wizards input commands using any combination of options. You should be encouraged to customize the command options in any way

Figure 4.10 *A typical AutoCAD dialog box; shown here is the dialog box generated by the **Ddrename** command.*

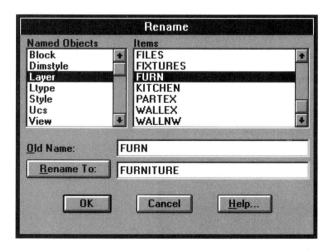

IN-HOUSE CUSTOMIZATION: A GUIDED STRATEGY

that enhances productivity. Some aspects of command input, however, must reflect officewide standards for several reasons.

- Many customization tools and other timesaving routines may be programmed to work by using a particular command input method. A routine initiated by tapping an icon on a digitizing tablet, for example, may not be accessible by a keyboard macro or line command. If one workstation has a digitizing pad or 40-button mouse, and others do not, then the ability to share Scripts, AutoLISP routines, and other tools dependent upon these input devices throughout the office is reduced and certainly requires additional modification to work.

- Some means of input are cheaper, at least in terms of the cost of equipment, than others. If you rely on a standard PC mouse, pull-down menus, and icons, you employ a far cheaper configuration than if you rely on a digitizing tablet and puck, which cost several hundred dollars per workstation (not to mention the time you'll need to spend editing and plotting a customized tablet layout).

- With highly individualized AutoCAD configurations other users will need to spend considerable time getting up to speed. If your workstations are dedicated to specific users, different configurations are not usually an issue. But if people in your office move around a lot (and don't bring their computers with them), or if you hire short-term CAD staff regularly, you will find that they can be more productive if they can use standard AutoCAD commands for routine functions.

There are several ways that your office can develop the optimal balance of office-standard versus individual customization. You can establish office-wide defaults (documented in an office CAD manual) that should not be overwritten; individuals can feel free to perform additional customization as long as it does not override the office defaults. You can establish alternative configurations, which are saved in separate AutoCAD menu files (identified with the .mnu extension). The office standard can be initiated by specifying the Office.mnu; a personalized configuration can be initiated by selecting Bobsmith.mnu, for example.

✔ **Step 5.** **Customizing Your AutoCAD Drawing Environment**

One of the keys to AutoCAD productivity is the establishment of prototypical drawings and libraries. Before these can be produced, however, you must establish defaults for the drawing environment that should be embedded in every drawing file that is created subsequently.

It is essential to understand that most of these defaults will *not* be systemwide and therefore cannot be easily specified through the procedures taken in Step 4. Rather, they are *drawing file specific*. So new drawing files will share customized defaults only if they are cloned from AutoCAD *prototypical* drawing files (the only alternative is to use AutoLISP routines or Scripts, but this requires more time and skill). You can set up new drawing files by using the **Mvsetup** command, but again you have to respecify basic parameters every time you create a new drawing file. If you don't save your customized drawing defaults in a prototypical drawing file, you are back to square one every time you start a new project.

Because the defaults that are programmed in your prototypical drawing file will be cloned into dozens of other office standards, it is essential to get these defaults as final as possible before the prototypical drawing files are developed. It is better to spend several days or weeks on these defaults (and the work covered in Step 6) up front than to spend weeks modifying hundreds of office templates that prove unsatisfactory because you didn't pay enough attention to type styles, pen weights, and other features from the beginning.

The following defaults and options should be reviewed by all who have some say in the drafting process and in the resulting output.

Drawing Units

The first default to be customized is the drawing units, using the **Ddunits** command to open the **Units Control** dialog box (Figure 4.11). AutoCAD defaults to decimal units; you will usually want architectural units (this sets one drawing unit equal to one inch), or possibly metric units. You will also want to select your preferred format for presenting angles and the default direction of the AutoCAD coordinate system, which is normally counterclockwise. For both units and angles, the default level of precision is important (see Chapter 7). You want to avoid both rounding errors and extreme levels of precision.

Drawing Limits

Once the default units are specified, you can set your drawing limits. These limits in effect establish the boundaries of your electronic drafting board. AutoCAD offers no on-screen indicators such as points or a border, but you can determine the current drawing limits by turning the drawing **Grid** on. The grid never exceeds the boundaries set by the drawing limits.

The AutoCAD drawing universe is a 3D positive-negative X,Y,Z coordinate system, just as you used in high-school trigonometry. Theoretically your drawing area can be infinite; for practical drawing purposes, however,

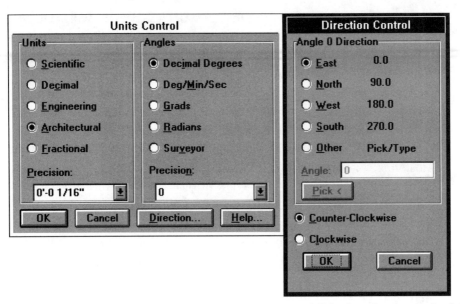

*Figure 4.11 The AutoCAD **Units Control** dialog box, showing the Angle **Direction Control** sub-dialog box*

most CAD users specify drawing limits which relate to traditional drawing sheet or drafting table sizes. As Appendix A shows, your drawing limits are determined not only by the physical dimensions of your sheet but also, because designs are created in "true" scale, by the drawing scale at which your drawing will plot. Therefore, a drawing of an office floor plan to be plotted at 1/8" scale onto a 48" long × 36" wide sheet of vellum will not have upper drawing limits of 48 and 36 inches (assuming the lower limit is 0,0) but more likely of 384 and 288 feet, respectively.

Although AutoCAD operates the same whether you work in the positive or negative quadrants, most offices set up 2D drawings limits to be in the positive X,Y quadrant (you cannot specify limits for the Z-axis) with the lower limit set at the AutoCAD default of 0,0 and the upper limit set at point which represents the upper right corner of the drawing sheet or border. Working in the positive quadrant offers several advantages:

- Most CAD users find positive coordinates easier to work with when specifying points in terms of distances, angles, and relative coordinate points. Many users find working with negative coordinates or combinations of positive and negative ones confusing.

- When the lower limit of 0,0 is used, linking together different drawings or inserting one drawing into another as a block becomes easier to manage. Since 0,0 is widely used by many CAD drafters as both the lower drawing origin and as a typical block insertion point, you will find this origin facilitates exchange of drawings with other users.

- The relationship of your drawing area to your plotting area is more straightforward, especially if you have to move your plotting origin from the default of 0,0 (one of your options for specifying the plotting area is to specify the drawing limits).

A primary purpose of drawing limits is to ensure that the basic drawing area relates to the size of your plotter or printer media. You can go beyond the drawing limits to test blocks or to "doodle" ideas, and you can revise your drawing limits at any time. For routine construction drawing, however, working within the limits is preferable, and you can encourage this by activating the *Limcheck* variable, which produces an error message when you attempt to draw outside the drawing limits. Keeping your limits as small as necessary and drawing within the limits also reduce the extent of the drawing area which needs to be regenerated and this in turn saves time.

Scale Factors

The traditional approach to architectural drafting, when done by hand, is to draw designed objects at a defined *architectural ruler scale*, whether 1/8" = 1'-0" or 3" = 1'-0". All architectural targets, lines, dashes, and other notations are drawn at a *uniform* scale. Many CAD programs reflect this tradition by allowing you to specify a scale before you begin drafting. When you are ready to plot the drawing, you generally plot at fullscale unless you want an enlargement or reduction of your drawing.

AutoCAD does things in reverse. You draw design entities and buildable objects *in full scale* and then plot the drawing at the architectural scale you want it read at on paper. This means that construction notes and symbols, for example, that would plot as 1/2 inch high in a 1/8" scale drawing are actually drawn in the AutoCAD file as 4 feet high.

In order for drawing titles, architectural graphics, and text to read uniformly, however, you must scale them up or down accordingly. You do this with scale factors. Scale factors provide a ratio to scale objects uniformly and quickly for each scale of drawing you plot. The scale factor for a 1/8-inch drawing is 96 and is calculated as follows:

If 1/8" = 1' or 12'
then 1" = 8 x 12" = 96"

The scale factor is 48 for a 1/4" scale drawing, 24 for a 1/2" scale, and so forth (see Appendix A). You specify scale factors through a range of AutoCAD commands, including **Ltscale** to set the spacing of dots and dashes of noncontinuous linetypes; **Dimscale** for establishing the size of dimensions; and **Ddinsert** for inserting of blocks and external drawings at a particular scale.

You may not find AutoCAD's defaults for scale factors graphically acceptable. The dashed linetype at a scale factor of 96, for example, may be too tight or too spread out for your use on a 1/8" scale drawing; always test graphic standards with different scale factors on a given scale plot and refine them to conform to office graphic standards.

One decision that affects your use of scale factors as well as the layout of your plotting drawings is whether to use paper space in addition to model space. *Model Space* is AutoCAD's default working mode for creating, or modeling, designs in "true" scale. *Paper Space*, a relatively new AutoCAD feature, enables you to take objects drawn in Model Space and place them in a range of views or scales on an electronic sheet designed for plotting. Paper Space enables you to present a single design in many different scales and angles; alternatively, you can place multiple details, enlarged plans, and elevations—all designed to plot at different scales—onto a single sheet. In many respects, the relationship between Model Space and Paper Space is analogous to the relationship between a simple text processor and a sophisticated desktop publishing program, where the latter enables you to arrange graphics, ordinary text, and other elements in many different ways, all with a view to creating a more sophisticated document.

You can switch quickly between Model and Paper Space, and you can also plot drawings using only Model Space. However, you will have more control over the output of complex drawings if you use Paper Space, and you will have better results in a shorter amount of time if you plan your customization with Paper Space in mind. Doing so affects two particular aspects of customization. One is the scale factors you use to link Model Space to Paper Space. The other is how you plan prototypical drawing sheets and blocks, as described in Step 5. For additional information on Paper Space and Model Space, refer to Chapter 10, "Layout and Plotting" of the *AutoCAD Release 13 User's Guide*.

Linetypes

All drawn entities in AutoCAD have a linetype that you assign either through the **Linetype** dialog box (Figure 4.12) or the **Ltype** option of the **Layer Control** dialog box. Most entities are drawn with a continuous (solid) linetype, but you will need to specify linetypes to convey column lines, limits of work, soffits, over- and under-counter cabinets, and other standard architectural graphic line conventions.

You can select additional linetypes from the *Acad.lin* file or create your own. Your customized linetypes can create variants in spacing on the basic AutoCAD linetypes, or they can include special shapes, text screens, and

Figure 4.12
The AutoCAD *Select Linetype* dialog box

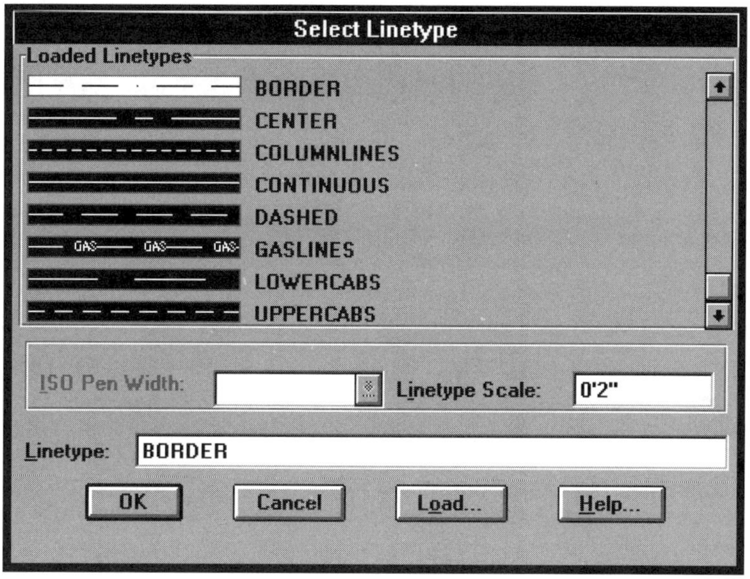

ISO standards. For example, you could create a new linetype that spells out at regular intervals "Limits of Work" to clarify your project's physical scope, or you could draw a linetype that represents wall insulation.

In Release 13, you can set a drawing-wide linetype scale through the **Ltscale** command or an individual linetype scale through the **Celtscale** command.

Since you have the option to create new linetypes or rename existing ones, consider naming your linetypes not for their appearance but rather for their projected use. Thus you can name linetypes for column bubbles, column centerlines, matchlines, and ceiling soffits. This feature helps drafters default to the appropriate linetype more easily.

Layer Standards

Layers are among the most important features of any CAD program of note, and particularly of AutoCAD. They enable you to generate an unlimited number of drawings from a single CAD file.

Layers are electronic versions of the traditional drawing overlay system. To explain further the layers concept, many CAD manuals and instructors use the analogy of clear acetate sheets, each of which contains text or graphics in different colors. Different combinations of these acetate sheets produce different drawing plots. While the acetate analogy is accurate, it touches only the tip of the iceberg in terms of the benefits of layers for production, global changes, and plotting. Unlike traditional overlays, CAD layers enable you to instantaneously change (or edit globally)

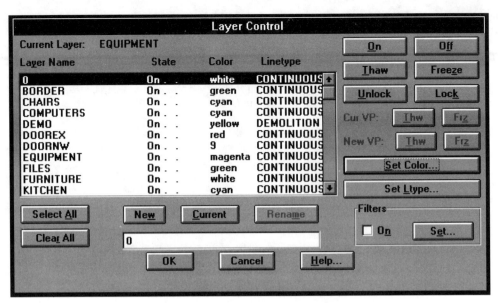

Figure 4.13 The AutoCAD *Layer Control* dialog box

properties of all objects on a particular layer; such properties include colors and pen weights, linetypes, and whether objects can be edited or locked (see Figure 4.13). (You can use other commands, such as the **Filter** command, to select objects by criteria, such as by size or object type, that are independent of the layers they belong to.)

Every design office should issue a set of CAD layer-naming standards. While layer names can be added or modified throughout the course of a project, using office standards minimizes start-up time and learning curves for new projects and assures greater consistency in drawing management and output. They also ensure that AutoLISP routines and other time-saving routines which incorporate layer names will work. Layer standards also have important implications for working with consultants and clients and even for marketing. Therefore, while CAD drafters work with layers on a daily basis, project managers and others must be involved in the establishment and development of an officewide layer-naming system. See Chapter 7 for guidelines on establishing officewide layer-naming standards.

AutoCAD permits you to create *unlimited* layers, each containing individual colors and linetype characteristics. While most design projects rarely require more than twenty layers, complex projects can use fifty or more. Re-creating layers for projects of this magnitude is time-consuming, so CAD-efficient design offices find ways to automate layer creation; this not only saves time but ensures officewide consistency. Your standard layering system should be embedded in a prototypical drawing, but you can also automate layer creation through Scripts and AutoLISP routines (see Step 8).

Many AutoCAD software add-ons load sets of layers upon installation. Carefully review and customize these layers as necessary.

Type Styles

The early versions of AutoCAD and other software offered limited ranges of typefaces, most of which were, well, graphically primitive computer typefaces. Generally, variants of Courier, these typefaces were obviously machine-produced. The early versions of AutoCAD used AutoCAD-specific typefaces that approximated but did not truly match other computer typefaces; AutoCAD typefaces have improved steadily in choice and range. With Release 13 for Windows, you can finally use Windows TrueType and PostScript fonts in addition to AutoCAD's fonts. AutoCAD also provides styles for non-Roman characters (you have to purchase add-ons for many foreign typefaces, however), ANSI-standard geometric symbols, and other figures.

With AutoCAD, you cannot create any text until you define a type style. To do this, you select a font file (similar to a typeface family), then specify height, compression, and other characteristics, and then assign this collection of variables a style name. Between the font files provided with AutoCAD (see the *AutoCAD Relaese 13 User's Guide*, Appendix 3) and the options for modifying them, you choose from an almost unlimited selection of typestyle options. For precisely this reason, you should select a limited number of styles early on; otherwise a motley crew of type styles will appear on your drawings, and the result will not only be unattractive but also confusing to clients and contractors.

Traditionally, many offices have accepted the limited, program-defined type styles of AutoCAD and developed their own type styles that had nothing to do with the typefaces they used for correspondence, specifications, and schedules, or for the firm name and logo. With Release 13, you can use the same Windows TrueType fonts that you use for other Windows applications; thus you can establish a uniform graphic standard for all of your documents.

Is Computer-Generated "Hand Lettering" Appropriate?

One consequence of expanded choices in computer-generated typefaces is, ironically, the availability of fonts that resemble handwriting. AutoCAD offers two such type styles (City and Country Blueprint), and you can access more through Windows TrueType fonts and third-party products. These "hand-lettering" styles have become ubiquitous, but are they appropriate for extensive use? Apart from the philosophical issue of whether it is appropriate to use machines to generate the appearance of handwriting, these typefaces are often difficult to read, particularly when used in very small sizes. Are they really suitable in design and bid documents for anything other than large titles?

As with linetypes, you can name your type styles not only according to their innate characteristics but also according to their intended use. Thus you can name styles of text for drawing titles, client names, detail and elevation titles, and construction notes and legends. This helps drafters select the appropriate type style for different situations.

> Most Windows applications use typefaces that specify type size in typesetter's units, that is, points, where 72 points equals 1 inch. AutoCAD's own type styles use the current CAD drawing's default units. If the units are architectural, the type size is expressed in inches. If you wish to have AutoCAD type styles match the height of typefaces used in other applications and the typical type size is 12 point, for example, you should specify that your AutoCAD text height (adjusted for scale factors) be 12/72" or .1666" (11/64") high; for 10-point type specify 10/72" or .1388"; and so forth.

Dimension Styles

AutoCAD provides you with almost unlimited options for dimension styles. The default variables, however, have not become standard among designers in the A/D community, and you will want to modify them to reflect your preferences. As shown in Figure 4.14, you can define your own dimension styles, including styles for straight lines, arcs, leaders, and dual systems of measurement. You can also add ticks and arrows of your own design and specify a wide range of characteristics from size of extension

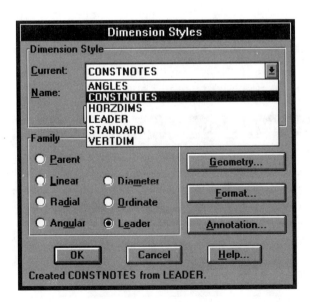

Figure 4.14
The AutoCAD ***Dimension Styles*** *dialog box, showing user-defined custom dimension styles*

lines to tolerance factors. AutoCAD Release 13 also lets you create dimension style families, which bundle together variants of a particular style.

As with linetypes and text styles, you can name dimension styles for their intended use rather than for their features.

Hatches

The AutoCAD **Hatch** command enables you to select an object (or set of objects) and hatch it. AutoCAD offers sixty-seven defined hatch patterns, many of which reflect ANSI- and ISO- standard patterns, and building materials (you can use hatches instead of the **Fill** mode to poché walls). You can also modify in infinite ways the scale and angle of defined hatches, thus producing an endless variety of hatch patterns. Not only can you purchase additional hatches from third-party developers, but you can also program your own.

By presetting hatch palettes you can create officewide standards for conveying materials, shading, and renderings. As with other elements, you can name custom hatches for their intended purposes (and also include the hatch pattern names in any materials legends that appear in your drawings). Documenting these customized hatch standards limits the temptation to modify the hatch scale or angle. Drafters can also use the **Bhatch** *Inherit Properties* option to select the appropriate hatch from an AutoCAD-based materials drawing legend and copy it to another object.

Multiline Styles

AutoCAD Release 13's new **Mline** command enables you to draw up to sixteen parallel lines simultaneously. As shown in Figure 4.15, you can specify and save your combinations of lines, joints, and colors as well as the use and shape of end caps.

For designers, **Mline** functions as a wall-drawing tool. Your customized line definitions can include standard wall types for exterior walls, core areas, and office partitions. Your naming policy should anticipate the use of each style; you can even key the defined multiline names and appearance to your office standard partition types as shown in Figure 4.15.

✔ *Step 6.* *Creating Prototypical AutoCAD Drawings, Templates, and Libraries*

To really make AutoCAD a productive work tool, CAD-efficient design firms establish libraries of templates or prototypical CAD drawing files that contain officewide drafting defaults (as developed in Step 5) and standard blocks and details.

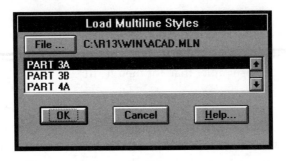

Figure 4.15
The *Load Multiline Styles* dialog box

Most design managers understand the need to establish standard CAD files; CAD standards are simply the electronic version of preprinted drawing vellum and drafting film sheets and notebooks of graphic details and standards. Electronic standards, offer even more features than paper ones. Only familiarity with AutoCAD, however, makes clear the range of these features, which must be customized to make prototypical files truly powerful.

Prototypical Drawings

The customization that results from Step 5 must be stored in a prototypical drawing. All standard drawing sheets, sketch sheets, and other templates are variants of this drawing. Once an initial sheet is designed, the others can be produced as you go along, since each new drawing is a variant of the previous one.

Developing a library of prototypical templates should be done in a systematic way to ensure that all are produced. Appendix A shows a prototypical checklist, which you can consult in developing the prototypical drawing files you need. Such a list also proves helpful when you need to change or update the library. Any computer spreadsheet or database program can generate such a form.

You do not have to—and may not want to—imbed prototypical items such as firm name, client information, and other information in your standard drawing files. You can instead link them using AutoCAD's External Reference (**Xrefs**) capabilities. When the client name or your own office address changes, you can instantaneously update all the project drawings to reflect the new information.

Prototypical production drawings can contain features to further automate drafting and reduce the time spent on repetitive tasks. Typical time-saving features include

1. grids and guidelines. Create a layer with a meaningful name, such as "guidelines," and that contains a base grid for organizing different components of your sheet. This is analogous to using grid paper under a sheet of drafting film or vellum. The grid layer, which is hidden during routine plots, should reflect an office-standard module. Many firms use the size of their detail boxes as the base module. You can go further and set up a smaller module grid. If you subscribe to the ConDoc methodology of drawing production (see Chapter 6), you can replicate their grid (including the alphanumeric coding) in your prototype drawing.*

2. drawing limits that anticipate the sheet size and scales that the drawing will plot at. The limits you specify must reflect not only the actual paper sheet size but also the constraints of your output devices; setting drawing limits also forces you to create margins (see Step 7).

3. a block containing attributes (data fields) for all information that goes into drawing title sheets. You can establish attributes for drawing names and numbers, scale, date, revisions, and so on. Tapping this attributed block—shown in Figure 4.16—encourages CAD users to enter and update title sheet information systematically.

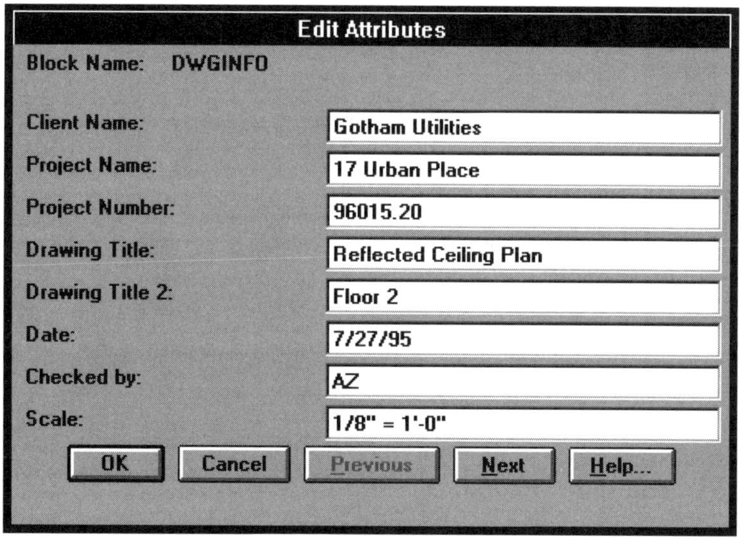

Figure 4.16
In your prototypical drawing sheets, you can use an attributed block to quickly enter and update base drawing sheet information.

---

*Through its AIA Master Systems Program, the American Institute of Architects distributes MASTERKEY for AutoCAD 2.0, a third-party add-on for AutoCAD which applies ConDoc methodology to CAD drawings. Contact AIA Master Systems at 1-(800)-424-5080.

4. a text box for the AutoCAD drawing file name (see Figure 4.17). This can be a component of the aforementioned title information box, an individual text string or separate attributed block. Whichever you use, the drawing file name is an essential component of your drawing plot; without it you'll have trouble determining which CAD file(s) produced which plots.

5. a plot information text line. This text line, which can be contained in an attributed block, should specify the date of your plots from the drawing file. Most firms place this text along the vertical left-hand drawing sheet border, at the binding edge. At the very least, this text should contain not only the plots' date but also their time and purpose as well as information on the set of layers used to produce them. You can write an AutoLISP routine to automatically stamp the time and date (make sure your computer's clock is properly set) before you run plots.

An AutoCAD block is a single object comprised of a collection of drawn entities. Through the **Block** command, these entities are locked together and cannot be separated or edited individually unless you apply the **Explode** command. Blocks can be used an infinite number of times within a single drawing file and if you don't like your first pass at your block design, you can **Redefine** the block, and all incidents of that block within a drawing file will immediately change to reflect your revisions.

In some ways, blocks are analogous to sheets of preprinted symbols and letters that you pull off and stick onto paper drawing sheets. However, AutoCAD blocks are far easier to manipulate than stick-ons; they can be

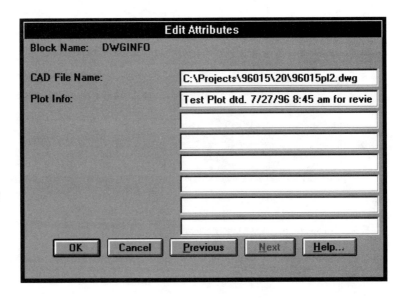

Figure 4.17
Every drawing plot should list the name of the CAD drawing file that produced the plot, the plot date and its purpose. The data attributes shown here belong to the same attributed block shown in Figure 4.16.

copied, resized, mirrored, rotated, and arrayed. They can be objects within one drawing file or they can be separate drawing files which can be inserted in other drawing files. Most AutoCAD symbol libraries are simply collections of drawing files intended to be used as blocks (in other CAD software, the equivalent of blocks may be called cells, symbols, or glossary items).

For serious drawing in AutoCAD, blocks offer many advantages over the use of separate graphic entities or of sets of editable entities which you can define through the **Group** command

- *Blocks are easier to use than bundles of separate entities.* As single objects, blocks are easier to place and move around in drawings. You do not have to worry about inadvertently missing a stray line when you reposition a block, whereas with a collection of many items you can inadvertently miss, erase, or move an object.

- *They resist tampering or cavalier editing or deletion.* Although blocks can be changed, you must explode or redefine them to make changes. This takes more conscious effort than if you can simply tap on an entity to edit it. Blocks protect important items from being carelessly changed. Moreover, even if one incident of a block is exploded or erased, you can still reinsert the block into your drawing as long as AutoCAD can reference the definition of that block.

- *They consume less drawing memory.* AutoCAD "memorizes" all the *incidents* of a block but inventories the block's *components* only once. As the size of your drawing expands to contain more and more information, such memory-saving features become increasingly important.

- *They can be redefined globally.* If you have 150 identical building columns that are drawn separately and not as blocks and you then decide to redesign the columns, you will have to change each one individually or else change one of them and replace the other 149 original items with a copy of the new column. If the column is a block, however, you redefine the block once and all 150 incidences of that block are instantly updated. Using blocks can save hours of drafting and coordination time as changes are made throughout a project.

- *Blocks contain data.* You can attach text and numeric attributes to blocks, as shown in Figures 4.16 and 4.17. Thus you can not only use the same block with different codes, but you can also link block attributes to external text files and databases. You can also query the attributes for specific values, sort them, and do calculations. Simple uses of attributed blocks include keys for partition types, column bubbles, room numbers, power and signal circuits, furniture, equipment, doors, and windows.

> **Should Blocks Be Created on Layer 0?**
>
> Experienced AutoCAD drafters fall into two camps on the question of whether blocks should be created on layer 0, which allows the blocks to assume the characteristics of any other layer they are moved to, or whether they should be created on the layer they are most likely to be placed upon, in which case they may have to be redefined before their layer definitions can be changed. Whatever conclusion is reached in your office on this debate, apply it consistently when forming your block libraries.

Design firms generally organize blocks into block libraries. Typical libraries include but are not limited to (see Appendix A for a sample checklist of items for block libraries):

1. partition plan symbols
2. reflected ceiling plan symbols
3. power and signal plan symbols
4. finish plan symbols
5. plumbing plan symbols and fixtures
6. mechanical symbols and fixtures
7. furniture and equipment symbols and fixtures
8. handicapped person graphics
9. enlarged plans of standard layouts for stairs, toilet rooms, and other spaces that must adhere to many codes and regulations

Blocks can contain other blocks; blocks of blocks are called *nested blocks*. Typical nested blocks include individual drawing symbol blocks contained within a drawing symbols legend block. When inserted into your drawings, these nested blocks provide at once both the legend graphic and block definitions for the legend symbols.

An important part of CAD customization is *documenting* blocks. Many firms print out all their blocks on separate 8 1/2" x 11" preformatted sheets. As shown in Figure 4.18, these sheets can provide more than just the block's name, insertion point, and location on the computer file directories; they also can contain information on the origin of the block's design, projects in which the block was used, and other pertinent information. Such documentation forms an essential reference library for all CAD users and designers who want to work from existing office CAD standards and draw inspiration from past projects.

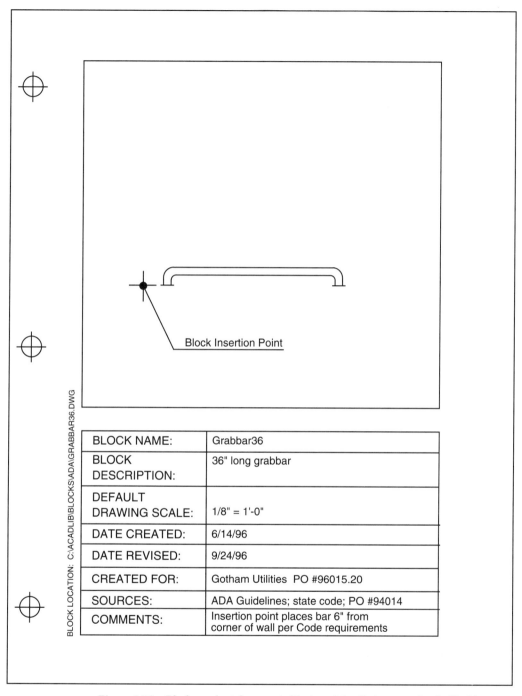

Figure 4.18 Blocks are best documented by templates that summarize the block's contents. This template is itself an AutoCAD drawing.

When designing blocks with text attributes, design the block to work with the maximum number of digits you are likely to use. For a door number symbol block, for example, even if the standard door number is three digits, plan the space so that the symbol block text could fit comfortably even when it is five digits.

Detail Libraries

Every established design office maintains binders of standard office design details. These include everything from partition types to window and door details. Unless you are committed to doing all detailing by hand, you would be wise to build an electronic version of your detail library (see Appendix A for a checklist for assembling a detail library).

Details are a variation on blocks, except that they are designed to be *exploded* (unblocked) and modified according to the project's needs. Some firms store each detail in a separate drawing file; others combine similar details into standard detail sheets, and then they pick and choose what they need for particular projects. Others draw details from scratch; still others purchase and modify third-party add-ons that reflect industry standards.

- As with handling hand-drawn details, organization and documentation of CAD-generated details is key. Document each detail on paper and keep all details in binders in your office library. If your office creates details both by hand and on CAD, merge the masters together, so designers can review all available details.

- Review all standard details carefully. Office standards can contain embarrassing typos, outdated standards, and design faux pas. Quality control should be of particular concern with regard to CAD-generated details, which tend to appear more perfect and final than they actually may be. Ideally, standard details should be carefully reviewed before they are used on a project and then again when a project is completed and ready to be archived. As discussed in Chapter 6, reviewing details should be part of an ongoing evaluation process.

- As with drawing sheet templates, imbed an implicit grid into your standard detail template. Grids help designers to locate and size dimensions consistently, to use text and leaders in a logical format, and to place the designed components properly.

- For details that can be modified in various ways, you can put on a hidden layer additional information to guide drafters in their editing process (not unlike with the AIA MASTERSPEC guidelines for specifications). On a partition detail, for example, you could list typical situations that require the two-hour rated variant.

- Details come in many scales. Various techniques are available for adjusting scale factors; some move between Paper and Model Space; others draw details as separate drawings and insert them or use the **Xref** command to link them to detail sheets. A preferred approach to accommodating different scales should be established before the detail libraries are developed.

Although basic office details should be converted to CAD as early as possible, building a comprehensive detail library is an ongoing process. Working on the detail library is a perfect assignment to give CAD users to occupy them during lulls. It also serves as a useful means of training junior designers and drafters, provided senior staff can review the details for oversights. Completed projects form an automatic source of additional details.

Slide Libraries and Tablet Overlays

As blocks and detail libraries proliferate, locating the item you want can become tedious, even with the best documentation. One way to facilitate finding what you need, without resorting to paper copy, is to create slide libraries using an external AutoCAD utility called *Slidelib.exe*. This program helps you create pull-down menus that present graphic facsimiles of the blocks, details, or other tools you want to use. Typical items featured in slide libraries are doors and windows, electrical symbols, and office seating.

If you use a digitizing tablet, you can use unallocated cells on the tablet for icons that provide access to your block libraries or to open drawings that contain the standards you use the most. You can also purchase third-party software specifically designed to file and retrieve blocks. General computer-based file managers can also help you manage block libraries.

✔ **Step 7.** ***Customizing Your Output Devices***

AutoCAD Release 13 enables you to support twenty-nine output devices, such as printers and plotters. These devices must be defined in the **Config** menu. Device definitions contain not only the machine's manufacturer and model information, but also plotting defaults, such as plot origin; maximum paper size; and plotted scale factor. Since most firms rarely use

> ### Does CAD Impede Design Training?
>
> Some designers worry that highly automated CAD systems will impede training of young designers. If the newer members of the design professions never have to draw a standard architectural graphic, drawing sheet, partition wall type, or column grid from scratch, such people argue, how can they really understand what good drafting and design are? This concern is heightened by the concern that CAD's more uniform output might obscure a drafter's level of competence.
>
> This concern is valid, as truly automated CAD does limit the need for stellar drafting skills, in the same way that spell checkers limit the need to be a first-rate speller. On the other hand, CAD can allow designers to advance more quickly past mundane drafting tasks and to move onto more interesting issues. For example, the time which previously was spent manually drafting the same ceiling detail over and over can now be spent researching alternate products and assemblies. Like all computer-generated work, CAD *input* does require intelligence and professional judgment. And as with any form of work, CAD *output* must be checked and verified.
>
> Although senior members of the design professions may worry about the design skills of future generations, industry surveys show that design firms are increasingly unwilling to provide any form of training for their employees. For such firms, CAD becomes both a substitute for training and a training tool itself.

more than half a dozen different output devices, multiple definitions can be created for each device; this helps to automate plotting procedures for the various permutations of sheets, ink, sheet sizes, and drawing scales you would typically print or plot, as Figure 4.19 shows.

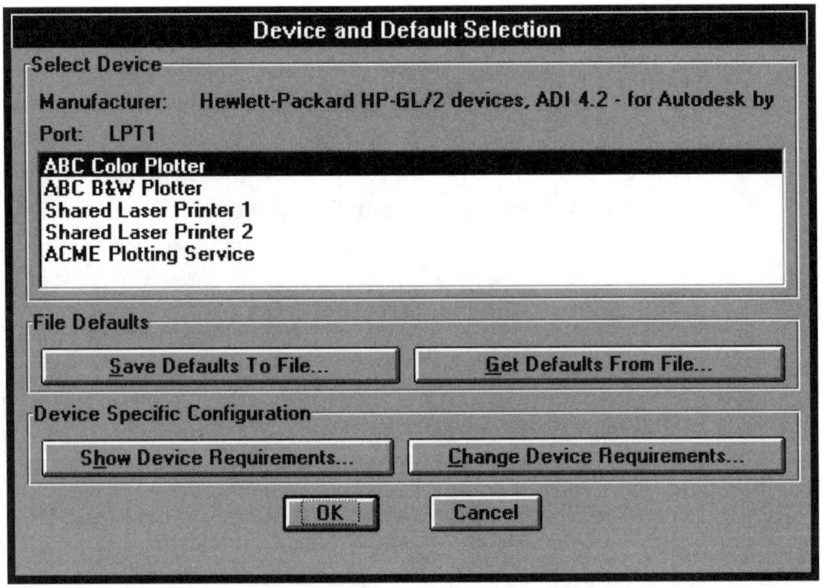

Figure 4.19
The AutoCAD Device and Default Selection box

An essential but sometimes confusing step in output configuration is specifying the pen settings in the **Pen Assignments** subdialog box. As Figure 4.20 shows, each pen (which is assigned a color number that correlates to on-screen colors) can have a specific pen number, linetype, speed, and width (the specification options vary according to the type of printer or plotter you use). Aggressive use of all these options can break down the relationship between what you see on-screen in terms of line characteristics and what plots, but the purpose of these options is not to confuse but rather to permit greater flexibility in producing output. To use pen assignments effectively, you must plan and document pen settings before plotting (see Chapter 8). Working with office-standard pen menus is helpful not only for plotting but also for specifying colors for new drawing layers and objects; CAD users should be encouraged to assign colors not arbitrarily but rather with a view to output.

Specifying paper sizes for your output is also important. The **Paper Size** dialog box (Figure 4.21) contains standard sizes, but paper size is *not* necessarily identical to plotting area size. If your plotting origin relates to a particular point in your drawing file, you can specify a custom user sheet size to automate plotting of that particular file.

In establishing your prototypical drawing sheets with a view to plotting, you must decide whether the 0,0 lower limit represents the actual corner of a plotted sheet of paper or the lower left-hand corner of your plottable area.

Color	Pen No.	Linetype	Speed	Pen Width
1	7	0		0.010
2	7	0		0.025
3	7	0		0.025
4	7	0		0.035
5	7	0		0.025
6	7	0		0.025
7	7	0		0.050
8	3	0		0.025
9	7	0		0.025
10	7	0		0.025

Modify Values — Color: 7 (white); Pen: 7; Ltype: 0; Speed: ; Width: 0.050

Figure 4.20 The **Pen Assignments** *subdialog box*

Figure 4.21
*The **Paper Size** dialog box*

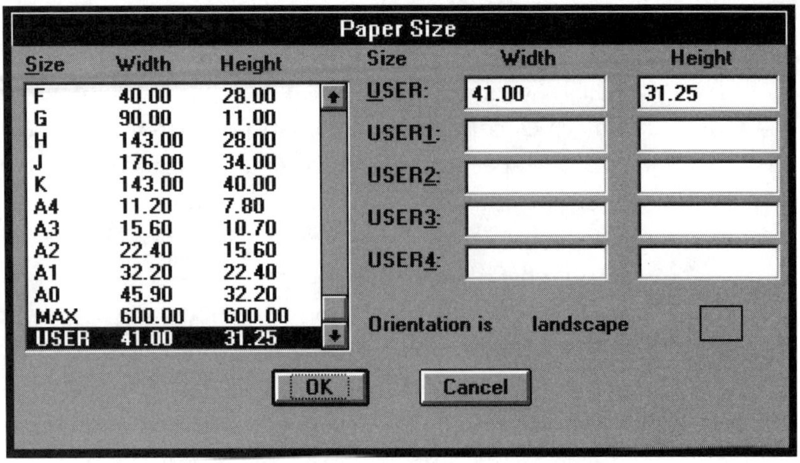

All plot configurations should be thoroughly tested *before* you face a serious plotting crunch. Anytime you upgrade to a new release of AutoCAD, or add or reconfigure an output device, you should schedule time to retest your configurations. If you rely on plotting services, you should have them run (complimentary) test plots as well. While the AutoCAD plot preview options are useful, they are not the same as hard copy. You really don't want to miss an important project deadline because no one had the opportunity to run test plots and check for such basics as pen widths and plotting limits.

Save all your output configuration definitions to individual ASCII text files with a *.pcp* (plot configuration) extension; if the defaults are changed or overwritten, you can retrieve them by using the **Get Defaults from File** option in the **Device and Default Selection** subdialog box. You can also clone and edit *.pcp* files to create variations on your standard configurations. When you send drawing or plot files to plotting services or consultants, be sure to include a copy of the pertinent *.pcp* files for their reference.

Along with AutoCAD itself, you can also customize the software that resides in the plotter or printer. Many high-powered plotters, for example, can be programmed to maximize speed, quality, queuing, and other output characteristics. Any adjustments to output devices must be thoroughly tested, and for software that plots or prints to them—not merely for AutoCAD.

✔ **Step 8.** **Customizing Special AutoCAD Applications and Links to Other Software**

Steps 1 through 7 of your customization process are typical steps all firms must take just to bring AutoCAD up to a reasonably productive level. If you wish to use AutoCAD's more powerful capabilities, plan to customize those as well.

Scripts

Scripts are text files that summarize a series of keyboard-generated commands. They can instantly complete keyboard routines that otherwise could take you many minutes to implement. In addition to saving time, Scripts help force consistency in various operations and limit the likelihood that some essential command will be overlooked by an inexperienced or harried CAD user.

Scripts, which are used in many other software packages and which function like macros, can be used for many purposes:

- to set up a new drawing file,
- to turn on specific combinations of layers and views for editing and plotting,
- to set up files for plotting, and
- to generate slide shows of CAD drawings.

Script files are identified by their .scr file extension, and they can be edited in any text processor that reads ASCII files. Every office should develop a library of standard Script files to automate frequently performed functions.

AutoLISP Routines

AutoLISP is not a speech impediment but a programming language developed by Autodesk to run in AutoCAD. It is an AutoCAD-specific variant of the Common LISP programming development language. Unlike many compiled programming languages, AutoLISP is relatively easy to learn. Even without formal training, you can easily edit existing AutoLISP routines with any word processor that reads and writes ASCII text files.

Conceptually, AutoLISP routines resemble the Scripts and macros that are used in other software programs (see Figure 4.22). AutoLISP routines, however, are more powerful than Scripts; they automate commands that cannot be accessed from the keyboard, create dialog boxes, perform complex calculations; and produce drawing results that otherwise might take

Figure 4.22 A portion of an AutoLISP file—here the AutoCAD Attredef.lsp file—viewed in a text editor. Note that the header of this file contains a description of the routine's purpose and results. You can add the author's name, copyright information, and other pertinent data.

```
;;; DESCRIPTION
;;;
;;; This program allows you to redefine a Block and update the
;;; Attributes associated with any previous insertions of that Block.
;;; All new Attributes are added to the old Blocks and given their
;;; default values. All old Attributes with equal tag values to the new
;;; Attributes are redefined but retain their old value. And all old
;;; Attributes not included in the new Block are deleted.
;;;
;;; Note that if handles are enabled, new handles will be assigned to
;;; each redefined block.
;;;
;;; -----------------------------------------------------------------;
;;;
;;; Oldatts sets "old_al" (OLD_Attribute_List) to the list of old Attributes
;;; for each Block. The list does not include constant Attributes.
;;;
(defun oldatts (/ e_name e_list cont)
   (setq oa_ctr 0
         cont T
         e_name b1
   )
   (while cont
      (if (setq e_name (entnext e_name))
         (progn
            (setq e_list (entget e_name))
```

many key commands and mouse movements. Many basic commands in AutoCAD are actually AutoLISP routines.

When you launch AutoCAD, you automatically load certain AutoLISP routines. You can load, or unload, other routines through the **Appload** function; additional AutoLISP routines can be found in the AutoCAD sample files and (if you have a CD-ROM drive) in the bonus CD-ROM that comes with AutoCAD software. Many third-party software packages written for AutoCAD use AutoLISP routines, and others often appear in *CADalyst*, *CADENCE*, and other AutoCAD-oriented magazines. AutoCAD provides you with the tools and documentation to enable you to write your own routines; you can also edit AutoLISP routines that come with AutoCAD software or third-party applications.

Sample AutoLISP routines that you can write or acquire generate dialog boxes to set up new drawing files, to load office-standard layers, or to plot an entire set of construction floor plans from one file. A very exciting application of AutoLISP routines is for interactive design routines, as discussed in Chapter 6. You can use AutoLISP routines to design everything from accessible toilets to ornately patterned brick building facades. You can also automate routine drafting functions, such as pochéing walls, adding headers and numbers to doors, and performing area calculations.

If you seek more programming power or connections to other Windows software, such as word-processing, spreadsheet, or database programs, you may wish to write routines using the AutoCAD Development System (ADS) rather than as AutoLISP routines. ADS requires some knowledge of C language or other high-level computer languages. ADS works in AutoCAD much the same way that AutoLISP routines do, but ADS routines require more time and programming skills to develop. As AutoCAD becomes even more Windows-based, you should anticipate the introduction of more powerful programming tools which link CAD files to other applications.

Color

If you have the output devices to take advantage of AutoCAD's color capabilities, you can create beautiful renderings and useful schematic diagrams (see Chapter 7). Choosing among 256 colors (more when rendering in 3D AutoCAD) can be time-consuming.

Most design offices have already developed signature color palettes for work done with pencils and markers. These can be easily mimicked in your AutoCAD drawings. The quickest way to develop palettes is to plot out the AutoCAD file *chroma.dwg*, included in your AutoCAD **Sample** File directory and select palettes from that. You can program your plotting configurations to reflect your color standards by saving the defaults for pen colors to a *.pcp* file (see Step 7).

Programmable Dialog Boxes

You can customize your working AutoCAD interface and facilitate the use of customized routines by programming your own dialog boxes. You can do so by writing ASCII files in AutoCAD's Dialog Control Language (DCL); you must have AutoLISP or ADS programming skills to use DCL effectively. Guidelines for creating and formatting dialog boxes are given in the *AutoCAD Release 13 Customization Guide*.

3D Parameters

If you use AutoCAD's built-in 3D capabilities routinely, you can save time by setting up standard 3D templates. Although every project has special requirements and 3D offers unlimited options, you can build in some time savings by customizing the following variables:

- view layouts to show typical viewports for both drawing and plotting
- User Coordinate System (UCS) and viewport presets, to set the viewing angles
- defaults and palettes for rendering with colors, surface materials, light, shade, and camera angles
- Scripts or AutoLISP routines for automating plots and slide shows of 3D images

Database and OLE Links

AutoCAD offers several ways to connect entities in drawings to external databases (see Chapters 6 and 10) so that you can link drawn entities to data for reports as well as help coordinate drawings with specifications, schedules, and other construction documents.

Whether you use the AutoCAD SQL (Structured Query Language) feature to connect drawing files to database software or OLE (Object Linking and Embedding) to establish links to Microsoft Word, Microsoft Excel, and other Windows software, setting up databases takes time, so it is worthwhile to develop standard template linkages for typical applications.

With OLE you can imbed into your drawings spreadsheets that you have created in Excel or Lotus 1-2-3, or you can import construction notes written in Word or WordPerfect. To do so effectively, however, requires that you set up the parameters and graphic standards in *both* AutoCAD and the other software applications. Once you have accomplished the setup, you have a complete set of linked prototypical construction documents.

File Conversion Routines

If you regularly import and export AutoCAD drawing files to and from other file formats (see Chapter 10), you can save time by automating the process. For example, you can write an AutoLISP routine that sets up your AutoCAD drawing files so that your consultants get only the information they need and in the format that helps them the most. If you have clients who request *.dxf* files that use their layer names, you can write a Script or AutoLISP routine to automate the process of renaming your CAD layers to conform with the clients'.

Protecting Your Customization

Given the time and money you will spend customizing your AutoCAD system, the last thing you want to risk is losing your work or having someone override your defaults by mistake. So before the customized configuration is authorized for general distribution, protect it by taking the following steps:

- Back up your customized computer configuration electronically. Ideally, you should make two backups of it on tape or CD-ROM. One backup copy should be in the office for easy retrieval; the other should be stored off-site, ideally in a safe-deposit box or other fire- and moisture-resistant storage place.

- Document on paper what you have done (this documentation will be part of your office CAD manual, as discussed in Chapter 9). This documentation should include both written guidelines for the work as well as sample printouts of prototypical drawing sheets, blocks, details, and the digitizing template. You can also use your word processor to print out all ASCII-based support files that you have created or modified; these include files with *.mnu*, *.pcp*, *.lsp*, and *.scr* extensions.

Implementing Your Customization

Your customized configuration normally is programmed on a single AutoCAD workstation. Once complete, the customized setup will need to be transferred to other CAD workstations (this is not necessary if you run a network version of AutoCAD with the software installed on a file server). Installing or reinstalling and configuring the software is tedious and may interrupt routine computer operations. Several techniques can be used to speed up the process:

- You can purchase software (originally designed to link laptop computers to regular workstations) to copy installations from one machine to another (all computers must be relatively similar for such software to work).

- You can carefully select all the files you have created and modified and replace existing ones in other machines with the new ones. You must be careful to edit the DOS and Windows *Config.sys* and *Acad.bat* files when you do this.

- If you use a network, your network software may enable you to copy or mirror a hard drive from one machine onto another.

> **Who Should Perform CAD Customization?**
>
> Contrary to what some designers might imagine, AutoCAD does not have to be customized by a CAD wizard. Most of the procedures discussed in this chapter can be followed by anyone with a basic knowledge of AutoCAD and the ability and willingness to expand on that knowledge as required. And CAD wizardry often indicates a general fascination with CAD that can interfere with the more focused requirements of customization.
>
> Other skills may be more important than high-powered computer skills, including knowledge of architectural drafting, construction, and the general design process; a good sense of architectural graphics; the ability to analyze and synthesize different opinions about customization issues; aggressiveness in pursuing answers to technical mysteries; good communication and organizational skills; and the ability to complete projects.

Feedback and Subsequent Revisions

A carefully executed, comprehensive customization will work for you for many years. Even the best customization, however, is only ever 95 percent complete. No sooner are your new standards issued than someone changes his or her mind, a contractor in the field has a comment, or a new version of AutoCAD appears.

So a normal part of customization is receiving feedback and incorporating it through periodic revisions. To make sure that CAD managers or drafters don't get continually caught up in making minor changes, schedule periodic reviews—perhaps every six months at minimum—to implement whatever changes are necessary. For major projects, the project close-out phase, when both electronic and paper drawings must be archived, often is an opportune time to review and revise your customization.

5 Making AutoCAD a Team Player: The CAD-Confident Project Manager

You can hire the most highly skilled designers and the fastest drafters to carry out your projects, but despite such talent, your projects may never reach their full potential—whether from a design or a management perspective—unless you also provide skilled project management. A project's success can ultimately depend on its project managers. The larger and more complex a project team is, the more essential good project management is.

As shown in Figure 5.1, the project manager plays the central role in the evolution of the project. Besides monitoring fees and schedules, the project manager coordinates all the parties involved in the project: clients, contractors, consultants, regulatory entities, and, of course, the design team. The project manager not only manages people, delegates tasks, and keeps an eye on the fee but also protects the project's design quality; when the project is efficiently run, the design team can spend more time on design issues and less time on troubleshooting. As design becomes the province of more and more specialists and regulators, the project manager's role in design projects becomes more complex and essential.

Project management is as important for CAD-based projects as for traditionally drafted jobs, and the general scope of responsibilities remains

Figure 5.1
The project manager serves as the clearinghouse and coordinator for design projects.

the same. The technicalities change, however, mainly because the project phase, sequence, and time requirements shift. In addition, when used effectively, CAD itself becomes a team member and therefore has to be managed, although not in quite the same fashion as one manages humans.

Managing CAD projects often perplexes even the most experienced and skilled project managers. A primary reason is that the average project manager today completed his/her design education before CAD was integrated into standard design school curricula. Most project managers in design firms worked their way up through the ranks, accumulating experience in traditional design methods and drawing production; few as yet are experienced CAD users. Thus many of them must manage a process they do not fully comprehend. This situation persists in part because CAD-proficient employees often get trapped in CAD production (see Chapter 9) and therefore do not get exposure to the range of design experience that would make them eligible to become project managers.

This project management dilemma can be mitigated to a large degree if project managers, and their superiors, take a *proactive* approach to CAD and examine how it affects the following management tasks:

- scheduling CAD-based projects
- staffing projects
- refining the project scope
- delegating tasks
- managing team communications and coordination
- implementing quality control
- scheduling output

Is CAD Faster?

When project managers schedule projects on CAD, their primary concern often seems to be whether CAD is *faster*. To compare CAD with hand drafting in terms of the speed of output, however, is counterproductive and misses the point.

The two drafting processes are not strictly comparable. Hand drafting often *appears* faster initially, especially when only lines and shapes are being drawn. Proper CAD drafting requires giving those entities some *intelligence*, such as actual dimensions or attributes; moreover, the initial input should be as accurate as possible so that revisions later on can be made quickly. As shown in Figure 5.2, the CAD process requires more input time initially but less time overall when the project is looked at as a whole.

Ultimately a well-configured CAD system produces clearer and more consistent drawings, provides a greater degree of accuracy, and improves project coordination. Thus, when debating whether to use CAD, perhaps the real issue you should address is whether CAD can do the job *better*, not merely faster.

Scheduling CAD-based Projects

Experienced project managers and operations managers in design firms usually have a good grasp of scheduling *hand*-drafted projects because of years of experience as well as their own training and capabilities in drafting. With AutoCAD and other computer drafting programs, however, project scheduling can appear mysterious. Only with experience and some basic understanding of CAD can project managers schedule most routine

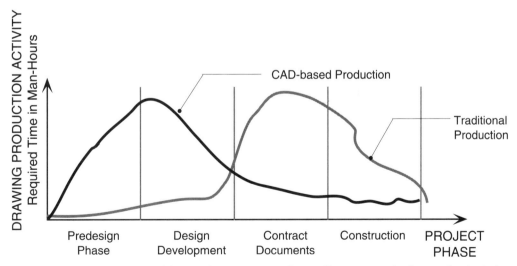

Figure 5.2 CAD requires more person-hours of input at a project's onset, but in the later phases and overall, CAD takes less time. The more complex and large a project's scope, the greater the potential benefits of CAD.

CAD projects with the same confidence they would apply to hand-drafted work. Experience also provides understanding of the variables that can affect a project's schedule.

Recent History

For firms that are already CAD-proficient, the best scheduling guideline is experience. Reviewing recent projects of similar size, scope, and billable time can give you a base schedule to work with.

Don't rely solely on the project documentation, however. Talk to the project team members, especially the CAD users. Find out whether they produced their CAD documents efficiently and at an acceptable level of quality; whether the project schedule proved realistic, and whether deadlines were met without undue overtime or stress. Did the team produce CAD documents that you could clone for new projects? Since the project's completion, has the office CAD system undergone changes that could affect the project schedule positively, such as the installation of faster computers or plotting devices? Alternatively, is anything currently happening to the office computer system, such as the installation of software upgrades, that could temporarily slow work down?

Never assume that because your CAD-based projects have met increasingly tighter project schedules that you can continue to accelerate schedules infinitely. CAD does speed up work but only to a limit; moreover, its speed frequently depends on the availability of information and decisions, which ultimately hinge on human participation and therefore subject to finite possibilities for acceleration. Moreover, unpleasantly tight project schedules limit the opportunity to perform the fun tasks of design, thereby raising the issue of why one ever chose to practice in the design professions at all.

Availability of CAD Skills

A design team that possesses a high level of AutoCAD proficiency definitely can meet tighter project schedules than a team with limited CAD skills. General AutoCAD skills, however, are not quite as valuable as familiarity with *your* customized office CAD configuration. New hires who are well trained in generic AutoCAD normally take at least *one day* minimum to learn basics such as file directory organization, plotting routines, and so forth. So if you have to bring in a new member to the team, budget sufficient time to train him or her on your system.

Computer/Drafter Ratio

CAD project scheduling is simplified if the ratio of CAD drafters to computers is 1:1. Then you only need to schedule the drafters (assuming some slack time for computer downtime); the computers come attached.

If your office has fewer CAD stations than CAD users, however, you have to schedule each CAD station as a separate resource. You must allow for slack time owing to computer bottlenecks. You may have to schedule work in shifts. You also need to have a clear understanding of the skills and relative rate of productivity of each team member and of the speed and power of each computer. Working in shifts can be productive for teams whose CAD users also fulfill non-CAD responsibilities and when the project schedule is flexible and fairly generous; it is *not* recommended for fast-track, tightly scheduled, CAD-intensive projects that are prone to sudden changes in deadlines and scope, since thereby you subject your CAD drafters to continual stress.

If you can't guarantee a 1:1 ratio of CAD stations to CAD users, then you should assume that CAD-based drafting time requirements will increase by at least 25 percent.

Availability of CAD Standards and a Customized CAD System

If your office has customized its CAD system as described in Chapter 4, then your project will move along faster than if you have either no CAD standards to work from or only unsatisfactory ones. A well-planned, customized CAD system that is already actively in use can reduce production time by at least 20 to 30 percent.

Modification of Office Standards

Even if you have a workable customized CAD system, you may work on projects for which you have to modify or even ignore your office standards. This often occurs with government and corporate projects where the client requires you to adhere to its own CAD standards for layers, blocks, and colors. If you provide external services to facilities management departments, expect to be asked to abide by their standards as a matter of course. With long-term contracts during which you may have the opportunity to customize alternate CAD standards, you will eventually get up to speed; in the short run, however, adjusting to new CAD standards will slow you down.

Hybrid Production

Although the tendency is declining, many firms produce drawing sets using *both* CAD and hand drafting. For example, floor plates might be drawn on CAD, and then notes and dimensions would be added by hand. This approach was justified with early versions of CAD that offered limited text and dimensioning capabilities; now, however, you can enter most information as quickly in CAD as by hand.

Although hybrid production may initially seem like a time-saver, it usually ends up taking more time overall. CAD drafters often have to duplicate handwork, especially at a project close-out when they have to generate CAD "as-builts." Hybrid approaches also create problems when the CAD portion has to be substantially reworked and replotted; for every new plot, the hand drafter has to be able to replicate the handwork done on the previous plot. The benefits of CAD as a coordination and quality-control tool thus may be reduced. Moreover, you must insure that both CAD and hand drafting skills are available to make the revisions. If you must rely on hybrid production, try to limit handwork to specifically designated tasks and completely separate drawing sheets, such as details or building elevations and sections; the designer can draw these on CAD-generated (reverse-read) sheets bearing the project title and border.

If you must produce some drawings or portions of drawings without AutoCAD, consider various means to get the information into AutoCAD as efficiently as possible when time permits. If necessary, you can scan and digitize images, or cut and paste and import various files from other software (see Chapter 10).

Precedents for Similar CAD Projects

If your firm's specialty is restaurants and your new project is a trendy bistro, you will be able to produce the design faster than if you are embarking on your first venture into health care facilities. Similarly, if you've traditionally specialized in small single-floor buildings or tenant fit ups, you will find that sort of project moves along faster than when you're tackling your first forty-four-story office building. Major differences in project type, scope, and complexity will require you to modify your existing CAD standards; for example, you may have to create symbols for new types of equipment and fixtures, revise your layering system, and even change your drawing sheet size and format. But you can modify an existing CAD customized setup faster than you can create a new setup from scratch. And what you learn from new types of projects can, in turn, be worked into templates for use on future projects of a similar nature.

Project Complexity

The more complex a project, the longer a schedule it usually requires. Intelligent use of CAD, however, allows you to deal with these complexities faster than you could with traditional production methods.

Typical features that add to a project's complexities include

- graphic standards and drawing formats *not* based on your office standards
- drawing sets containing unusual sets of layers or large quantities of layers
- constant and ongoing changes in project scope
- multiple building sites or floor plates with few common elements
- project sites or building floor plates which are too large to fit onto a single drawing sheet
- few opportunities for standardization and use of blocks
- few opportunities for variations on standard blocks and elements
- lack of an existing and accurate CAD file for the base site or building
- major discrepancies between drawings or CAD files and field conditions
- projects requiring coordination of many consultants and specialists
- large drawing sets
- CAD drawings that take longer to open, regenerate, and plot because of their large file size
- CAD drawings that, either because of their content or because of your plotting capabilities, take exceptionally long to plot

Computer Speed

Although an ultrafast computer cannot speed up an inept CAD drafter, CAD *meisters* definitely benefit from working with faster machines. AutoCAD Release 13 on a 386 PC with 16MB of RAM will move almost half as fast as a Pentium procesor with 32MB of RAM and a graphics accelerator. With CAD, a high-powered computer is a necessity, not a luxury. The more additional computer power you can give your CAD drafters, the faster they will produce.

Plotting Capabilities

Your production turnaround time is directly tied to your ability to produce plots within a given time. If you use a fast inkjet or LED plotter or send plot files to a nearby and reliable plotting service, your plotting time requirements will be minimal compared to the time you would spend using an older pen plotter or working with services that have to messenger drawings from across town to you at rush hour. See Chapter 8 for specifics on scheduling plots.

Plotting Quality

If you use a state-of-the-art laser or inkjet plotter, you can run plots with your choice of media and quality of pen lines. Most of the newest plotters permit you to select print quality ranging from draft to final to enhanced quality and to print on either vellum, bond, or film. An enhanced-quality plot on film, however, can take almost twice as long as a draft-quality plot on vellum. If *draft* quality plots will suffice for certain drawing sets, you will save plotting time, not to mention ink and plotter wear and tear, if you accept this quality of output.

Plotting Schedules

With the recent improvements that both AutoCAD and manufacturers of plotters have made, CAD plotting has become more reliable, faster, and more versatile. Nonetheless, writing plot files, setting up a plotter with the proper paper and pens or ink, and monitoring its output takes time. This is time that otherwise might be spent doing drafting, design, or specification tasks. Careful scheduling of plots or making do with a quick 8 1/2" × 11" laser plot instead of full-size drawing sheets reduces drafting interruptions and speeds up production overall.

Staffing Projects

Part of project scheduling is project staffing. You can consider a well-configured CAD station to be almost the equivalent of an additional team member. Knowing how and when CAD can substitute for additional human effort is essential. The more your office "mainstreams" CAD into its overall production, to the point where CAD's role is "transparent," and blurs the distinction between drafting and design, however, the less time you will need to spend scheduling CAD alongside other staff.

Project Management Software

An invaluable aid in scheduling projects is a quality project management software package. These products enable you to track all your resources (including both people and machines), schedule major and minor deadlines, keep track of project fees and expenses, and build in time for schedule changes, employee vacations and illnesses, and bad weather. You can use project management software to divide tasks into dependent and independent variables, to test different project scenarios, and to input schedule changes as they occur. The software tells you when you have too much or too little staff and when you have gone over budget. Typically, it also generates a range of reports and graphics, including budgets, schedules, task lists, calendars, and Gantt and PERT charts.

Some project managers, such as Microsoft Project and Claris's MacProject Pro, are generic packages, designed to work in all types of industries and situations. Others, such as SureTrak Project Manager by Primavera Systems, have defaults that respect A/E/C conventions and can even create charts within AutoCAD, although they do not require AutoCAD to run. Still others, such as AutoProject by Research Engineers, Inc., run inside AutoCAD; these packages not only use AutoCAD to generate graphics but may also address specific issues of CAD production, such as plotting time and splitting work between 2D and 3D.

As with other computer tools, project management software is most productive if you develop templates for use with typical projects. Such templates can speed up project setup by providing defaults for the formats and contents of office standard reports and graphs. You can key in standard project phases and design tasks, as well as staff names and standard billable rates. As with all templates, you must carefully review the underlying assumptions, revise them to reflect individual project requirements, and see if any data—such as wages and billable rates—in the template file is obsolete.

Even if they have no internal inclination to use project management software, design firms increasingly find that their clients and contractors use them. Particularly on larger, more complex projects, designers may be required to use project management software or to provide data in a format which other parties can incorporate into their own electronic project managers. Thus, it behooves designers to at least gain some familiarity with this kind of software.

As with hand drawing, effective CAD production is more a question of the *quality* of the drafters than the *quantity*. A single well-trained, strategically oriented, and design-sensitive CAD user will prove more productive than four slow CAD novices. Moreover, skilled CAD users can effectively combine design and drafting functions. Thus, with the right CAD staff, you should be able to reduce your production staff while improving project quality.

Many small to medium-sized design projects can be completed comfortably by *one* CAD drafter, who ideally would see the project from concept through to completion. This arrangement is easier to manage and also provides continuity for the drafter. Given an opportunity to "own" their work, CAD drafters, like designers in general, often are more alert to what the project requires and work more intelligently.

While a single drafter is often highly effective, you must have backup for when the drafter is ill, on vacation, or otherwise unavailable. Factors that help a CAD "understudy" step in quickly include adherence to office CAD standards, a wide knowledge of CAD among design team members, and good documentation of what has been accomplished in CAD for a particular project to date.

If multiple drafters are scheduled to work on one project, they prove most effective if their tasks are apportioned in certain ways, as will be described further. Too many "hands in the pot" are counterproductive and frustrating, and if the organization and separation of work are not well delineated and documented, multiple CAD users can actually duplicate or undo each other's work.

Refining the Project Scope

At the time when a client awards you a design contract, you usually have an understanding of the project scope in the broad sense. Fine-tuning the project schedule involves visualizing the actual products—drawings, specifications, and other output—that your team will produce and then communicating those requirements to the team. The sooner the team has an understanding of the project requirements, the sooner it can be productive. With CAD, you can move ahead on certain standards while the project is still in the predesign phase (see Figure 6.2), thus reducing the time needed for production. CAD users can be expected to move ahead on certain portions of the project even while some design decisions are still evolving. For example, even while you are refining the partition plan, the drafter can scope out partition types, set up preliminary sheets of details and elevations, and do a first pass at the reflected ceiling plan.

For many project managers, developing a cartoon set or miniature mock-up of the projected set is a classic approach. CAD allows you to generate cartoon sets on standard letter-size paper very quickly. These sets can include all the officewide CAD standards, ranging from sheet formats to standard details, which might prove relevant to the project; with an ordinary office laser printer, these sets are highly legible. The cartoon sets enable you to delineate the project scope quickly by editing existing templates rather than conceiving items from scratch. The CAD drafter can then set up all the CAD drawing files while you consider other project issues.

Another useful tool is the drawing matrix, shown in Figure 5.3. This matrix, which can be easily produced in a spreadsheet application such as Microsoft Excel (used here) or in a project management software program, previews the drawing contents, the CAD templates that produced them, and other output. It tells the design team how the CAD portion of the project will be produced, down to the exact file names, and what prototypical CAD files will be used. The matrix can be augmented by including

ABC Design Inc.
Gotham Place Renovation
Project 9643.00   Date: 2/22/96

Sheet No.	Sheet Name	CAD Prototype File	CAD Filename	Script File	Plotted Scale	Target % Check Set Completion		Comments
						4/14/96 50% Check Set	6/16/96 90% Check Set	
	Cover Sheet	Cover.Dwg	9643Covr.Dwg	N/A	N/A	80%	85%	
A001	Symbols & Abbreviations	A001.Dwg	9643A001.Dwg	N/A	N/A	95%	95%	
D101	Demolition Plan - Floor 1	Plan.Dwg	9643PL1.Dwg	Demo.Scr	1/8"=1'-0"	65%	95%	
D102	Demolition Plan - Floor 2	Plan.Dwg	9643PL2.Dwg	Demo.Scr	1/8"=1'-0"	65%	95%	
A101	Partition Plan - Floor 1	Plan.Dwg	9643PL1.Dwg	Part.Scr	1/8"=1'-0"	55%	90%	
A102	Partition Plan - Floor 2	Plan.Dwg	9643PL2.Dwg	Part.Scr	1/8"=1'-0"	55%	90%	
A201	Reflected Ceiling Plan - Floor 1	Plan.Dwg	9643PL1.Dwg	RCP.Scr	1/8"=1'-0"	50%	90%	
A202	Reflected Ceiling Plan - Floor 2	Plan.Dwg	9643PL2.Dwg	RCP.Scr	1/8"=1'-0"	50%	90%	
A301	Power & Signal Plan - Floor 1	Plan.Dwg	9643PL1.Dwg	P&S.Scr	1/8"=1'-0"	66%	90%	
A302	Power & Signal Plan - Floor 2	Plan.Dwg	9643PL2.Dwg	P&S.Scr	1/8"=1'-0"	66%	90%	
A401	Finish Plan - Floor 1	Plan.Dwg	9643PL1.Dwg	Fin.Scr	1/8"=1'-0"	66%	90%	
A402	Finish Plan - Floor 2	Plan.Dwg	9643PL2.Dwg	Fin.Scr	1/8"=1'-0"	66%	90%	
A701	Enlarged Plans - Floor 1	A701.Dwg	9643PL1.Dwg	N/A	1/8"=1'-0"	66%	90%	X-Ref base from 9643PL1.Dwg
A702	Enlarged Plans - Floor 2	A701.Dwg	9643PL2.Dwg	N/A	1/8"=1'-0"	66%	90%	X-Ref base from 9643PL2.Dwg
A801	Elevations	A801.Dwg	9643A801.Dwg	N/A	3/8"=1'-0"	55%	90%	
A802	Elevations	A802.Dwg	9643A802.Dwg	N/A	3/8"=1'-0"	55%	90%	
A901	Details	Details.Dwg	9643A901.Dwg	N/A	Varies	60%	90%	
A902	Details	Details.Dwg	9643A902.Dwg	N/A	Varies	50%	90%	
A903	Details	Details.Dwg	9643A903.Dwg	N/A	Varies	30%	90%	May not be required; by hand if necessary
FP101	Furniture & Equipment Plan - Floor 1	Plan.Dwg	9643PL1.Dwg	FURN.Scr	1/8"=1'-0"; 1/4"=1'-0"	45%	85%	Issued Separately from General Bid Documents
FP102	Furniture & Equipment Plan - Floor 2	Plan.Dwg	9643PL2.Dwg	FURN.Scr	1/8"=1'-0"; 1/4"=1'-0"	45%	85%	Issued Separately from General Bid Documents

Figure 5.3 *A CAD drawing matrix*

information such as which drafter is assigned responsibility for a particular drawing, and various contingencies that could affect the project. You can also include drawings to be provided by your consultants. If your office has already created a prototype of such a matrix, you do not need to depend heavily on CAD users to do a first pass at developing a matrix. If such a matrix is kept updated throughout a project, it can prove invaluable not only for planning but also for ongoing team management and communications.

Delegating Tasks

The larger and more complex your project team, the more important task delegation becomes. This is just as true with CAD as with hand drafting. You need to assign tasks to the team members who can perform them most efficiently; at the same time, you should delegate tasks so that young drafters can be exposed to new learning experiences.

Effective task delegation requires recognizing that people work best when they can take responsibility for their own work, receive feedback, and understand the project in its overall context. Ineffective task delegation can diminish an employee's morale and willingness to be responsible while increasing the likelihood that things will fall through the cracks.

CAD task delegation depends on several factors. One is the ratio of CAD machines and CAD-proficient drafter/designers to the overall staff; if CAD has permeated your office, you can spend less time dividing work according to employees' suitability for CAD and more time focusing on matching tasks to the abilities and interests of particular people. If CAD skills and equipment are at a premium in your office, you will need to spend more time determining if a task is at all suitable for CAD.

Another key factor in delegating work is the configuration of your computer system. If you use a centralized network file server, CAD users can simultaneously share the same electronic base, such as a building floor plate. Using AutoCAD's External References **(Xrefs)**, you can "see"—but not affect—current changes to other parts of a project while working on your part of the project. With this setup, you can tell instantly when the reflected ceiling plan you are laying out needs to be modified to reflect the ceiling diffuser that the engineers have just added. Figure 5.4 shows diagrammatically how four CAD users can share a base floor plan as they complete their own portion of the work. If you don't have a network that permits the use of **Xrefs**, you must rely on other communications, such as e-mail or paper memoranda, to issue updates. You must also ensure that

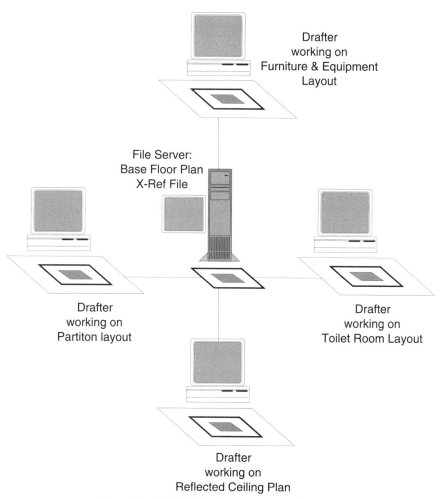

Figure 5.4 Multiple drafters can use **Xrefs** to share the base floor plan while each develops a different aspect of the project design.

your CAD filing system reflects and distributes updates without inadvertently overriding anyone's work.

Design teams delegate CAD work in several ways. The more CAD users you have, the more formal you need to be about splitting the work, and wherever possible you should use your CAD system to help coordinate these users (see Chapter 7 for more information on various approaches to working with multiple floors and buildings).

- Dividing work by layers (content). With this approach, one drafter enters reflected ceiling plan information, and another works on floor finishes. This approach only works well if the CAD users can simultaneously reference an electronic base drawing by using AutoCAD's **Xrefs** capabilities. Otherwise, you have to have drafters working on copies of the same plan. You also lose the benefits of ongoing drawing coordination that skillful drafters can provide when they enter additional data into a drawing which is always automatically updated to reflect changes. Dividing CAD files (as distinct from dividing *information*) by layers is not recommended when working on floor plans unless you can apply a disciplined approach to using **Xrefs**.

- Dividing work by floor. As with dividing work by layers, dividing work by floors is not recommended unless you can use External References to gain access to a current reference base of exterior walls, columns, and core; otherwise, multiple drafters have to continually track changes to shared building elements. Separating work by floors only works well when the floors are genuinely different, share few common building elements, or if you have a base plan that is unlikely to change.

- Dividing work by drawing type. Unless you're using **Xrefs**, most teams split up work by type of drawings, such as plans, elevations, and details. This is the easiest way to ensure continuity and to minimize confusion. You nonetheless must develop a means of tracking changes that affects cross-referencing, so that elevations and detail reference numbers, for example, will immediately be reflected in both plans and all other drawings where these references appear. You can do this through the traditional means of creating check sets or by the use of electronic CAD-links, such as **Xrefs**, SQL database links, and OLE (see Chapter 10), or by using AutoLISP-based queries.

- Dividing work by project stage or task. Splitting up work by project phase is another common way to delegate work. Here, one person may work on schematic layouts, another on design development, and another on production drawings. With this approach, you want to ensure that the person(s) working on the latter stages of a project isn't completely dependent on the person responsible for the initial work. Communicating a clear scope of the project and target results helps enormously in this regard.

In addition to delegating work among CAD drafters, you must also clarify what work, if any, will be done by hand or by another computer program. You can add this sort of information to a matrix such as that pictured in Figure 5.3. Note that when you assign tasks that involve entering information in other computer programs, your staff will need to know what sorts of files can be shared with CAD files (see Chapter 10); otherwise, you'll find that someone spends hours typing into CAD, for example, lengthy construction notes and specifications that are already contained in a word-processing file and that could have just been pasted directly into the CAD drawing.

Managing Team Communications and Coordination

Team communications and coordination are among the project manager's most important responsibilities. The larger and more complex the team, the more effort the project manager must make in this regard. Indeed, many project managers often feel that they rarely perform any other function.

CAD must be addressed in all project communications and coordination. Often, however, CAD users may be left out of the "information loop" as it were, by not being informed of the project's general progress. Managers may assume that CAD users neither want nor need to know all project developments, when in fact lack of up-to-date information limits their ability to be proactive, independent, and able to use CAD to help anticipate and reconcile discrepancies. You will see better work from your CAD staff if you include them completely in all aspects of the team operations and give them the same attention you give other team members. You should extend the same treatment to CAD users that you would to other team members:

- Make sure they receive routine communications, including memos, updates, and bulletins.
- Include them in regular project meetings.
- Introduce them to the project parties, including the client, contractors, and engineers.
- Arrange for them to come along on site visits.

A key element that facilitates team communications and coordination is the physical setup of the project team. Projects generally go more smoothly if all team members physically work in the same pod; communications, both informal and formal, can take place faster, and project managers worry less about communicating information formally because the odds increase that it will circulate informally. Physically separating CAD drafters from project teams and placing them in a CAD ghetto (see Chapter 9) slows down the pace of communications and dialogue and also isolates CAD users. Moreover, when CAD users are physically isolated, CAD remains a mysterious process to those who don't use it, while CAD users have less of an opportunity to broaden their knowledge of

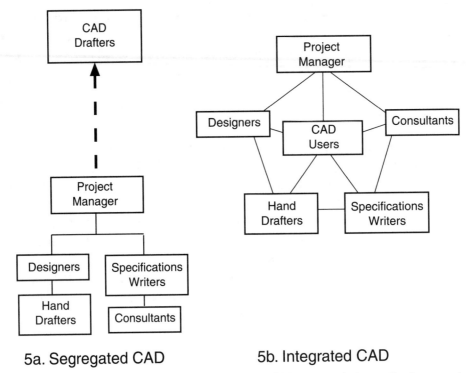

5a. Segregated CAD **5b. Integrated CAD**

Figure 5.5 *The ideal physical location of CAD users relative to other team members is shown in diagram b.*

design. Therefore, the ideal physical relationship for CAD users vis-à-vis design pods, shown in Figure 5.5b, is a workspace arrangement that eliminates any vestiges of a separate CAD area.

Implementing Quality Control

CAD drawings often appear so finished and precise that you might imagine that their contents and accuracy match their appearance. Not so. The same checks and quality-control measures traditionally used with hand-drafted documents should be applied to CAD-produced work, perhaps more so because CAD output looks final at a very early stage in the design process. Psychologically, many reviewers of CAD drawings (as well as other computer-generated documents) overlook obvious errors because the overall results look so pristine.

To build in quality control, a classical coordination method is to schedule regular printing and review of project check sets. Check sets should be scheduled to be generated on particular dates and for target percentages of completion as shown in Figure 5.3. Check sets present a way to cross-check and integrate CAD work and non-CAD work. They form the basis for making formal changes and major redlines, so they should not be superseded by verbal comments and interim plots. Many project managers put a moratorium on subsequent plots or changes to CAD files until check sets are reviewed and changes approved.

As described further in Chapter 6, you can use a range of CAD-based techniques to electronically check and coordinate drawing sets and specifications. You can "redline" electronically as well as on paper, and you can run a series of AutoLISP routines to check for inconsistencies and omissions.

While well-planned CAD use can speed up a project, aid coordination, and improve drawing quality, it does require meticulous monitoring. Changes made in electronically based media are much more difficult to track than in paper-based projects. Therefore, you must have the means to leave a "paper trail" of electronic changes. The date and purpose of all CAD plots must be carefully noted, and you should establish and follow procedures for approving and implementing changes to CAD files. To do so, you may want to use a range of AutoCAD-based document management software or write an AutoLISP routine that records when changes to a CAD file were made, why, and by whom. Software add-ons such as Autodesk's WorkCenter file management software, for example, can actually compare two versions of the same drawing and graphically highlight the differences between the two.

Scheduling Output

One major project management task is scheduling production output. With traditional drafting, this often means ensuring that a diazo machine and staff person will be free at a designated time to print drawing sets. With CAD, you also need to schedule plotting time. If you use in-house plotters, you should inform the rest of the office of the days and times when your team will need to use the plotter, and you should ensure that adequate plotter paper and plotter pens or ink cartridges can be found in the supply closet. If you use plotting services, you should inform them at least one or two days in advance of a major deadline and make sure that their machines are operating smoothly. Most plotting services have their own delivery service for getting completed plots to you, but arranging for a backup messenger can be good insurance against Murphy's Law.

Although plotting time requirements vary by project, you should schedule at least *one* day for plotting, including allowance for plot errors and failures. Therefore, you should plan to complete your plots one day prior to your deadline. Familiarity with plotting speed and capabilities may allow you to be more precise in estimating the time requirements. For example, if you have to plot sixty 42" × 30" sheets on an inkjet plotter that generally plots final-quality plots at 15 minutes per sheet, you need to budget a minimum of 15 hours (or at least two days)—assuming that you encounter no plotting problems—for plotting, and add time for writing plot files, reloading plotter paper, and resending plots because at the last minute an error in a drawing file needed to be corrected or because an ink cartridge ran dry. A table such as the one shown in Figure 8.6 is a useful way to trade all the information necessary to schedule plots.

What Do Project Managers Need to Know About CAD?

Only rarely are project managers also CAD wizards. Nor do they need to be. Indeed, too much technical knowledge can distract project managers from their fundamental responsibilities. To run CAD-based projects, however, they must demonstrate a general knowledge of how CAD works, its strengths and limitations, and an ability to schedule it. Ideally, they should have a basic understanding of the following CAD commands and procedures:

- basic AutoCAD file menu commands, such as **Open**, **Exit**, **Save**, **Print**, and **Search**
- blocks, their uses, and how to design them so that they can be easily redefined to generate global changes later in the project (see Chapters 4 and 6)
- layers, including what entities appear on which layer and what layer combinations generate various drawings. Familiarity with the office-standard layering system is essential, as is understanding the implications of switching to nonstandard layering systems (see chapter 7).
- time requirements for completing basic CAD routines (see chapters 5 and 8)
- modeling, including its scheduling requirements and its relationship to 2D CAD
- office-standard CAD prototypes for drawing sheets, block libraries, and details (see Chapter 4)
- techniques for dividing and delegating CAD-based tasks (see Chapters 5 and 7)
- office-standard electronic filing system and the office file-naming system (see Chapter 10)
- the major file formats, such as *.dxf*, used in file conversion routines, the time requirements for converting files, and each file format's capabilities (see Chapter 10)
- the office's output devices, including printers and plotters, their speed, media options, and output quality (see Chapter 8)
- service bureaus and plotting services, including their range of services, turnaround time for service, and rates. (See Chapter 8)
- general knowledge of the CAD issues that affect the professional practice of design and an awareness of the design issues that can affect the use of CAD (all chapters, but especially Chapters 5, 7, 9, and 10)

The same scheduling considerations for plotting hold for any other output, or input, your project requires. Other forms of production you have to budget time for include having items scanned in or translating AutoCAD files into *.dxf* files.

In some firms, a CAD manager may oversee the mechanics of production; in others, a designated CAD-proficient member of the project team assumes this responsibility. Regardless of who makes the actual arrangements, the project manager must keep all parties up-to-date on production deadlines.

As described in Chapter 9, project managers have a range of sources for obtaining knowledge about CAD-related issues. The best option—because it is particular to your office—is in-house training by a CAD manager, senior CAD drafter, or other CAD-proficient colleague. Various professional associations also produce seminars on CAD management. An introductory course in CAD can also be useful, although such courses don't usually address the attendant management issues.

6 AutoCAD in the Design Process

The design professions generally describe their work as a *linear* and *sequential* process consisting of nine phases, more or less. As shown in Figure 6.1, designers usually relegate CAD use to the "Document Production" phase of this process. In its 1993 survey of the architectural profession, the American Institute of Architects (AIA) found that only 33 percent of its CAD-using membership used CAD for conceptual design; CAD still was used primarily for construction drawings.*

Using CAD only during the construction documents phase gives you access to only 25 percent of CAD's features. Moreover, if you restrict CAD's role to the drawing production process, you are violating the number one principle of effective computer use: *Don't reinvent the wheel.* If you wait until the project design is complete before you create your CAD files,

---

*The design phases presented in Figure 6.1 loosely parallel the eight project phases that the AIA presents in *Article 1.1* "Schedule of Designated Services" of *The AIA Standard Form of Agreement between Owner and Architect for Designated Services* (AIA Document B163, 1993 edition). Design firms use variants on these phases depending on the nature and scope of their practice; no industrywide standard for either phase, scope, or nomenclature yet quite exists. Note that you can use CAD for almost all of the fifty-three designated services, as well for the thirty supplemental services, described in AIA Document B163 (see Chapter 7).

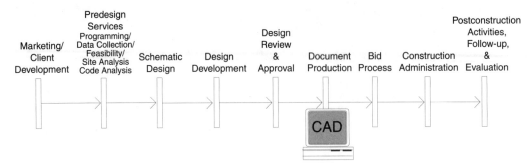

Figure 6.1 *The traditional approach to using CAD in the design process*

you have already drawn the same base information over and over. Moreover, if you didn't rely on CAD's database capabilities during the initial phases of programming and schematic design, you've not only reworked the visual design; you've lost opportunities to reuse and to analyze your data to create a better, more accurate solution to your clients' needs.

Once you incorporate CAD into the overall design process, you'll see instant benefits in terms of time savings, project coordination, information retrieval, decision making, and possibly even design quality. You'll also start to look at the design process differently. Instead of a linear process, CAD will seem like a circular process, with your CAD system as the central spoke, as shown in Figure 6.2. You can view your work not as a rigid sequential

Figure 6.2
The progressive and profitable approach to using CAD in the design process

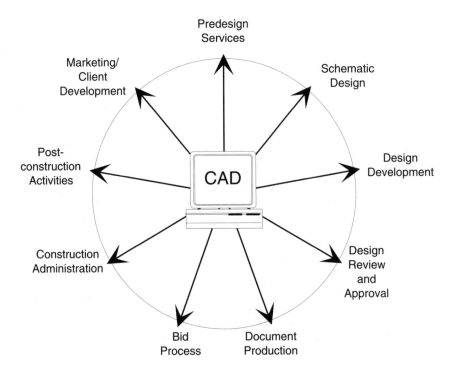

process, but rather as a flexible dialogue in which you'll go back and forth between broad schematic issues and details, between concepts and facts. And you no longer need to feel compelled to collect data in a specific sequence or format; regardless of when and how you obtained your client and project data, CAD can store it for you and present it as you need it.

A holistic approach to applying CAD to the entire design process not only reshapes your sequence of work but changes the *relative* time requirements for each phase, as shown in Figure 5.2. You'll find that the initial stages of client development and programming require more time than the later stages; most of the CAD work involves building your project database, which serves as the foundation of the rest of the project. Once your database is organized and the base information has been input in the computer, the later design work accelerates because nothing ever again has to be developed from scratch. Instead, your design is merely modified and refined as the project develops.

This chapter describes how you can use AutoCAD in all phases of design projects, with the goal of both improving your work and speeding it up. The stages presented here reflect the standard design phases. Many projects, of course, do not include all these phases, and others may not follow any "standard" sequence. A properly customized CAD system and an open-minded attitude about CAD, however, will enable you to use CAD effectively for projects of a wide range of sizes, scopes, and schedules.

Getting the Project: Marketing and Client Development

To get CAD-based projects, design firms naturally must first have clients. AutoCAD can serve as a powerful tool to help develop and expand your client base, and it can shape your marketing program in several valuable ways. You can incorporate AutoCAD in your marketing strategy by

- marketing AutoCAD-based services,
- incorporating current and completed AutoCAD-based projects into your existing marketing materials, and
- using AutoCAD to produce new marketing materials.

Marketing AutoCAD-based Services

In the early days of AutoCAD and other PC-based CAD systems, CAD was so unique (and so expensive) that its use in almost any context justified charging special fees. CAD itself is now no longer so exotic, and design firms cannot market it as a special service for routine production drawing. You can, nonetheless, market AutoCAD for special services, which are divided below into design services, client markets, and geographical markets.

By Special Services

Although routine drawing production on CAD does not warrant special marketing, other services do. Typical special services include

- building inventories and site documentation
- CAD database creation
- external drafting and plotting services
- external document management services
- facilities management outsourcing services
- design of models and mock-ups
- training services
- visualization studies, including 2D and 3D views and audiovisual presentations
- third-party software development

Note that AIA Document B163: *The AIA Standard Form of Agreement between Owner and Architect for Designated Services* (1993) includes in its "Schedule of Designated Services" thirteen categories of "Computer Applications" (Article 2.4.78), ranging from programming to construction cost accounting. Computer services are listed under these categories, along with descriptions and guidelines for offering these services, but the information should be applied with caution, as standard agreements and forms for dynamic fields such as computer-related services tend to become outdated very rapidly.

By Client Markets

Traditionally, many design firms have established their reputations by being specialists in various building types, such as health care facilities, labs, and commercial renovations. CAD not only helps you maintain your area of expertise in specialties; it can give you a competitive edge. You can tailor not only CAD but all your computer applications to suit the particular issues that concern your target markets. You can easily develop symbol libraries, AutoLISP routines, and templates that reflect current industry standards for building types.

In a highly specialized world, specialized applications of CAD exist. Government agencies, such as the Department of Defense, as well as the facilities departments of larger corporations, universities, and other institutions now rely heavily on CAD as a matter of course. These organizations issue detailed guidelines on CAD use to the designers they hire. These guidelines reflect not only the owners' goal of ensuring computer compatibility, but also their need to comply with national and even international guidelines for electronic compliance and document management as mandated by the Occupational Safety and Health Administration (OSHA), the International Standards Organization (ISO), and other organizations. As these requirements become more complex and widespread, potential clients will pay increasing attention to design firms that demonstrate knowledge of, and ability to work within, these guidelines.

By Geographical Markets

During the past decade, an increasing number of American design firms have sought to go beyond their local geographical area and to develop a national or even international clientele. Thanks to today's electronic data transfer capabilities, design firms are now less geographically restricted in their practice than previously. Although proximity to a building site during construction is always necessary, you can now conduct much of your work by modem and fax. CAD, which is both a national and an international computer language, allows you to expand your geographical range more easily and logically (see Chapter 7).

Many design firms have acquired or established partnerships with firms in other cities and countries in order to service clients who themselves own or lease properties in many different locations. Having an efficient national or international computer network to back up this service approach is an invaluable selling tool for many designers and is worth investing in if you seek this type of clientele.

Incorporating AutoCAD-based Projects into Your Marketing Materials

The marketing staff of most design firms is well aware of the need to use CAD-generated drawings in the firm's marketing program. They are not always familiar, however, with the various ways AutoCAD files can be imported into desktop publishing and presentation software programs (see Chapter 7). Familiarity with file translation and transfer techniques (see Chapter 10) can save turnaround time and expense in the preparation of marketing materials.

You can use AutoCAD files to create a number of typical marketing materials.

Presentation Boards

You can use AutoCAD to generate presentation boards of any size and scale—up to the maximum sheet size your plotter can handle. Depending on your plotting capabilities, your plots can be rendered in black-and-white or in up to 256 colors. You can also cut and paste drawing elements from AutoCAD into other software programs for use in multimedia presentations. An AutoCAD-produced floor plan or building elevation, for example, can be combined with photographs and spreadsheet charts in a desktop publishing file. Alternatively, you can "clip" a statistical chart or graph from Microsoft Excel or PowerPoint and paste it into an oversized AutoCAD-generated sheet.

Marketing Publications and Brochures

You can import drawings from AutoCAD into project sheets and marketing brochures that you produce using QuarkXPress, Adobe PageMaker, or any number of desktop publishing and layout programs. You do not need to send AutoCAD drawings out to reprographics services to develop camera-ready art or provide extensive reductions; use your file exchange capabilities to save time and money preparing artwork for these publications.

Proposals

You can import AutoCAD files into word-processing and page layout programs when you respond to requests for proposals (RFPs), issue letters of inquiry, and prepare standard qualifiers. You can use your existing library of AutoCAD images for proposal graphics ranging from building sites to client logos. Images that aren't already in the computer can be quickly scanned in (see Chapter 10). Working with existing computer graphics not only saves enormous amounts of time, stress, and money but also ensures consistency in the images you produce.

Slide Shows and Animated Presentations

Using AutoCAD's **Mslide**, **Vslide**, and **Script** commands, you can put together slide shows of images for dynamic client presentations. You can use 3D Studio and other software programs to present animated 3D walk-throughs and flyovers. If you have a computer with a sound board, you can even combine AutoCAD images with sound. You can also import AutoCAD images into programs such as Microsoft Powerpoint or Novell Presentations to create slide shows, charts, overheads, and handouts.

Be careful about going to special efforts to develop 3D presentations and other specialty materials that you cannot easily produce with your current staff and software. Marketing such presentations can create a level of expectation on the client's part that you may find difficult and costly to meet.

Office Tours

When potential clients seriously consider hiring a design firm, they often visit the firm's office and receive an office tour. These days, part of the tour inevitably includes appraisals of the office's computer system, including terminals, plotters, and other devices.

In the early days of CAD, just having a CAD station and a plotter was impressive. Nowadays, prospective corporate and institutional clients not only are rather jaded about basic computer capabilities but also tend to be knowledgeable and sophisticated in their assessment of your system. They may ask about, and expect to see, sophisticated network configurations, a complete range of state-of-the-art input and output devices, and evidence of a customized CAD setup. They may ask you questions about your CAD budget, hardware acquisition strategy, the version of AutoCAD you use, backup procedures, and so forth. Even though marketing directors do not oversee CAD operations, they should have at their fingertips current and accurate statistics on the office's computer resources.

In preparing for office tours, make sure that your CAD manager and other CAD-knowledgeable staff will be on hand to answer technical questions and to demonstrate the firm's capabilities. If you want CAD stations to demonstrate a certain program during the tour, plan this activity in advance and review the projected display yourself. Nothing is as embarrassing or potentially damaging as telling clients about a special CAD project and then not being able to pull it up on the screen.

 Windows contains a collection of "screen savers" that temporarily hide the computer screen when the computer is inactive. You can customize your screen savers to include your own firm name or logo or other special design that can run during office tours. During tours it is advisable to disable screen savers that are ugly, frivolous, or downright offensive.

Marketing Libraries

Every design office with a marketing function maintains libraries of photographs and slides of their projects. These images, while essential components of slide shows and publications, can be unwieldy to catalogue, store, and retrieve. With recent software developments, however, marketing staff can not only catalogue these materials more easily but also in ways that make retrieving images less time-consuming and more foolproof.

For AutoCAD files alone, one form of electronic slide library can be developed through the AutoCAD **MSlide** command as described above. You can also use an external AutoCAD utility called *Slidelib.exe* to create slide libraries; these can be accessed by AutoCAD pull-down menus.

You can also develop comprehensive marketing databases using database programs, such as Claris's Filemaker Pro or Microsoft Access, that can store both text and graphics. You can create project profiles that both list the project statistics and contain all your electronic graphic images of the project. You can use the query capabilities of these programs to retrieve project records that are most similar in type to the prospect you are developing marketing materials for. You can also use databases to summarize firm statistics, such as the number of active clients, the amount of square footage under design and construction, and the percentage breakdown of clients by type of projects.

Marketing managers may also find helpful the growing number of software packages devoted to documentation management and storage. Programs such as AutoCAD WorkCenter track and preview files on your computer system. Other applications enable you to build electronic slide libraries, optically stored on CD-ROM, so that instead of spending hours thumbing through multiple binders and boxes of slides and photographs you need only access a single electronic library. Using film scanners you can build a digital version of your existing library using slides and photographs.

For marketing managers to use AutoCAD-generated materials effectively, they must be in close communication with their CAD-proficient colleagues. Marketing staff should be familiar with the file import capabilities of AutoCAD, as described in Chapter 10, and must establish procedures for collecting, documenting, and storing CAD files that can be used in marketing presentations. They should review and, as appropriate, receive copies of CAD files for completed projects before the projects are archived. And they should consider the firm's computer inventory part of their own database of firm resources and qualifications (as recommended in Chapter 10).

Predesign Services: Programming and Data Collection

Although predesign services, such as programming and site analysis, are not formally considered design, they form the foundation of designs. The process of collecting and analyzing data about a client and actual or potential project sites produces the design "recipe" that shapes all subsequent design phases. The more comprehensive and accurate you can be at the predesign stage, the more smoothly and competently you can execute the rest of the project and the better chance you have of reducing costly errors and changes.

CAD is a powerful tool for making the most of the predesign phase. It can serve as an expandable and infinitely flexible database that stores all the information you have about a client. This information can be graphic, such as information on site topography or building envelopes, or data-oriented, such as dimensions and text and numerical attributes; it can be about buildings, communities, people, or codes and regulations. Provided that it is electronically accessible, the information can be in almost any format and scale. Because it can be "hidden" until you actually need it, information does not have to be input at a specific time or in a particular sequence; it can be entered into the CAD database as soon it is available and verified, and then it can be displayed only when it is relevant.

In keeping with the CAD principle of entering data into a computer only once, you want to review and analyze the data you collect carefully before you enter it into your CAD database; errors and omissions will come back to haunt you; correct data will serve you well throughout the project.

Programming and Database Creation

The programming process is essentially a database-creation process. As discussed above, CAD databases incorporate both *graphical* and other forms of data. The graphical information automatically includes not only an object's visual configuration but also its dimensional information; for every line, arc, polygon, and other entity, AutoCAD automatically tracks the length, angle, perimeter, and area. This is a powerful feature not only for dimensioning but for calculating areas and distances—tasks that are essential during the programming phase.

In addition to graphical information, AutoCAD stores other types of data. Any entity or combination of entities defined as a block can have attached data attributes. As shown in Figure 6.3, for example, you can create a block in the form of a rectangle that represents a typical employee workspace. To each incidence of this block, you can attach unique attributes or fields that contain information on the user's name, personnel status, department, division, even telephone extension. This information can be accessed through the AutoCAD drawing simply by selecting the attributed block or by using the **Ddattext** command to export the data as an editable ASCII text file, which can then be converted to a database or spreadsheet.

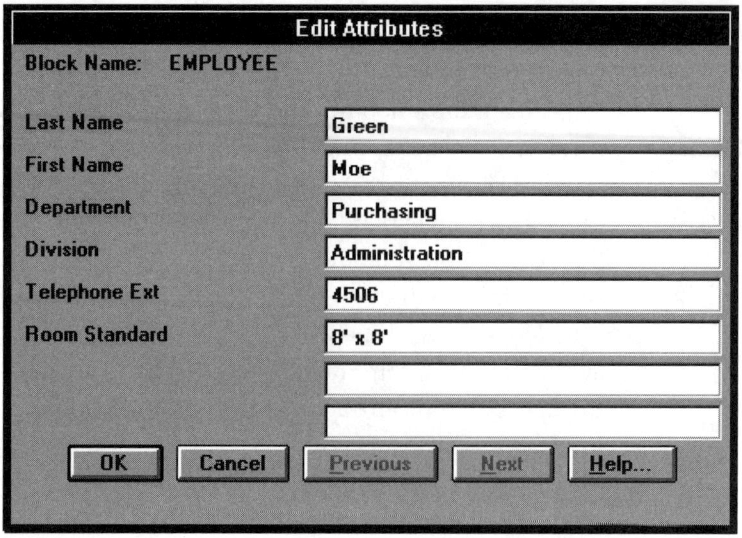

Figure 6.3 A graphic entity, or AutoCAD block, can contain data attributes that link images to external databases. Shown here are typical data fields in an attributed AutoCAD block for an employee workstation. The information typed into these fields can then be used to create databases and reports.

You can further increase the power of block attributes by linking attributed blocks to external databases using AutoCAD's Structured Query Language (SQL) capabilities. The SQL feature connects the data contained in each incident of a block to databases created in standard relational database software such as dBase, ORACLE, Borland Paradox, and Microsoft Access. You can also purchase a range of third-party add-ons that load preformatted databases for typical architectural applications.

Especially for long, large, and complex projects, the SQL database link is valuable on several levels:

- You can expand or contract the database's features as the project evolves. Such flexibility is helpful considering that these days project scopes are rarely static.

- You can input data in any sequence. You can define the scope of the data you'll need to collect early on but wait until it is available or can be verified before you enter it into AutoCAD. You can "hide" empty data fields in AutoCAD while you are waiting to receive the information.

- You can coordinate drawings with programming statements and other documents. SQL links help ensure that every space, person, and other documentable entity in your program statement also appears in your drawings. If an element in the program is changed, deleted, or added to, this can be reflected automatically in both drawings and written reports. Thus, you reduce the possibility of overlooking an end-user or piece of equipment when you develop your initial design schemes or when clients make changes to the program.

- You can "input" your data once and "output" it in an infinite number of formats. With AutoCAD alone, you can present diagrams and drawings in an infinite number of sizes and formats, and with countless combinations of layers. With AutoCAD in conjunction with SQL you can also generate a variety of reports according to your client's needs.

- You can control the presentation of information to reflect the needs of particular project phases. At the start of a project, for example, you might collect a complete inventory of your client's employees, equipment, and furniture. Showing such detail on drawings during the schematic design stage, however, might be both distracting and precipitous. So you can hide what would be premature at this stage and present only the information that allows you to address the issues at hand. Yet later you can use that inventory to ensure that your schematic designs accurately reflect specific program requirements.

With AutoCAD Release 13 for Windows, you can augment AutoCAD's SQL capabilities with file exchange procedures and enhanced Object Linking and Embedding (OLE) features, and you can copy and paste objects across Windows applications (see Chapter 10). With these capabilities, you need never input information from scratch more than once in any Windows application. Rather, you "recycle" the data throughout your various software.

Feasibility Studies

An important function of the predesign stage is assuring that the proposed project is actually *feasible* in terms of space utilization and availability, budgets, and compliance with building codes, zoning requirements, the Americans with Disabilities Act (ADA), and all other applicable standards and requirements. With AutoCAD, you can use templates and AutoLISP routines to analyze and test regulations and requirements against a client program and site. For a building renovation that must be upgraded to comply with ADA and new fire code regulations, for example, you can overlay templates of compliant toilet rooms, egress stairs, and exits onto a CAD floor plan to review what compliance would require in terms of using and reconfiguring existing space. With SQL Databases and third-party software, you can link these potential renovations to cost estimation databases. With these tools, you can quickly determine whether square footage and funds are indeed available to make the project viable.

In addition to analyzing requirements, you can use AutoCAD to test different ways to meet these requirements. For a company that is looking to lease new office space, for example, you can test its program in a range of spaces. You can use CAD-based floor plates of actual buildings that are potential lease sites. If your client hasn't yet begun looking at sites, you can nonetheless test a program of employee space standards, equipment, and circulation requirements against typical building core configurations and column grids, thereby having a good estimation in advance of the minimum space configuration and square footage that your client actually requires.

Merging Drawing Information

Data collection and site analysis often may require merging different sets of graphical information. You might need to superimpose the plan of an existing building drawn at 1/8" scale with a 1:20" scale engineering site survey and utility plan. In AutoCAD, you can easily merge these two plans and plot the results at the scale of your choice. If you separate the information from these plans onto different layers, you can generate infinite drawings from a single CAD file. As long as you can get the information into your computer, the scale, orientation, and other features of the information are irrelevant.

Schematic Design

The information you collect in the programmatic phase forms the foundation for your schematic designs. If this information is accurate and properly organized in a computer database, you can *reuse* it during the schematic design phase to produce schemes that are accurate and logical extensions of your predesign work. Being able to reuse this information not only saves an enormous amount of time but also reduces the possibility of introducing inconsistencies or duplications, or omitting necessary elements.

Flowcharts and Diagrams

Flowcharts and other diagrams are standard tools used in schematic work to convey information and design concepts graphically. Typical diagrams include

- adjacency matrices
- bubble diagrams
- blocking diagrams
- stacking diagrams
- flowcharts

How Do Clients Use CAD?

Increasingly, institutions, corporations, and government agencies are becoming as CAD-proficient as the design consultants they hire. Over the past decade, a growing number of organizations have come to recognize the value of using Computer-Aided Facilities Management (CAFM) to manage their properties more efficiently. Even clients that don't own real estate find CAFM useful in planning office moves and comparing leasing options, for instance.

As discussed in Chapter 7, clients benefit enormously from AutoCAD add-ons and other CAFM software that enables them to link CAD-generated building floor plans to databases. A well-constructed CAFM database enables users to track and manage their space more accurately and efficiently, thereby allowing them to realize substantial cost savings over time. In addition, building owners have less need to hire architects and space planners to draw their property on paper or CAD from scratch; they can simply hire designers for other services and provide a CAD template at the start of the project.

Another important use of CAD-related tools pertains to document management. Organizations find that computers, instead of helping to realize the goal of a paperless society, have simply loaded offices with even more paper trails to organize and track. The costs of mismanaging paper are high, so many corporations, institutions, and government agencies seek tools which will reduce or at least make it easier to manage the flow of paper. For facilities managers and others who must reference large architectural and engineering drawings in the course of their work, storing such sheets on site in tubes or flat files is unwieldy and an ineffective use of expensive office space. Many offices thus are able to justify the costs of scanning paper drawings into CAD and then using Total Document Management (TDM) software to retrieve and view electronic versions of their drawings as needed. Since many firms may not have the in-house staff or equipment to convert their hard-copy documents to electronic files, they often hire qualified design firms or service bureaus to do the work.

For designers, CAD-literate clients can seem a mixed blessing. On the one hand, with CAD files you and your client can exchange information and ideas more quickly. On the other hand, CAD gives clients the ability to review not only your design results, but also your production process. A growing number of clients expect their designers to use the clients' CAD standards, including layers and graphic standards; willingness to conform to these standards may determine whether you get a project to begin with. Clients who have established an efficient and productive in-house CAD system also tend to assume that the external designers they hire will respond as quickly to project changes or additions as they themselves can.

These graphics are used to present project schedules, analyses of client work flows, and preliminary space layouts. A range of software applications, including AutoCAD add-ons, are available specifically for the purpose of creating visual presentations of this nature; most will save your diagrams in file formats that AutoCAD and other standard Windows applications can exchange.

You can also diagram flowcharts directly in AutoCAD; this capability is particularly useful if you need to produce charts on large, architectural-size sheets. You can create libraries of typical chart components by making a set of attributed blocks, and you can also produce sets of drawing templates that store typical shapes, typefaces, and pen weights; thus you can ensure greater consistency in your chart presentations. One benefit of using AutoCAD for graphics such as block diagrams is that AutoCAD's area measurement capabilities allow you to size components to reflect their actual or relative space allocation. You can easily modify these shapes to reflect revised area requirements.

Adding and Editing Information

If your schemes evolve and focus, you can use AutoCAD's block redefinition capabilities to reflect these developments. Rather than redrawing and adding more information to hundreds of individual shapes, you can take existing blocks and use the **Redefine** command; each affected block will immediately assume the revised characteristics. As Figure 6.4 shows, for example, you can redefine a block so that it evolves from a simple rectangle that represents a typical manager's 7'-0" × 7'-0" systems workstation to a

Figure 6.4
*AutoCAD's block **Redefine** command enables you to quickly and globally convert schematic graphics into detailed, dimensionally precise spaces.*

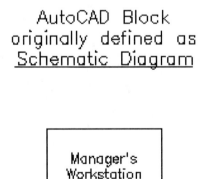
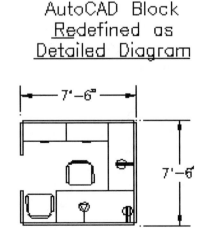

layout that reflects a proposed layout of panels and furniture; you can even include an initial scope-out of electrical and telephone outlets for your consultants and the client's in-house facilities department to ponder.

If you have in-house programming skills or acquire one of a number of third-party software packages designed for facilities management, you can feed your program into an existing CAD-based floor plan or building site. The CAFM programs (see Chapter 7) automatically create blocking and stacking diagrams based on the CAD plans and employee data you supply. You can test an infinite variety of "what-if" scenarios by tweaking certain program parameters; provided your CAD information is valid, the resulting diagrams will automatically and accurately reflect the changes. Although this application is used primarily for space management in larger organizations, you can use such software for other types of schematic design, ranging from layouts of museum exhibits to major urban planning schemes.

Tracking Schemes

During the schematic design phase, design ideas proliferate. CAD can help you not only to develop different schemes, but also to track and store them. By storing your drawings electronically, you'll use up far less space in your drawing flat files. At the same time, provided you follow a policy of keeping all project work until project completion, you can easily retrieve schemes that you had put aside but that later seem more appealing. You can use Copy and Paste functions to merge components of different schemes to produce the final design. If later in the design development stage—or even during the document production or construction phases—a discarded scheme suddenly seems more viable, you can easily retrieve the CAD file and paste the scheme into your working drawing.

CAD designers track schemes in various ways. One method is to create a separate file for each alternative under consideration. This method works best for very large files, but you risk harboring inconsistencies should the base information common to all the schemes change; making the common information an AutoCAD External Reference can help reduce inconsistencies of this nature. Another common method, which works well for small, simple designs, is to create all schemes in one drawing file; each design is drawn on a separate layer or series of layers, and the various layers are turned on and off as necessary to produce different plots.

Regardless of which method you use to test different schemes, CAD file management is key. CAD files must be properly named, and the plots they produce must contain, and should always print out with, the file name, date, and purpose. Notating which schemes have been approved and why is also helpful for future reference.

Presenting Multiple Schemes

Except with very simple, fast-track jobs, during the design development process *multiple* schemes will be tested and reviewed. If you want your client to review all your ideas objectively and fairly, take care to present the schemes in the same form, using identical scales, paper sizes, and production processes. Clients tend to respond differently to CAD drawings than to hand sketches. Even the most embryonic idea takes on a finished appearance in CAD.

If only half your schemes are CAD-generated, you may get more balanced feedback if you plot them and trace over the plot to produce a hand-done drawing that matches the other schemes in form if not in content. You can also plot your CAD files in gray or very thin black lines and go over the plots with markers or colored pencils. Alternatively, you can purchase an AutoCAD add-on that will "squiggle" the lines of your plots so that they look hand drawn.

Design Development

Once you've tested your client's program in various schemes and gotten approval for a particular design direction, you can then develop your design. During this phase, you look at your design in detail and make decisions that will eventually form the foundation of your construction documents.

Just as you can use AutoCAD to convert your programming data into design schemes, you can use AutoCAD to speed up the process of turning a loose scheme into a detailed, specific design. At this stage, you may need to refer to programming information that was not relevant during the schematic design phase. For example, you might start addressing specifics about employee and equipment requirements. If you had previously entered this information into your AutoCAD database as soon as it was available, you can now incorporate it more quickly into the development of your design.

For design firms that see CAD primarily as a construction documentation drafting tool, CAD at this point serves merely to express graphically what the designer wants to do. Used effectively, however, CAD can help you not merely graphically express your ideas but actually produce more effective design work.

Does Using CAD Preempt Full Exploration of Design Possibilities?

Many CAD skeptics believe that the structured and precise environment of CAD drafting forces design decisions to be finalized too early in the design process. They argue that CAD prevents designers from being "loose" and open to different possibilities, and that CAD data entry forces designers to take a premature concern with dimensions and other data of a precise nature.

Another concern of CAD skeptics is CAD's apparent preference for linear shapes over curved ones. This concern reflects the fact that AutoCAD as well as most other CAD software traditionally has had problems calculating arcs, ellipses, splines, and other curved shapes as accurately as it can lines, polygons, and circles. (With Release 13, AutoCAD introduced the **Spline** command, which creates nonuniform rational B-splines (NURBS), an improvement over AutoCAD polylines; Release 13 also improved AutoCAD's **Ellipse** command.)

This CAD skepticism does have some validity. But designers should remember that computers can't—and aren't meant to—replace creativity that, despite the many advances in artificial intelligence, still only originates from the minds of people. Nor should CAD eliminate the need for, or value of, sketches on tracing paper, doodles on the backs of envelopes, or ideas generated during freewheeling brainstorming sessions. Rather, CAD's role in the schematics and design development stages should be to collect information and ideas as they become available and to store them for future reference. In the hands of skilled designers who aggressively push the parameters of the software (rather than letting CAD set the limits), CAD and certain "sketch" programs can test ideas with greater accuracy and offer additional options for consideration. The successful use of CAD, however, depends on designers' individual skills, their attitude toward computers, and their ability to recognize when CAD is restricting them.

CAD skeptics should also bear in mind that the traditional concept of the schematic design phase as a loose, creative brainstorming process doesn't always dovetail with the needs of many projects today, which must comply with extremely specific requirements for employees, equipment, and code compliance, usually under tight design and construction deadlines. Clients often cannot afford to consider schemes, however innovative, that cannot be guaranteed to address their program requirements down to 1/32". Anyone with experience designing hospitals, factories, laboratories, and other complex buildings confronts significant design constraints at the start of the project and often finds that CAD is essential for tracking complex requirements. With intelligent use of CAD, all designers can reach a sound blend of reality and creativity in their design schemes.

Evolving from Schematics to Developed Designs

As previously stressed, the underlying principle of all computer-aided designing is never to enter data more than once but rather to edit what you have. Therefore, if you have used CAD to create your schematic designs, you should be able to convert them into developed designs by *modifying* them rather than by drawing the designs entirely by scratch.

> ### Can CAD Produce Better Designs?
>
> Since CAD's inception, designers have been involved in an ongoing debate about CAD's impact on design. CAD skeptics argue that CAD's impact on design is either negligible or nonexistent; that it imposes rigidity in the creation and organization of shapes; that it encourages repetition rather than variance; and that it forces premature design solutions. The greatest insult such skeptics can give a building is to suggest that "it looks like it was designed on CAD."
>
> CAD proponents rebut this skepticism by observing that with CAD they can make design elements much more precise, they can achieve greater geometrical complexities, and they can test more options.
>
> As with most debates, both viewpoints have some validity. And as with most software, CAD is only as good as its users. CAD programs resemble word processors in that they can, indeed, produce a lot of drivel, but they can also encourage talented designers to refine their work, express ideas more cogently, and meet clients' needs more precisely. In the high-pressure design environment of the 1990s, CAD may be the only means many designers have to achieve under unrealistic deadlines what previous generations did with more time.

As previously discussed, AutoCAD's block **Redefine** command can convert general graphic entities to more developed schematic blocks. During design development, you can, in turn, take the schematic blocks and *redefine* them as detailed layouts that reflect actual products and assemblies. At the design development stage the manager's workstation shown in Figure 6.4, for example, can be redefined into an even more detailed layout that reflects actual products. Similarly, a block representing a 24" × 24" column that was originally drawn as a simple polyline can be redefined as a dimensionally accurate American standard S18 × 17 steel flange with fireproofing and two layers of gypsum wallboard.

Note that in addition to *redefining* blocks, you can *substitute* other blocks for existing ones using the **Ddinsert** command. In the previous example of the columns, you can substitute the schematic column with a block from your library of standard column types.

Since new entities are often introduced to a drawing during the design development phase, you can't redefine or substitute blocks for every entity. But you can apply the block redefinition concept to all additional blocks you create or take from your CAD library. This allows you to change repetitive elements quickly and globally and to revise drawings at the last minute when changes occur.

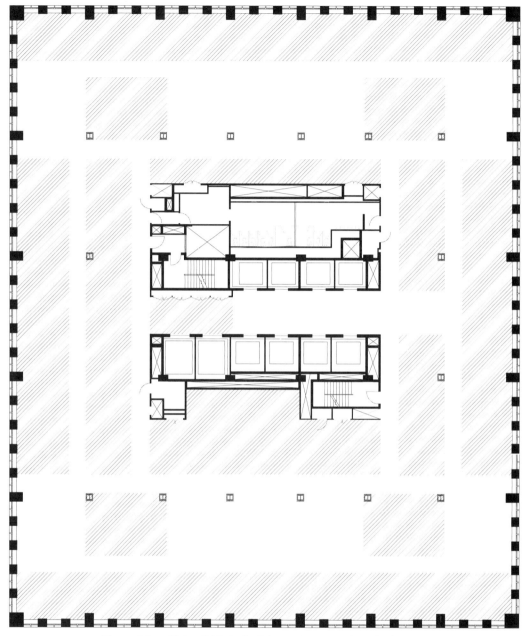

Figure 6.5a *Using CAD in the Schematic Design Phase facilitates the conversion to developed design and, eventually, to construction documents. For a multifloor corporate tenant, the design firm of Griswold, Heckel & Kelly, Inc. developed an AutoCAD-based prototypical schematic layout, based on the existing building floorplate for all the floors. . . .*

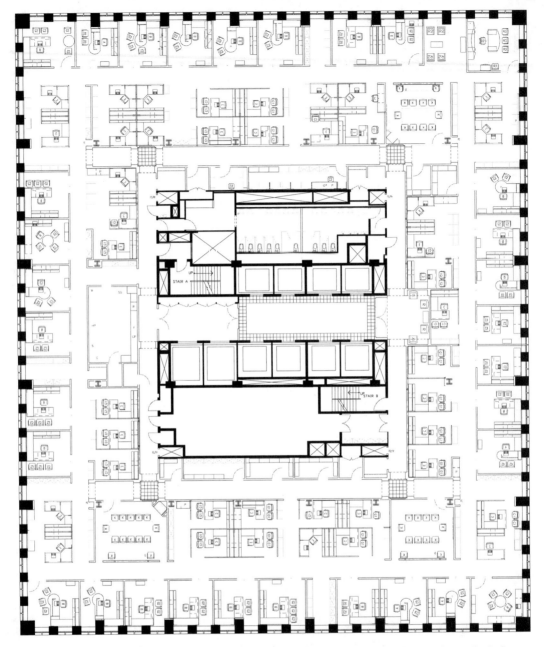

Figure 6.5b *As the program for each floor and department, along with standards for office and workstation layouts and furniture, evolved, the CAD base developed in schematics was reworked to incorporate the program particulars....*

Figure 6.5c *Although the final design layout varied from floor to floor, all layouts share common design elements and an overall approach to zoning and circulation. For complex projects such as these, CAD can help designers retain their original design concept while addressing the particular requirements of specific spaces.* AutoCAD files provided courtesy of Griswold, Heckel & Kelly, Inc., Chicago, IL.

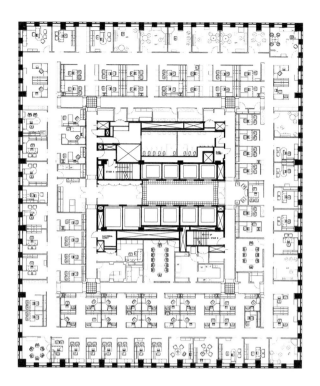

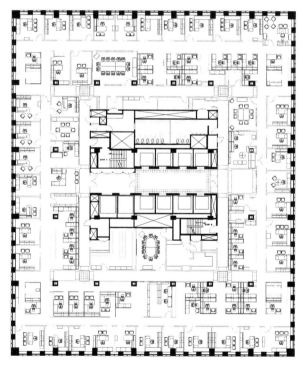

DESIGN DEVELOPMENT

Interactive Design Routines

One of the most powerful ways to design with CAD is through the use of interactive design routines. With such routines, dialog boxes or the AutoCAD command line (see Chapter 4) prompts you for information on design requirements and preferences, and on the basis of your responses the program produces a "first pass" at a design issue. You can write the design routines yourself using Scripts or AutoLISP, or you can purchase a range of third-party add-ons to do the job.

A typical design routine might address staircases, one of the most mathematically complex and highly regulated yet potentially exciting architectural elements. Through a series of commands or a dialog box, such a routine would prompt you to supply certain design parameters including

- the type of staircase you wish to design (such as straight, L-shaped, or spiral)
- the floor-to-floor height(s) in the space where the staircase will be located
- the dimensions, construction, and finishes of the walls of the stair enclosure
- the dimensions and configuration of the stair landings
- the tread depth and riser heights (the defaults can reflect local building codes)
- the height and configuration of handrails (again, reflecting codes)

Based on the information you provide, AutoCAD can generate in either 2D or 3D a staircase that meets these criteria. You can then further refine the plan until your stair design satisfies you both functionally and aesthetically.

In addition to stairs, you can program or acquire design routines for many other typical design elements and tasks, including

acoustical designs

cabinetry and millwork

ceiling grids and light fixtures

column grids

doors, windows, and other openings

elevators and escalators

exterior fenestration

lighting calculations and layout

parking spaces

roofs

walls and floor types

Naturally, design routines produce results that are only as good as the program used to create the plan; never rely solely on these results without further *manual* study and analysis. But these routines can be great time-savers, especially for elements that require complex calculations. They also serve as an electronic checklist of all the decisions you have to address and the codes and other requirements with which you must comply.

Products and Components

A major goal of design development is ensuring that you can actually build your design. For projects of ordinary scope and budget, this usually means determining whether you can build your design with existing products and standard assemblies. For unusual or custom projects, you have to be able to present complex ideas clearly enough so that a fabricator or installer can determine the design's feasibility and, if possible, provide a budget price for it.

Many of the tools you develop during the AutoCAD customization process (see Chapter 4) can aid you in this. Your libraries of blocks (symbols) can be used to lay out dimensionally "realistic" parking spaces, plumbing, and elevators. The wall types that you develop with the **Mline** command can help you determine whether your diagrammatic floor plan will work using standard partition types.

In addition to your own libraries of blocks and details, you can purchase AutoCAD add-ons from manufacturers, which essentially function as electronic product catalogues. These add-ons provide plan, elevation, and detail drawings of actual doors, windows, furniture, and other items that you typically need to specify (see Chapter 7). While such libraries should not be relied upon exclusively, since products usually must be bid out to different manufacturers or suppliers, these product libraries do give you a head start not only in your graphic layout but also in cost estimation, as many of these manufacturers' add-ons can be linked to specifications and cost estimation programs. In addition to providing CAD-based blocks, many manufacturers also often generate CAD-based drawings themselves for products ranging from auditorium seating to window details.

In situations where specific products cannot be considered, you can nonetheless often benefit from working with generic products that meet standards and guidelines particular to a given industry. Like individual manufacturers, a growing number of manufacturers' associations now sell or distribute for free electronic versions of their industry standard guidelines; some have developed programs to help you perform calculations and analyses. Some of these packages are not AutoCAD add-ons, but you can link the data to CAD drawings; others are add-ons. The American Institute of Steel Construction (AISC), for example, sells libraries of blocks for AISC standard steel shapes. Such packages are particularly useful for products for which you need to write performance standard specifications rather than proprietary ones.

Models and Mock-ups

Although models are also useful in schematic design, they are often essential in design development. Models enable you to test your concepts in plan, elevation, section, and 3D views. You can also test color schemes, lighting, traffic circulation, and other issues of particular concern.

As explained in Chapter 7, AutoCAD can be a useful assistant model maker, whether you are producing traditional physical models or electronic 2D and 3D presentations and visualizations. Modeling in AutoCAD offers several advantages:

- You can quickly develop an unlimited number of design options. You can store all of them on your computer system rather than in overflowing flat files. And in addition to creating these options, you can document—on a hidden drawing layer—the rationale behind each one.

- Your models can be extremely accurate. While a physical model created at 1/16" scale is not usually reviewed for its dimensional precision, with CAD you do have the comfort of knowing that your model is dimensionally accurate and consistent and therefore feasible.

- You can produce the models in an infinite array of scales. While your designs will be drawn at "true scale," AutoCAD enables you to present them at any scale. Thus you can plan your model to fit whatever scale is most useful for the purposes of your presentation, available media, even the size that will work best with your chosen method of transportation to the client.

- You can control the level of detail. With AutoCAD's layers, you can select the level of detail that is appropriate for the scale of your model and stage of design.

- You can analyze many more variables. Particularly with Autovision, 3D Studio, and other software applications, you can review AutoCAD designs in terms of colors, light, shadows, materials, and (in a non-tactile way) textures. To replicate all these variables with traditional physical models would require an enormous inventory of rendering supplies and skills in stage set design and construction. CAD is a much more compact process in this regard.

In many instances, a small-scale model is not useful for addressing design issues. When clients deem it worthwhile, they often commission mock-ups of complex design elements that must be seen in full scale to be assessed accurately. Mock-ups typically are commissioned for door or window units, office and workstation configurations, ceiling and lighting installations, and millwork. Although generally manufacturers and fabricators produce the mock-ups, often you are responsible for coordination and specification. You can use your existing AutoCAD designs to produce the documents for the mock-ups. If the mock-up is approved in part or in its entirety, you can then incorporate the CAD file into your construction documents.

Geometry and Math

At the heart of much great architecture lies a strong mathematical and geometrical base. Most designed objects are essentially solid pieces of geometry, and AutoCAD offers powerful capabilities that help you to select, develop, analyze, and document these shapes.

As shown in Figure 6.6, AutoCAD's basic tools include commands for creating basic 2D shapes such as circles, ellipses, arcs, rectangles, and polygons, and solid or hollow 3D primitive shapes such as cubes, pyramids, and spheres (see Figure 6.7). With most all these commands, you can create shapes through various methods, depending on what you need to achieve and the parameters you want to set for the shapes. To create an arc, for example, you can select from ten approaches, depending on whether

Figure 6.6
The AutoCAD toolbars for 2D and 3D shapes

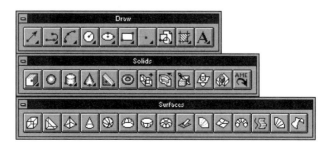

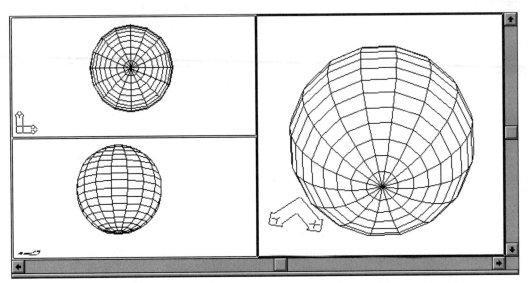

Figure 6.7 You can take advantage of AutoCAD's powerful geometric tools to instantaneously create and document complex geometric shapes in both 2D and 3D.

your arc specifications are driven by specific points on the arc itself, by the arc radius, or by an arc chord length. You can design shapes according to particular dimensions, angles, and rotations or according to their physical relationship to other objects. And rather than spending hours calculating how to fit a twenty-five-sided polygon into a 16'-0" diameter circle, for example, you can use the **Polygon** tool to instantaneously create the polygon by specifying the number of sides and the polygon's relationship to the radius of a circle.

In addition to creating powerful geometries, AutoCAD makes it easier to document them. You no longer need to spend hours calculating the angle or center point of the arc of a curved wall which you've designed. Instead, you simply tap on the arc using the **List** or **Ddim** command. And you can easily tweak your shapes through commands such as **Ddmodify**. Not only do you save time producing these dimensions and measurements, but contractors and fabricators will appreciate the information and greater level of accuracy and consistency.

Apart from creating and documenting geometric shapes, many basic AutoCAD drawing and editing commands allow you to promote and control your selection and placement of drawing elements so that your ultimate design is more logical and ordered (if that is what you strive for). Traditional design training also encourages the use of spatial concepts such as hierarchy, symmetry, balance, space and voids, grids and axes,

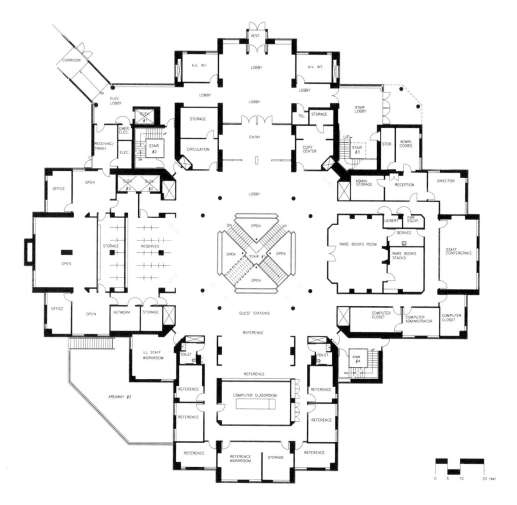

Figure 6.8 *Even in the final construction documents phase, this design for a new law school library retains a strong geometric symmetry while accommodating a complex client program. Typical AutoCAD commands such as **Array**, **Mirror**, and **Rotate** help ensure that the original geometric organization is retained as the design evolves.* AutoCAD drawing of the Boston College Law School Library provided courtesy of Earl R. Flansburgh + Associates, Inc., Boston, MA.

space, and color and light. You can use AutoCAD to experiment with, and then implement, these concepts. Commands such as **Array**, **Rotate**, **Mirror**, **Osnap**, as well as the aforementioned 2D and 3D shape tools all serve to help you execute these classic concepts (see Figure 6.8).

DESIGN DEVELOPMENT

Architectural and interior design schools **finally** are discovering that, ironically, state-of-the-art CAD is a powerful tool for teaching traditional design concepts. Increasingly, design studios are becoming "electronic," and more and more newly minted design graduates are including CAD-generated projects in their student portfolios. Design schools have also begun to show more support for CAD-generated thesis projects.

A useful tool embedded in all these design commands is the AutoCAD **Cal** command. An AutoCAD calculator that you can use in conjunction with most drawing commands, **Cal** enables you to create new entities by specifying them in terms of numbers, vectors, and relationships among various "object snaps" (osnaps) of other entities and by specific drawing coordinates as you create new objects. Thus you can construct objects precisely without having to touch specific points of objects already on the computer screen. Another benefit of **Cal** is the ability to establish a consistent mathematical relationship between objects. For example, you could specify that the length of every new entity you draw is a mathematical derivative of a previously drawn entity. You could use **Cal** to embed the golden mean and other special proportions into your designs.

Design Checklists

Ideally, design development involves considering every decision clients and designers must make before the project is approved for construction documents. As designs become more complex, designers have a harder time remembering all the issues they need to address. Many therefore rely on an office-standard design checklist to ensure that important design issues aren't overlooked.*

AutoCAD can apply the checklist concept to your CAD files in several ways. One is through the use of templates, as described in Chapter 4, which contain all typical building elements configured in a standard way. You can overlay these templates onto your design scheme and compare the two to detect omissions, discrepancies, and conflicts. You can also use AutoLISP routines to develop interactive design routines, as described above, and search commands to look for designated entities and alert you if they are not present or not up to code. For example, if you code all attributed blocks with the character ? to indicate an unresolved issue, you

---

*For a useful prototype of a design checklist, see the American Institute of Architects, *Architect's Handbook of Professional Practice,* Vol. 2, (Washington, D.C.: AIA Press, 1994), 646–48.

can then run a search for all blocks containing attributes with ? in them. SQL and OLE links can provide similar functions. You can also use the **Purge** command to locate empty layers and unused blocks; these often are clues to possible omissions.

Finish Color Boards

During design development, designers usually introduce specific materials, finishes, and color schemes for client review. Perhaps because of their colorful and tactile nature, materials and finish boards often are overlooked as a potential application for CAD. CAD, however, is a logical tool for producing background sheets for finish boards. The boards often are as large as typical architectural drawing sheets, so you can use a CAD plotter to print board titles and product descriptions, thereby saving hours at the office lettering machine. You can also plot—ideally, in a light halftone—the outlines of your finish samples, so you can affix the samples to the board squarely and neatly. You can then mount the plotted sheet on foam board or a similar substrate and paste on your samples.

With the recent development of more affordable color plotters and printers, as well as advancements in paper, you can plot colors, patterns, and textures directly onto a range of paper colors and textures. Plotter-generated texture and material samples are more efficient than dimensional samples that are glued onto boards (and are likely to become detached from the boards). And rather than pasting on pictures of proposed furniture, light fixtures, and other components, you can clip images from manufacturers' libraries or scanned artwork and paste them into a desktop publishing application along with elements from an AutoCAD file. With AutoCAD's rendering tools and with other Autodesk software such as AutoVision and 3D Studio, you can access and edit a growing library of standard building materials to render your designs in the finishes you are proposing. While CAD-generated images of materials are no substitute for actual samples, they do help clarify where the proposed colors and finishes will be used.

Design Review and Approval

Once you've completed the design development stage, you should be ready to submit your design for review and (ideally) approval. At this point, AutoCAD becomes a powerful and convincing presentation tool. Presentation boards, renderings, and animated studies all can help convince clients that the developed design should be approved.

Often at this stage your client is not the only party who must review and approve the design. Depending on the nature of the project, you may need to seek the approval of building and zoning authorities, preservation groups, financing agencies, neighborhood associations, and end users who may not necessarily be the owners. Many such groups may have criteria for the formats and materials presented for approval. AutoCAD's flexibility can help enormously not only in producing finished boards, slides, and other presentations but also in refining the contents to meet the interests of your audience. If you are skilled in 3D, for example, you can create animated shadow studies to demonstrate to concerned residents how your new office high-rise will affect the level of sunlight in their neighborhood. You can present your new building storefront in its urban context to show preservation associations that you considered the existing adjacent buildings when you developed your design. In addition to presenting the drawings, you can annotate them with facts and other notations of particular concern to your intended audience.

You can also use AutoCAD to generate drawing sets to obtain permits from local and state building departments. In addition to the usual plans, you can use AutoCAD to create code-compliance diagrams, which very succinctly demonstrate how your design meets all applicable codes for egress, fire protection, and other building functions that you must address.

Document Production

The overwhelming majority of design firms use AutoCAD solely or primarily for the production of construction drawings. As stated before, in doing so they are using only a quarter of the software's power. And even during the document production stage, many firms do not take advantage of all the CAD tools available to produce a comprehensive, correct, and consistent drawing set. Intelligent CAD use can greatly improve the quality of your construction documents, which in turns helps your clients obtain better construction bids and gives you more control during the construction process.

Coordination and Quality Control

One of AutoCAD's most useful functions is project coordination. As a single database of unlimited capacity, AutoCAD stores all the information you collect at the beginning of the project and then publishes it in the form of bid documents. CAD also surveys the data and helps spot conflicts and unresolved design and specification issues.

As is often observed, CAD is merely the electronic version of the classical design coordination approach of overlay drafting, wherein you place different sheets one on top of another to review alignment of different trades and construction specialties. CAD, however, is much faster and more flexible in generating overlays, and the number and combinations of sheets that can be overlaid is infinite, limited only by your input. If you work in 3D, you take the coordination possibilities one step further.

When reviewing drafts of CAD-generated drawings, many designers plot the same combinations of layers that they would if they were running the final plots that form their drawing sets. You can gain more, however, if you plot unusual layer combinations that combine trades that are bid out separately yet together determine the success of your final design. In interiors, for example, you can more accurately assess your lighting layout by overlaying it on top of furniture, panel systems, and equipment; this composite layering helps highlight areas that remain underlit or will suffer from glare. You can superimpose electrical/data layers on the construction layers to ensure that you have not placed a light switch on top of a drinking fountain. You can use CAD files from your engineers or consultants to further check for any conflicts or inconsistencies. This overlay approach is particularly useful for projects that don't warrant elevations for every wall.

AutoCAD offers you other ways to check your work. As discussed previously, you can perform queries on attributed blocks to search for particular strings of text or numerical values. You can use the new **Spell** command to check text for errors. You can also use the **Change** or **Change All** option in **Spell** to globally revise a specific text string, if for example you wanted to change all or some of your references to wall finishes from wood ("WD") to paint ("PT").

If you link your drawings to external databases, you can use SQL links to highlight discrepancies. With a floor plan that is linked to a door and hardware schedule, for example, you can double-check that all additions, changes, and deletions to doors are reflected in both your schedule and your plans. Although American designers don't usually provide product and materials quantities in their documents, you can run totals to check against your budget; drawings linked to cost estimation databases can show you if your design is still within budget.

If you use Windows's OLE and clipboard capabilities to embed text, schedules, and other data from Windows applications into AutoCAD drawings, you should also take advantage of the quality-control tools that those applications provide to check your work. For example, prior to exporting a large body of text from your word processor, you can use its spelling, grammar, and hyphenation tools to check for errors. You can double-verify formulas and quantities in spreadsheets and databases by using the programs' query and macro capabilities.

Check Sets

Although AutoCAD is a powerful coordination tool, you still need hard-copy plotted check sets. Many designers seem to assume that computer-generated work is inherently correct; perhaps the finished appearance of most computer output promotes this assumption. But there are several good reasons for scheduling periodic check sets to review a project's progress:

- Mistakes do happen. Although some people seem convinced that computers are infallible, the fact is that CAD-generated drawings are as prone to errors as hand-drawn ones. Sometimes the software causes a glitch, but more often ordinary human error is the root of the problem. Although some CAD users are skillful at identifying and correcting their own mistakes, a second and more objective eye is often helpful. And many mistakes do not result from incorrect use of CAD commands but rather because the information which the CAD users worked from was wrong to begin with.

- No one can truly review the drawings accurately by staring at a computer monitor. Even very large computer screens, for example, cannot present at full plotting scale the entire contents of a 42" × 30" drawing. Only a properly scaled hard-copy plot can give you an indication of how the final drawing will appear in its entirety.

- Drawing coordination and quality control requires simultaneous review of all the drawings in a set, often in conjunction with written specifications. Basic AutoCAD allows you to open only one drawing at a time. The AutoCAD **Browse** option (Figure 10.6) and various software add-ons enable you to see snapshots of multiple drawings, but not with the clarity that hard copy provides. Even with the use of various quality control tools, as described in this chapter, trying to review and coordinate multiple drawings solely by computer is a risky business.

- AutoCAD's approach to displaying pen weights does not permit an accurate on-screen rendition of line weights. Unlike many paint and publishing programs, AutoCAD does not show actual line thicknesses on-screen; you specify them through your plotter settings. Even thick polylines are misleading; their ultimate thickness depends on the designated plotter pen setting. Although the use of office standards can give you an intuitive sense of the plotted pen weight, the image on the screen can still be misleading.

- Humans respond differently to screen monitors than they do to hard copy. Many CAD users find it difficult to edit work they've done on computer without stepping back and looking at their work on paper. Others are distracted by the glare of the screen and the need to bounce around and zoom in and out of the drawing file.

In addition to these reasons, check sets are essential for projects where some drafting is done by hand, as well as for coordinating drawing sets with written specifications and other components of the bid documents.

Note that check sets of CAD-produced drawings are most useful if they are planned and timed according to a schedule of target check-set dates and at certain key stages of completion specified at the onset of the project; a distributed schedule, as shown in Figure 5.3, helps communicate the goals and purposes of check sets. Scheduled check sets not only help the project team focus on production priorities; they also reduce the impulse to constantly plot drawings for the sake of plotting. No matter how fast your plotter operates, constantly running plots distracts your team from drafting tasks and quickly clutters up your flat files with interim drafts that are never as informative as formal check sets that reflect major changes. In addition, the more plots you have hanging around, the greater the possibility that the wrong draft will be referenced.

Although occasional full-scale check sets are recommended, many firms increasingly rely on smaller-scale check plots for interim reviews. Many designers run small-scale check sets on the largest size paper their regular office laser or inkjet printer will handle. Others plot check sets at 50 percent of the final size. The high quality of today's inkjet and laser plotters and printers often makes these miniature plot versions as legible as their full-size counterparts. Apart from saving space and paper, smaller-scale check sets are easier to store, copy, and distribute (especially with fax machines). Increasingly, clients, contractors, and consultants also prefer the smaller plots for routine project communications. Many also find that small-scale plots help "punch out" the essence of a design, which is useful when you're concerned with broader issues rather than with details.

Although scheduled check sets should be sufficient means for reviewing the progress of a project, many project managers continue to feel uneasy if they cannot obtain daily updates. As mentioned above, running continual plots—except with *extremely* fast-track projects—is counterproductive and potentially confusing. Several other techniques can be used to increase the comfort level of managers. The project team can run quick laser plots on 8 1/2" × 11" sheets at the end of each workday; the project manager can then file them in a project book. The project manager can view the working CAD file(s) directly on the computer every day. As described in Chapter 3, project managers can use one of a growing number of electronic file/document managers to monitor the progress of project drawings. They can also use AutoCAD Release 13's **Open** command to *preview* or *read* drawing files without worrying about inadvertently changing them. The more project managers can learn about using CAD themselves (see Chapter 5), the more confidently and self-sufficiently they can monitor the development of their projects.

Redlines

Although the principle and importance of redlines remain the same for CAD as for hand drawing, some technical differences do manifest themselves. Here are a few tips:

- Don't hide mistakes with correction fluid or glued-on sketch revisions. If direct redlines won't read clearly, tape redlined pieces of tracing paper or bond paper over the area to be revised. With CAD you don't usually need to completely *erase* the area and then redraw it. More often, you merely need to move, stretch, or otherwise *change* the work. Completely covering up incorrect work makes it harder for CAD drafters and project managers to assess what needs to be done.

- When you are indicating that something should be removed from a drawing, clarify whether it is to be absolutely *deleted* from the project—and hence the drawing file—or whether it should merely be *moved* to another layer (and perhaps hidden).

- Use whatever CAD knowledge you have to help drafters—particularly newly trained ones—to enter redlines more efficiently. If, for example, you want to change the graphic appearance of a standard electrical outlet block that has been inserted into the drawing fifty times, point out that the block for the outlet needs only to be *redefined*, not redrawn. If a curved wall needs to be reshaped, provide the geometric parameters that will enable the drafter to select the **Arc** command option that will calculate your revised wall most accurately.

Although most redlining is still done by hand, you *can* redline electronically. A simple method is to create a layer called **Redlines**, make it the current layer, and lock all other layers so that nothing else will be touched. You can then enter comments, clouds, and entities on the **Redlines** layer, and the CAD drafter can then incorporate these changes as appropriate.

You can also purchase AutoCAD add-ons that offer more high-powered redlining capabilities. Autodesk itself offers a redlining capability through its new WorkCenter and View document management software. Such applications may be overkill for small projects run by individuals or small teams, but they have the potential to be extremely valuable for large, complex projects; by using these programs you can create an electronic "paper trail" of who made revisions and when. They can also help you plan and monitor changes to complex projects, such as CAD drawings that are essentially compilations of External References **(Xrefs)** and Windows OLE–generated files and/or of the efforts of consultants from many different offices.

Links to Specifications and Cost Estimation Software

As the design development section has described, drawings and data can be linked by SQL or OLE features. During the documents phase, these linking capabilities can be used for keeping track of specifications and cost estimates. At this stage, database links are particularly valuable because you must monitor frequent changes in designs and specifications. You may want to test "what-if" scenarios about products and assemblies. In addition, you usually must watch the construction budget and, as necessary, seek more cost-effective solutions to design issues.

For most architects and interior designers, the easiest way to link drawings to specifications and cost estimation software is to purchase third-party software that can read AutoCAD files. These packages automatically load built-in blocks and AutoLISP routines to generate the links. Note, however, that you often have to use their tools from the outset of the project in order to get the benefits of these links. For example, you may have to draw all your partitions and doors with special wall and door tools in order to connect these elements to schedules and budgets.

As with all databases, the results of using a CAD-based database are only as useful as the information that goes into it. A cost estimation database that provides you with cost information, such as R. S. Means data, should always be double-checked to ensure that price information is current and reflects regional and local price indexes. Similarly, you should make sure that your electronic product catalogues do not contain obsolete products. You also must be aware when these programs calculate costs differently than a contractor would. Most cost estimation programs simply calculate construction costs by multiplying unit materials and labor cost by the quantities to be built. Contractors, however, often build into their fees allowances for fixed overhead costs and savings on bulk quantities, and unit costs cannot accurately reflect this price.

Many designers base their specifications on the AIA MASTERSPEC master specifications format, which is available as a word-processing add-on from AIA Systems.* At present you cannot directly link MASTERSPEC to AutoCAD drawing files for internal coordination purposes, but you can import your MASTERSPEC text files into drawing files. If you incorporate manufacturers' specifications into your own documents you can also import electronic versions of them, which many manufacturers provide, into drawings and specifications.

Cross-Referencing

A major challenge that construction documents present is the process of cross-referencing the different drawing components. Designers must ensure that references to detail and elevation on plans match the numbers on the detail and elevation drawings. Numbers for doors, windows, and construction notes should match schedules and notes. References to products and finishes should coincide with specifications.

A project manager's dream is to be able to link these references so that when one element is changed everything else immediately reflects that change. In CAD, automated cross-referencing is not yet built in, but you can use various features of AutoCAD and Windows OLE as well as AutoCAD software add-ons to approximate automated cross-referencing. You can, for example, use SQL to create database links between attributed drawing blocks, schedules, and lists. You can use AutoCAD External References to link together items that are common to multiple drawings.

Although future versions of AutoCAD for Windows are likely to improve in this regard, the current "plain vanilla" AutoCAD is incapable of providing airtight coordination. So often the best bet is to purchase a third-party overlay, as described in Chapter 3, that anticipates and automates the cross-referencing process. If members of your office have skills in creating AutoLISP routines and SQL database links, you can develop your own coordination overlay.

---

*For a complete product overview of MASTERSPEC and related products, contact AIA Master Systems at (800) 424-5080.

Many architects are familiar with the ConDoc methodology of drawing production and coordination. Developed by two architects, ConDoc seeks to improve both the appearance and internal consistency of bid documents by providing standards for the formats, organization, and naming of architectural drawings and a system of linking keynotes on drawings to specifications. ConDoc has been a major proponent of a five-digit keynoting system, based on the Construction Specification Institute (CSI) divisions for linking architectural symbols to a list of master notes and specifications. As have many attempts to develop industry wide standards for architects and interior designers, ConDoc has encountered resistance; but it serves as a useful template from which to develop your own methodology. AIA Master Systems now markets an electronic version of ConDoc, called MASTERKEY, that can work with AutoCAD for Windows to "manage" drawing notations and help ensure quality control; the software is designed to be modified to suit particular office needs (see Figure 6.9). If you use ConDoc,

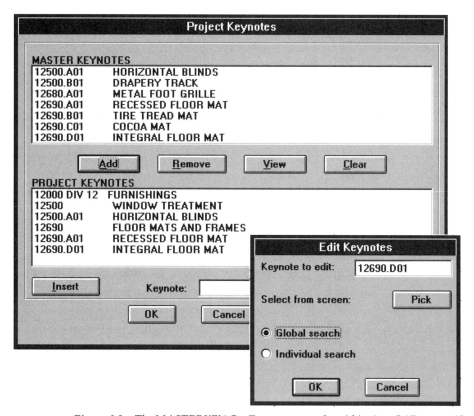

Figure 6.9 The MASTERKEY ConDoc system works within AutoCAD to provide a measure of quality control and consistency for construction documents. Image courtesy of McCarty Architects, PA, Tupelo, MS.

or if you are considering doing so, you may find the computer-based version useful.*

As previously mentioned, you can purchase an electronic version of the AIA MASTERSPEC system; although it does not provide electronic links to AutoCAD, it does include various tools for drawing coordination, including a quality-control checklist. You can use this checklist as a source of ideas for creating routines that you can run inside AutoCAD.

Coordination with Consultants

Usually from a project's start, designers are responsible for supervising a team of consultants. Most commonly, these are engineers, but they may also include experts in historic preservation, ADA accessibility, security systems, audiovisual products, kitchen equipment, and an endless array of other specialties. For CAD-based projects, CAD is a logical medium for communicating with, and coordinating, the work of all these specialists.

Ideally, you should be using CAD to communicate with your team of consultants as early as possible in the project. The document production phase, however, often involves the most intense CAD exchange. Many project managers assume CAD coordination is automatic and requires little guidance; in fact, someone has to coordinate the distribution of CAD information. To ensure smooth CAD communications throughout the process, you might do the following:

- Whether it is you or a colleague, someone on your team should serve as the project's CAD job captain. This person oversees all CAD exchanges and makes sure that all parties who use CAD are electronically compatible.

- Establish criteria for distributing CAD files to consultants. You'll need to issue updates to CAD files whenever a significant change is made to your design. For most projects, sending daily updates is neither necessary nor helpful, but you must ensure that your consultants are not using an obsolete CAD base to do their portion of the design. Experienced CAD drafters can often anticipate when design changes warrant updates, but you should not take it for granted that they will issue updates without instruction from you.

---

*MASTERKEY for AutoCAD software is sold through AIA Master Systems. Call (800) 424-5080 for more information. For a justification of the use of ConDoc in CAD production by one of its creators, see James Freehof, "ConDoc and Its Applications," *A/E/C SYSTEMS Computer Solutions*, (July/August 1993) 37–39.

- Ensure that you and your consultants are electronically compatible. Advances in file-exchange capabilities mean that almost any CAD program can read AutoCAD drawing or *.dxf* file formats (see Chapter 10). Compatibility, however, involves more than just file exchange. You need to know what operating platform consultants use, which disk formats their computers can read, whether they have access to a modem to receive and send files, and so forth.

- Anticipate the turnaround time for CAD file exchange. If you can exchange files by modem, the delivery time (depending on the modem speed and file size) is virtually only minutes. If CAD files have to be delivered on disk, you have to calculate the time for messengers, mail, or overnight express services.

- Clarify with both your staff and your consultants what kind of information will appear in shared CAD files. For reasons of both computer efficiency and liability (see Chapter 10), you should share only as much electronic information as is necessary. You usually don't need to send information about floor finishes, for example, to structural engineers. You will save your consultants both time and hard drive space if you distribute only those components of your files that are pertinent to their work.

- Automate your CAD exchange process. If you have to perform a series of tasks to create CAD files suitable for distribution, write Script or AutoLISP files to automate these tasks. You can write files for routines ranging from converting to *.dxf* files to binding External References **(Xrefs)** to your drawing. Once you automate the process, you will produce these modified files not only more quickly but also more consistently, an improvement your consultants will appreciate.

- Distribute information about your customized CAD standards and CAD drawing file setup. Even though your system for layers, blocks, drawing extents, and other CAD basics may seem perfectly clear to everyone in your office, it can appear mysterious to outside users. Many consultants will appreciate receiving a written explanation of your drawing setup; such a description can be cloned directly from your office CAD standards manual and kept in the office library of templates.

- If you use third-party software add-ons, special typefaces, custom menus, or other features that are not provided with "plain vanilla" AutoCAD, your consultants may have difficulties opening or using

> ### How Do Engineers Use CAD?
>
> Engineers are among the most aggressive users of CAD in general and AutoCAD in particular. AutoCAD is widely used in all engineering specialties, including civil, electrical, structural, mechanical, as well as HVAC, fire protection, and plumbing.
>
> In general, engineers seem to be ahead of their counterparts in architecture and interior design in terms of using CAD for functions beyond routine drafting. While much of their work does require using CAD to create layouts using symbols, stock components, and schedules, what is often more valuable to engineers is CAD's ability to help with complex calculations.
>
> CAD and other computer programs can perform and diagram in minutes calculations that, if done manually, would take hours or days. Not only does CAD make these calculations more precise, but the time savings it provides allows engineers to test a greater number of design options.
>
> Working with AutoCAD add-ons, engineers can automate and improve upon the typical design processes for their specialties. Using sound files, traffic planners review the noise levels that a new highway will produce. Lighting engineers map out lighting schemes that will reflect criteria given for illumination levels, quantities of light fixtures, and energy usage. Mechanical engineers use visualization techniques to determine whether the equipment they are designing can be accessible for maintenance.
>
> Engineers also use CAD to generate cost estimates and specifications. They can tie their drawings to databases that contain price information for standard engineered components and assemblies; with this data they can run cost-benefit analyses for different design options.
>
> Without needing to become engineering experts, architects and interior designers can benefit from understanding how engineers use CAD. Such knowledge helps in CAD file exchange; it also helps designers when they interview, evaluate, and consider engineers and other consultants for a project.

your CAD files; their CAD configuration may not recognize customized features. Usually, consultants can address this problem by substituting their own type styles, menus, and so forth. This process, however, is not only time-consuming but produces unexpected graphic results. You can minimize potential snafus by testing and editing your files in a plain, uncustomized version of AutoCAD before you distribute the files.

The Bid Process

When projects go out to bid, design teams often look forward to a momentary breather. Often, however, designers must instead answer queries from bidders and issue addenda in the form of revised plans, specifications, and sketches. CAD can help not only to generate addenda but also to track them. The changes issued with each addenda can be placed on separate layers for each revision. You can use AutoLISP routines to generate revision clouds with attached revision symbols that prompt you to provide the revision number and date.

If you link CAD drawings to cost estimation software, you can use these links to analyze and compare bids. Although most sets of bid documents still are issued in the form of paper specifications, prints, and sepias, you can expect contractors to start requesting CAD files derived from your bid documents. CAD-skilled contractors and subcontractors can use your drawings to do area takeoffs, cost estimates, and shop drawings. You can also expect that portions of the bidding process, such as the issuance of invitations to bidders and the issuance of addenda, will be conducted on-line over the Internet.* Be careful, though, about distributing your bid documents electronically; computer files can be easily manipulated, and you will be left with little in the way of a paper trail. You always want to ensure that the bids you receive are based on the drawings *you* did.

One way to circulate CAD-based drawings while ensuring that your original files remain intact is to write your drawing files to a CD-ROM disk using a CD recorder. Others can copy your files to their computers, but they cannot manipulate the disk itself (after all, ROM stands for "read-only memory").

The traditional paper-generated family of bid documents relies on formats that often have the propensity to be misleading or incomplete. With CAD in general, and 3D CAD in particular, you can easily generate designs in a wide variety of formats that clarify your design intent and allow contractors to dimension and price your work more accurately. Designers may find that presenting their work in isometric or other 3D formats results in better installations as well as more favorable bids, as contractors feel more comfortable with the base they are estimating from. For custom designs that cannot be accurately conveyed through written specifications, you can use visualization methods to convey the end results you have in mind.

---

*The American Institute of Architects already offers its own on-line service, AIA Online, which includes in its services the *Commerce Business Daily* and *CMD Early Planning Reports*, both of which can provide leads on jobs. See Chapter 10.

For many designers, the construction process is an invaluable source of feedback on the clarity and effectiveness of their drawings and specifications. Reviewing bids, answering questions from confused contractors, and reviewing shop drawings are just some of the ways the process of bidding and construction can show you how effective your work is. Don't forget to put this feedback to work in future projects. One of the best ways is to incorporate it into revisions to your customized CAD system (see Chapter 4); this way, the improvements will be available for future projects.

Construction Administration

Documentation

Throughout the construction process, designers must monitor the project's progress and issue design changes in the form of proposal requests, change orders, and sketches. You can use AutoCAD to produce sketches and drawing revisions and to log field changes and other developments. One primary advantage of using CAD to help track construction is that your CAD files will be fairly current and accurate when the project is complete; with a very fast turnaround, you can give your client an accurate as-built CAD file without having to do extensive field work or to account for undocumented changes.

Shop Drawings and Fabrication

The process of reviewing shop drawings and other submittals often requires designers to issue sketches that clarify the design intent or add an additional level of detail. You can embellish your CAD-generated details from the bid documents. If you accept any changes proposed in the shop drawings, you can update your details to reflect them, thereby documenting the changes for the record.

Increasingly, fabricators, ranging from stone carvers to millworkers to signage manufacturers, use CAD to cut materials, designs, and text. Many can use *your* CAD files as actual templates for cutting, scoring, or painting; this capability enhances the possibility that you will get the exact product and design results you specified, especially if you used custom components in your work. You may find in some instances that your ability to share your design intent electronically may reduce costs, because fabricators may feel that less risk is involved in producing a satisfactory product;

so you can help make custom work more affordable. Make sure, however, that giving CAD files to fabricators does not make you liable for any aspect of production or fabrication that falls outside your normal scope of responsibilities and expertise. And always ask for a shop drawing based on any CAD files you provided.

How Do Manufacturers and Fabricators Use CAD?

Given AutoCAD's predominance among industrial designers who use CAD, the odds are that an increasing number of products and assemblies you specify will be drawn and even fabricated using AutoCAD. The software is used in all areas of industrial design and manufacturing; the process is often called Computer-Integrated Manufacturing (CIM).

Industrial designers rely on the 3D capabilities of AutoCAD, AutoSurf, 3D Studio, and other software to design and analyze parts and assemblies. With CAD, designers can model solids and surfaces and can develop prototypes as well as production drawings that specify precise dimensions and tolerances. These designers also value AutoCAD's dimensioning tools, particularly its Geometric Dimensioning and Tolerancing (GDT) capabilities, which enable designers to meet ANSI, ISO, and other national and international standards for calling out acceptable tolerances in size and shape and for specifying material conditions and datum references.

Industrial designers also rely on Finite Element Analysis (FEA) software AutoCAD add-ons to test their products' performance under a range of simulated conditions. FEA packages are mathematical models that evaluate performance under variable conditions such as different types of stress, temperatures, electrical charges, and material characteristics. FEA is only a theoretical model and must be supplemented by other forms of modeling, but it can help manufacturers move toward the optimal shape, size, and materials for their products.

As with other AutoCAD applications, industrial designers and fabricators can link their drawing files to cost estimation databases. With these programs, they can evaluate the production break-even point for a product that sells at a certain price and is composed of a given set of materials and pieces. They can also perform "what-if" tests to compare different options.

In addition to using AutoCAD for design, manufacturers also use it for fabrication. Special AutoCAD plotting software instructs machines to create pieces from metal, plastic, fabric, wood, and other materials. This not only saves interim steps in transferring designs from the drawing board to the shop floor but also can reduce material waste; for example, CAD designers can calculate the optimal number of components that can be cut from a standard sheet of material.

In the near future, architects and interior designers can expect to see their own CAD designs directly fabricated by CAD-driven machines and will need to learn more about these applications in order to control the output better. Many notable architectural designers have branched out into areas of industrial design, producing chairs, tables, teapots, plates, and other products. If this work interests you, learn to produce your prototypes on CAD.

CAD in the Field

If you have a portable (laptop) computer, you can bring CAD to the field. If your laptop is not terribly powerful, you may want to install a minimal version of AutoCAD on it, or you might consider using AutoCAD LT for Windows. You can use an on-site CAD workstation to

- compare what you see on-site with what you've originally shown in your bid documents. You can immediately reconcile bid documents, construction drawings, and revisions,
- answer dimensional questions on-site, and
- create and document punch lists.

As mentioned below, contractors also use CAD to monitor and direct activities in the field. Many will install a CAD workstation on the construction site as their primary coordination tool. For large and complex projects, you may want to encourage your clients to request that a contractor install a CAD workstation as a precondition for hiring that contractor.

How Do Contractors Use CAD?

Construction is still considered a low-tech industry, since many construction trades do not rely on sophisticated electronics. The managers of construction firms, however, have often been at the forefront of computer usage. The more sophisticated construction firms were among the earliest and most aggressive users of computer programs that generated construction schedules and applied the Critical Path Method and other project planning tools. Today, construction firms still take the lead in encouraging the development of computer tools that will allow them to build more efficiently, profitably, safely, and quickly.

Contractors find that AutoCAD serves several useful functions. Linking CAD files to cost databases enhances contractors' abilities to generate bids and estimates more quickly and accurately. They can also price alternatives better.

Contractors themselves use AutoCAD to oversee and coordinate projects in the field. A growing number of contractors ask architects and engineers for CAD files, which they then "split off" and electronically bid out to different trades. Contractors may require that all subcontractors provide their shop drawings and other submittals on CAD, so that the contractor can superimpose the submittals on the original CAD files and compare them electronically.

CAD also enables contractors to track field changes and the effects of change orders and proposal requests. Not only can they track the cost impact of construction developments more accurately, but at the end of a project, they can produce CAD as-builts with a faster turnaround.

Move Coordination

Many design firms find a bread-and-butter source of income in coordinating moves for clients who are transferring their employees and furnishings from one building site to another. For such projects, designers provide both plans and schedules.

AutoCAD can facilitate move coordination. The most obvious use is providing installation plans that show coded symbols for offices, workstations, equipment, and other client property. CAD is even more useful when complicated moves are done in stages; you can compare the contents of various layers to ensure that all spaces are accounted for during each stage and to locate possible areas for staging items.

If your block symbols are attributed, you can link equipment and furnishings to a database. You can then use these databases to double-check that all pieces of furniture or equipment have a designated location and to verify their dimensions, weight, and power requirements. This information can help your client get favorable moving bids and avoid installation mix-ups.

Property owners who use Computer-Aided Facilities Management (CAFM) systems attach bar-code labels to their physical property and use bar-code readers to feed information into a CAFM database. This practice facilitates inventorying and tracking of property during complex moves. It can also reduce the need for extensive duplicate data entry.

Don't forget to use your CAD capabilities to produce construction signs. Your plotter will generate the requisite oversize sheet, and you can incorporate elevations and 3D rendered views in your sign. You can send a plot file to a reprographics or signage firm to produce camera-ready art. Your CAD-generated billboards not only provide required construction site information; they serve as an implicit marketing tool.

Punch Lists

You can use CAD in the very important process of completing a punch list for a project that is near completion. CAD can be used in several ways. You can use CAD files, whether you're working on-site or not, to track and record the status of outstanding tasks; if you've entered field changes and change orders into CAD all along, you can reconcile the status of the construction work with the paperwork more easily.

You can also use CAD, in conjunction with other software, to develop punch lists and to clarify work to be done. You can, for example, annotate and print portions of plans, elevations, and details that require additional work in the field, and then attach them to your punch list. If you find that work needs to be done on a series of doors, you can print individual sheets containing the door elevation along with information on finishes, door hardware, and signage; these can then be taped to the doors in question. If you have a well-integrated CAD database, you can simply retrieve the necessary punch list information rather than spend hours trying to compile it.

Postconstruction Activities, Follow-up, and Evaluation

Even when you have completed a punch list for your project, and all work is deemed substantially complete, you may still have additional tasks to complete. If you have kept your CAD-based work updated throughout the course of the project, your postconstruction work will go faster, and the final product you deliver to your client will be better.

As-Built Documents

Often the most important postconstruction task is the preparation of "as-built" or record drawings for the client. Increasingly, this task requires creating a CAD disk that shows *exactly* what was built, as opposed to what was originally bid on. If you documented construction changes in CAD as the work progressed, you can verify actual conditions more quickly. Otherwise, you should make sure that your fees cover the time required to document these conditions.

Preparing CAD as-builts often requires more than simply inputting field changes. You may be expected to merge the CAD-based work of your engineers and other project consultants and to review the consolidated files for discrepancies. Your client may expect you to edit your drawing files so that they comply with the client's CAD standards. At the same time, you will want to strip your files of data that the client doesn't want or should not have (see Chapter 10).

If you link your CAD files to specifications and other databases and simultaneously update the database portion to reflect as-built conditions, you can provide your client with a fully integrated as-built manual for future reference and maintenance purposes; this is an incredibly valuable facilities management tool, and you should, one way or another, be compensated for it.

Evaluation

The essence of the architecture and interior design professions is that our work is classified not as a trade but as a *practice*. Implicit in the term "practice" is the recognition that our work is always evolving and can always be improved. In order to know what aspects of our work require improvement, we must evaluate the work we have done to date.

Whether or not a client pays for a postevaluation study, you should take the time to review completed projects and to learn from them. While the primary purpose of review is for your overall practice, the conclusions should also indicate where refinements to your CAD-based operations are justified. Some of the issues that your evaluation should address include

- Did the project meet the deadlines, and if not, why?
- Was the time allocated for designing, drafting, and plotting adequate? Did your design fee cover it?
- Did the design meet the construction budget, and if not, why? Was it a question of the content of documents, such as unclear specifications or details that turned out to be unexpectedly costly? Could construction overruns have been reduced by more effective use of CAD?
- Did bidders and contractors find your drawings and specifications straightforward, or did you have to issue frequent clarifications and revisions?
- Were you able to incorporate changes to the project quickly and consistently?

Ideally, the evaluation process should be built into your routine project cycle as shown in Figure 6.10.

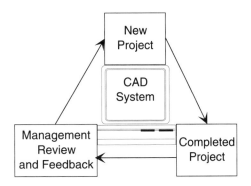

Figure 6.10
Feedback should be an integral part of the CAD-based design process.

Promotion

A successfully completed project can serve as a promotional tool to help you obtain additional projects, whether repeat business from an existing client or a project from a new client. At this stage, you have completed the cycle of design activities, as shown in Figure 6.2, and are ready to incorporate the results of your latest CAD projects into your marketing program.

In addition to marketing your own work, you can use CAD to help your clients market theirs. Clients such as real-estate developers can use your plans and renderings to market their new buildings to potential buyers and tenants. Nonprofit organizations such as schools, universities, and hospitals need presentation materials to make appeals to potential donors to raise the funds for the buildings you've designed for them; presenting building perspectives, elevations, and plans makes fund-raising appeals more effective. Government agencies will need visual aids to explain to taxpayers and the public in general where tax dollars designated for building projects are going. The more you can anticipate the promotional needs of your clients, the more you will be able to use your CAD system for activities beyond building design.

Although most promotional materials are still in the form of hard copy, such as printed brochures and magazine articles, you should expect to see your projects incorporated more and more into audiovisual and animated presentations and electronic media. You may even find yourself on the electronic highway. Recently, the New York Public Library created an Internet World Wide Web site to provide Internet access to its holdings in its new Science, Industry, and Business Library facility; the architect's 3D renderings of the library were used to introduce users to a visual electronic "virtual tour" of major library spaces.

7 Special AutoCAD Applications for Architects and Interior Designers

A major benefit of AutoCAD's "open architecture," that is, its ability to be customized, is that you can use AutoCAD for almost any computer-based work that requires high-powered drafting in conjunction with other tools. Since the software itself is so professionally generic, however, many architects and interior designers remain unaware of all the possible ways they can use AutoCAD. Moreover, although both Autodesk and many professional organizations distribute information about applications* of AutoCAD, the information is so overwhelming and so diffuse that few designers would ever have the time to collect and absorb all of what is available.

This chapter describes major AutoCAD applications that architects and interior designers can incorporate into their practice. Some of these require additional software or hardware, but others merely demand that users understand the full range of AutoCAD's capabilities and use them with the view of improving upon traditional means of practice.

---

*In this context the term "application" refers to a way to *apply* AutoCAD to various tasks; it does not necessarily refer to software applications.

The applications described here represent only a slice of what designers can do with powerful software like AutoCAD. More and more uses are constantly emerging. One way to keep abreast of developments is by reading AutoCAD-oriented publications, joining professional organizations and AutoCAD users groups, as well as by reviewing *The AutoCAD Resource Guide*. Another, simpler way is to use the power of analogy: if you can do something manually or in another computer application, you should be able to find an analogous AutoCAD command or application. Always look for ways to do things without erasers, without tape and clear adhesive sheets, without taking millions of measurements, and without trying to perform the same step more than once. With this approach, you will discover ways to get more from your AutoCAD investment.

Many third-party software packages for AutoCAD can help automate some of the applications described here. In addition, you can apply whatever in-house skills you have in AutoLISP routines, ADS, 3D, and other AutoCAD-related commands and utilities to develop your own computer-based approach to carrying out typical design tasks.

The Americans with Disabilities Act and Code Compliance

A primary professional responsibility of architects and interior designers is public safety and assuring every person's right to have access to public places. Designers spend much of their time ensuring compliance with federal, state, and local building codes and, since 1992, with passage of the Americans with Disabilities Act (ADA) of 1990, a civil rights law. Code compliance places constraints on design, but it also generates job opportunities: many design firms include ADA-compliance surveys, for example, among their services.

AutoCAD is a useful tool for ensuring that a project complies with the ADA and other building codes, whether at the beginning of the project or during the creation of construction documents. Most of the techniques listed below can be carried out using ordinary CAD drafting routines; you can also find third-party software add-ons that will accomplish similar ends.

- Develop an office-standard library of blocks for handicapped person (HP) symbols and fixtures for use in *both* floor plans and elevations. As the library expands, divide it into subdirectories for major groups of HP blocks, such as toilet fixtures, ramps, parking spaces, pay phones, and water fountains.

Design HP standard blocks so that their *insertion points* automatically place HP fixtures in the proper location relative to floors, walls, doors, and other building elements. For example, if the ADA calls for HP toilet grab bars to be located 6 inches from a wall corner, make the insertion point for a grab bar block not on the bar itself but rather 6 inches from the bar's end. When the blocks are inserted into drawings they will be positioned accurately relative to other building elements.

- Create AutoCAD *overlays* of typical plan layouts, sections, and elevations for all portions of a building or site that must reflect certain dimensions or arrangements of fixtures. These overlays should reflect required minimum or maximum dimensions. These requirements can then be blocked into a plan of an existing site or into a schematic layout to test the feasibility and legality of existing conditions as well as of proposed designs or alterations.

- Use AutoCAD templates to reconcile differences between the ADA and building codes. Although federal and state building codes are rapidly incorporating ADA requirements, discrepancies in requirements still exist, and, where they do, the more stringent requirement prevails. Overlaying AutoCAD templates can help identify conflicts and clarify which standards apply.

- Write AutoLISP routines to create interactive dialog boxes that test building elements for compliance. A sample AutoLISP routine, for example, would ask you to use the AutoCAD **Dist** command to measure all exit corridors to see if they meet the minimum width requirements or check if any dead-end corridors exceed 20 feet.

- Create attributed blocks that symbolically represent an actual or possible ADA or code violation. Attach these attributes to an external database software package that describes the violations, how to resolve them, and the potential costs of doing so. This data can be used to generate cost estimates for renovations that are necessary to ensure code compliance; estimating costs is particularly important where ADA is concerned as compliance with ADA in renovations is required only if resulting costs are considered "not disproportionate" to the overall project scope and budget.

- If you regularly produce ADA studies for clients, consider building a library of templates for studies, budgets, and drawings as well as blocks of HP fixtures, stairways, and other standard building elements.

THE AMERICANS WITH DISABILITIES ACT AND CODE COMPLIANCE

Table 7.1 Checklist of items for an AutoCAD-based library of blocks and templates that comply with the ADA and codes

Note: If you practice in several states or countries and therefore must comply with several building codes, copy and then modify your library blocks to reflect these variations. You can also convert your blocks to metric if necessary.

Topic	Sample AutoCAD Files
Building entrances	plan of entrance layout with required wheelchair turning radius and door size and swing orientation
Doors and doorways	plan of a typical 36"-wide door with 18"-wall at the door's pull side
Egress paths and exits	an AutoLISP routine that asks you to check the distance between building exits, the length of dead-end corridors, and the maximum travel distance from various building points to the nearest exit
Elevators	elevations of typical layouts for elevation control panels and railings
Furniture and millwork	elevations of counter heights for transaction counters, reception desks, and cabinet counters
Height and area limitations	an AutoLISP routine that, given certain parameters such as building structure, zoning requirements, occupancy type, maps out code-compliant plans and elevations on a given site
Kitchens	plan and elevation view of typical layouts for kitchen cabinets and appliances
Parking and drop-offs	plan view of typical parking lots with minimum widths for standard cars, and lift-equipped vans and the minimum number of required HP parking spaces
Places of assembly	plan view of an auditorium with code-compliant seat arrangements and quantities, HP seating areas, and exits
Ramps	plans and elevations of HP ramps showing minimum and maximum widths, railing configuration, and location and size of landings
Signage	elevation detail of ADA-compliant wall plaque at doorway entrances
Stairs and smoke-proof enclosures	an AutoLISP routine that, given certain parameters such as floor-to-floor heights and building occupancy, will lay out a stairway that meets code requirements for width, risers, treads, landings, and smoke-proof enclosures
Telephones, water fountains and other fixtures in public spaces	plans and elevations of ADA-compliant water fountains
Toilets and baths	plans and elevations of ADA-compliant private HP toilets, with toilet, sink, and grab bars in proper configuration
Turning circle	block of standard HP symbol of a wheelchair occupant within a 5-foot turning diameter
Walks, pathways and curb cuts	plans of proper dimensions of curb cuts at sidewalks

Area Takeoffs

For many clients, one of the most important services a design firm can provide is the preparation of area takeoffs or space calculations for new or existing buildings. Since area takeoffs have an immediate and major impact on business decisions, building owners consider this service essential. Indeed, many firms successfully market area takeoffs among their major services.

Area takeoffs are used to provide the base for many calculations, including

- the rentable space in commercial leases,
- owner percentage interests in condominium documents,
- overhead rates used in federal and state contracts and grants,
- construction estimates and material takeoffs, and
- property values and taxes.

Accurate takeoffs are essential to building owners because they form the basis of leases, contracts, taxes, and other arrangements that affect income and expenses; both the measurement of space and the way it is allocated significantly shape the bottom line. Area measurements also enable employers to more precisely allocate space to their employees, equipment, and other assets, and this in turn shapes the organization's efficiency and productivity.

Therefore, both building owners and managers as well as designers have strong incentives to use measurement tools that track space accurately, consistently, and quickly. AutoCAD is a natural tool, as it documents areas at "full scale," measuring to a level of precision impossible with a 1/8" scale hand-drawn plan and handheld architectural scale. When you are talking about large floor plates with square footage of five or six digits, a 3" difference in the location of corridor partitions or exterior walls—which a 1/8" ruler cannot accurately measure—can add up to hundreds and thousands of dollars to be accounted for.

Although more precise area calculations can justify an organization charging more for its spaces, they can also demonstrate when a building owner is charging for space it either doesn't own or is double-counting. Take heed that clients may not always welcome the results of your area calculations and may contest them.

Hand-done area calculations are inherently rough, especially when they are taken from small-scale floor plans and rounded to the nearest foot or tens of feet. Moreover, if the floor plates involved are complex and oddly shaped, you have to spend hours dividing the spaces manually into rectangles or other easily measurable polygons.

CAD programs such as AutoCAD allow you to calculate areas not only instantaneously but far more accurately, especially where irregular shapes are concerned. AutoCAD's **Area** command calculates areas—as well as perimeters—of any line, closed polyline, or polygon in inches and feet (or metric), to 1/256 of an inch, as shown in Figure 7.1. Unlike manual take-offs, AutoCAD easily calculates areas for complex, irregular shapes and can also add or subtract multiple areas as required. If the **Area** command is properly used, you don't have to worry so much about mathematical errors. And if the area being measured needs to be modified, you can simply reshape the area polylines, and the **Area** command will reflect the new square-foot measurements instantaneously.

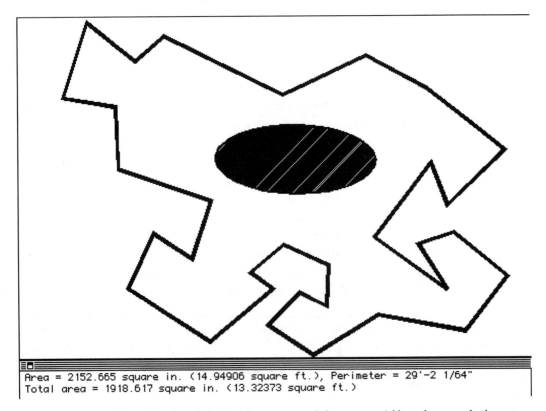

Figure 7.1 The AutoCAD **Area** command shows you quickly and accurately the area and perimeter for lines or closed polygons of any complexity. Here, the **Area** command has taken the area of a complex polyline and subtracted the ellipse (shown as shaded) in the polyline's center.

Area Takeoff Methodologies

In AutoCAD, calculating building areas is simple, but the underlying methodology is often extremely complex. There are several reasons for this complexity:

- The various users of area takeoffs define space differently and with different goals in mind: a contractor is interested in the *buildable* area; a flooring installer in the *carpetable* area; a real estate developer in the *rentable* area; a potential tenant in the *usable* area. Different uses require different takeoff methods.

- Area takeoff methodologies are not mandated. Although various methods have been introduced during the past two decades and several are now widely used, no method has been uniformly accepted. With the exception of government projects, which must adhere to the General Services Administration (GSA) standards, there is no legal or professional requirement stipulating that building owners use any particular takeoff method, and therefore there is no practical incentive to use any method that seems difficult or potentially unprofitable to implement.

- The various takeoff methods, while generally applicable to most buildings, cannot always address the specific needs and complexities of individual spaces. Many buildings, for example, have rooms that don't fit tidily into the space categories that typical takeoff methods promote. In particular, older buildings constructed prior to the introduction of formal takeoff methodologies are apt not to conform to standard space categories.

- Especially with regards to older buildings, you may not have sufficient information to apply standard methods properly. If, for example, floor plans for a building are not detailed enough to show the actual centerline of the exterior window glass or to distinguish between different types of nonusable space, building owners either must settle for less precise area takeoffs and classifications or pay to have their property accurately surveyed and building plans redrawn.

Over the past decades, several professional organizations have attempted to develop standard area takeoff methodologies. These methodologies tend to reflect the concerns of the organizations' membership, which most simply might be divided into owners and tenants or users.* The following organizations have developed some of the more popular guidelines:

---

*See Lawrence W. Vanderburgh, "By Whose Standards," *Interiors,* (February, 1994): 26–28. Vanderburgh also provides a useful graphic comparison of how five major space management methodologies allocate spaces among the major space categories.

- Building Owners and Managers Association (BOMA)
- General Services Administration (GSA)
- International Facility Management Association (IFMA)
- The Real Estate Board of New York, Inc. (REBNY)

In many contracts between owners and designers, one of these standard methods may be specified as the basis for the takeoffs the designer agrees to provide. Other agreements may reference a client-specific approach that the client's real estate or facilities management departments have developed and written guidelines for; append a copy of such guidelines to your design contract if you're expected to adhere to them.

Even when a standard or client-specific method is mandated, designers should always be conscientious about reviewing with their clients any areas that do not fall neatly into given space categories. These matters should be addressed early on because the area categories, such as rentable, usable, and common spaces, form the basis of often Byzantine calculations used to determine the square footage totals used in leases, grants, and other agreements. Once these calculations have been implemented, modifying them can prove difficult if not embarrassing. Typical building elements that may be open to interpretation include

- public corridors and corridor walls
- exterior walls
- exterior windows
- partitions between shared tenant or user spaces
- building service areas
- vertical penetrations (elevators, mechanical shafts, and so on)
- roofs, penthouses, and basements

Producing CAD-based area takeoffs is not the same as performing the actual, often complex, calculations required for establishing rental fees and other numbers. Make sure your contract clarifies what numbers you are to provide. Carefully document the basis of all calculations you make, in case they are called into question by your client or other parties. You should always be able to demonstrate how you arrived at your numbers.

A Suggested Procedure for Area Takeoffs

The first and perhaps most important step in developing a proper takeoff is to agree upon the floor plate that is to be measured. This floor plate may or may not reflect reality: with many older buildings that have undergone decades of undocumented renovations, an accurate base cannot be determined without either gutting the building or arranging for a professional survey. Either way, the floor plate that is accepted by all parties to be the working base for area takeoff calculations should be in effect "frozen." Any fundamental changes to this base may entail reworking takeoffs from scratch.

The second step is to review what you have in your CAD files, if any are available, and compare it to existing conditions. For new construction, you probably have a recent CAD file—possibly one you provided—and you just need to do cursory field checks. For older buildings it is less likely that you will have a CAD file already available, and if one exists, it probably contains severe and mystifying discrepancies. You may also find that your CAD base is not drawn in a way that enables you to comply with the client's area takeoff guidelines. For example, you will be asked to take areas from the centerline of the building's exterior glazing, yet the exterior perimeter conditions in your CAD file may be too inaccurately drawn to be meaningful; this is particularly a problem with CAD files created primarily for interior space, where the exterior wall and window conditions are considered a given and not included in the scope of work. Therefore they were not considered worth drawing in detail. To correct such defects can involve considerable staff time in field verifications and CAD work. Therefore, wherever possible, include in your fees sufficient staff hours to do the necessary work. Since CAD files of this nature may form the basis of calculations and negotiations for years to come, a correctly drawn floor plate can be well worth the fee.

The following procedure describes a typical way to document area takeoffs for a tenant space in a typical single floor of a commercial office building. Regardless of the takeoff method(s) used, the overall approach will work for most situations.

1. Open the drawing file that contains the proper floor plate. **Thaw** and turn on only layers that define the floor's base building conditions and partition layout; such layers include exterior walls and windows, new and existing partitions, core walls, elevators, and shafts. **Freeze** (hide) all other layers except those—such as the Room Name layer—that may help you identify spaces on the visible layers.

2. Make a new layer for the polylines used to create the takeoffs. This should have a logical name, such as Area_Takeoffs. Assign a color to this layer that will cause it to plot with a thick pen weight. As a side effect of using the **Make** command, your new layer becomes the current or active layer; lock all the other layers so that you can see but not change them.

3. Start drawing a 1-inch-thick Pline around a discrete area to be measured. Starting at the upper left-hand corner of the floor plate, wrap the polyline around the interior face of the exterior wall, along the window surrounds and face of window glass, and through the centerline of demising partitions, or as otherwise directed by your chosen methodology. Make sure that the final segment of the polyline meets the starting point of the first segment, thereby creating a *closed* polyline.

4. Repeat this process for all other distinct areas. This includes additional but physically separate tenant spaces, elevators, mechanical shafts, and stairwells.

5. Use the **Area** command to measure the area and perimeter of the polylines. Use the **Area Object** mode to measure a single polyline or polygon; use the **Add** and **Subtract** mode to add or subtract areas from a previous object. When you have accounted for all areas, hit the Enter key, and the screen will display the area in both square inches and feet, as shown in Figure 7.1. If the area totals required a lot of adding and subtracting, double-check your totals by applying the **Area** command to individual polylines or by using the **Boundary** command to create a single polyline from a set of several.

6. If appropriate, use the **Dtext** or **Mtext** command to create text lines that record the areas for each space you've measured.

7. If the areas are assigned to different categories, such as tenant spaces, shared building spaces, and overhead, create hatch patterns (using the **Hatch** command) to separate these spaces graphically and create a table that explains the hatch patterns and shows the resulting totals for each category.

8. Type in a description of the space, the purposes for the takeoff, the scale, date, and other pertinent information as appropriate. Use a type style that will be legible when plotted.

9. Print or plot the file onto paper. For area takeoffs, most clients prefer that takeoff plots appear on letter-, legal-, or tabloid-size paper that can be easily copied and bound with other report materials.

10. Create an archival copy (defined by a file extension such as *.arc*) of the file used to create the area takeoffs and store it in a safe place for

future reference. It is essential that you be able to retrieve the file in case the takeoff methodology is revised or the totals are questioned.

If you do area takeoffs routinely, take steps to automate the process. You can write AutoLISP routines to set up the layers as described above, hatch the polylines, and punch in text showing the areas for each space.

Area Takeoffs and Databases

Since AutoCAD-based area takeoffs are a means to connect graphic space with other sets of data and to perform and track often complex calculations, they are a logical application for AutoCAD's Structured Query Language (SQL) database feature. Indeed, many third-party applications for AutoCAD offer a built-in area takeoff feature for connecting floor plans to spreadsheets and relational databases. Most create "asset links" between the area polylines in CAD floor plans and data fields in a database program.

Most of these add-ons are priced at $1,000 or more on average, but you should consider investing in one if your firm aggressively markets area calculations as a specialty or if your clients include major corporations or institutions whose building projects require complex area calculations during the programming phase of design.

An alternative to purchasing an add-on for area calculations is to create your own SQL database link, working with the generic database packages for which AutoCAD provides SQL links. You'll need in-house programming skills, but with some work you can produce a unique and valuable tool that you can market, possibly earning a return on your own research and development.

Computer-Aided Facilities Management (CAFM)

The ability of AutoCAD's Structured Query Language (SQL) feature to create "intelligent" links between CAD drawings and other databases makes AutoCAD a natural for Computer-Aided Facilities Management (CAFM). Indeed, many of the most widely used CAFM software packages are AutoCAD add-ons (CAD-driven CAFM tools); others make a point of being able to read AutoCAD drawing or *.dxf* files but don't depend on AutoCAD to work.*

---

*For an informative comparison of five AutoCAD-based facilities management add-ons which include stacking and blocking features, see Eric W. Janson, "Manage Your Assets: Five Facilities Management Software That Work with AutoCAD," *CADENCE Magazine,* January 1995, 67–76.

CAFM software is marketed primarily to the facilities management departments of corporations, hospitals, educational institutions, government agencies, and other organizations for which real estate is a major activity and operating expense. By making connections between graphic renderings of physical assets and other property-related information, CAFM offers the following benefits:

- enables users to inventory their assets, including plant, equipment, and furniture as well as people, faster and more accurately
- reduces or eliminates duplicate data entry
- enables users to maintain and update links between real estate, employees, and financial projections and other data, and provides internal consistency among different sets of information
- enables users to generate more accurate forecasts of spatial needs in response to business decisions
- enables organizations to charge their space more precisely and fairly to users
- allows users to maintain building assets properly and to reduce deferred maintenance levels
- enables users to make better strategic management decisions about real estate

Although CAFM packages vary in their capabilities, most offer tools for carrying out the following activities:

- space planning, design, and management
- scenario testing, using adjacency, blocking, and stacking diagrams
- architecture and interior design services
- development of space standards for employees, rooms, furnishings, and equipment
- lease management
- maintenance schedules and operations
- construction management
- move coordination
- asset inventory management, including tracking of equipment, furniture, and other physical assets
- project budgeting and scheduling
- planning and management of basic building systems, such as electrical, communications, security, and fire protection systems

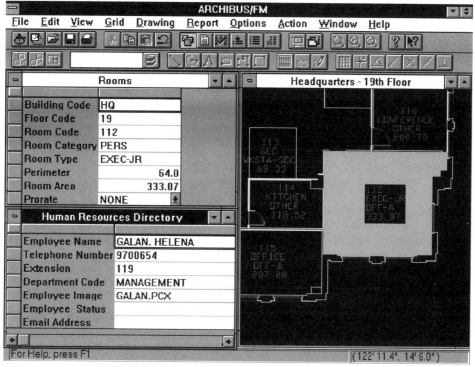

Figure 7.2 CAFM databases enable designers, building owners, and others to analyze, plan, and manage their real estate more efficiently. CAFM software such as ARCHIBUS/FM 10, shown here, link AutoCAD drawings to SQL databases, so that the impact of organizational developments—actual or projected—will be reflected in both drawings and other sets of information. Image provided courtesy of ARCHIBUS, Inc., Boston, MA.

- production of directories and reports, such as telephone lists and insurance inventories

CAFM and Design Practice

Although CAFM software packages are designed primarily for use by facilities departments, architects and interior designers would do well to familiarize themselves with these products and even consider using them in their practice for the following reasons:

CAFM affects your practice. As more and more corporate and institutional clients use CAFM, CAFM shapes how the clients determine their space needs and how they want your services delivered to them. CAFM can affect everything from CAD layer standards to programming. The more you understand how CAFM affects your clients, the more responsive you will be in delivering design services, and the more smoothly you will be able to communicate with your client.

CAFM can benefit your practice. CAFM can automate tasks that you perform routinely in design practice. With CAFM, you can develop and track client programs and test them in floor plates, create adjacency matrices, and generate blocking and stacking diagrams.

CAFM is a growing market for the A/D professions. Most recent surveys suggest that CAFM has penetrated less than 25 percent of its potential market of facilities management departments. This, combined with the trend in many organizations to downsize and subcontract some or all of their facilities functions, produces a potentially large market of clients who need CAFM-based services but can or will only acquire them through external sources—such as design firms.

The use of CAFM in-house has not been uniformly successful. Although many organizations have seen a return on their CAFM investment in terms of operating efficiency, others have not seen the purported benefits. Sometimes CAFM dissatisfaction results from improperly acquired or installed hardware and software; in many situations, it may be a reflection of inadequate training or ignorance about how to set up and operate a CAFM system, or the difficulties of data input. Regardless, dissatisfactions with CAFM create an opportunity for designers, who can provide *external* CAFM services that address *in-house* shortcomings.

CAFM overlooks some markets. In terms of costs, implementation, and basic features, most off-the-shelf CAFM packages seem directed to larger corporations, hospitals, universities, and other larger organizations with formal real estate departments. They are not products that a small nonprofit, restaurant, art gallery, residential landlord, or tiny but fast-growing entrepreneurship might consider relevant, much less affordable. Yet almost any organization with assets can benefit from a variation on CAFM that is affordable and addresses specific needs. If you have a particular industry specialty or type of client that could benefit from a CAFM package tailored to specific needs, you may be able to market your own variation on CAFM.

Selecting a CAFM System

If you choose to invest in an existing CAFM product, you should apply the same criteria as you would to any other AutoCAD add-on or complementary software package (Chapter 3). You should investigate CAFM packages with particular care for two reasons. First, they are generally fairly expensive packages—often more expensive than a single AutoCAD license. Second, they are rarely available from computer retail stores or mail-order houses; instead, they tend to be distributed exclusively by designated vendors who offer support and customization along with the software. These vendors frequently offer sleek product demonstrations but may be elusive in explaining the actual mechanics of the software. For CAFM programs that you are seriously considering, make sure you have an opportunity to actually use the program, not just view a demonstration disk or video. You might request that one of your own projects be used as a beta test. Ask for names of current clients as references and review the following CAFM features in depth with the vendor:

- price: list versus net. Are discounts available for multiple stations?
- network capabilities and licensing policies
- specific applications offered:

    communications and security management

    construction management

    equipment, furniture, and other asset management and tracking

    lease and real property management

    maintenance

    scenario, blocking, and stacking analyses

    space design and management

- features: Are the features fully integrated in one package? Must you buy separate modules? If the latter, how do they work together?
- customization features: Can you customize the software, or will the vendor do it? If the latter, are additional consulting fees involved?
- area takeoffs: Do they reflect the major area takeoff methods? Can you modify them or add your own space categories? Do they read and write in metric?
- reports: How many types of reports are available? What do they include? Do they meet typical needs? Can they be customized?
- number of records and fields allowed: Does the program have a ceiling on how much data it can manage?
- data verification: Does the program offer ways to verify the integrity of the database? Does it prevent duplicate entries? Does it resist casual or inadvertent data manipulations?

- query capabilities: Does the program impose limits on how you sort and retrieve information?
- platforms: Will the software run on Windows, DOS, Unix, Macintosh? Is cross-platform exchange feasible and reliable?
- relationship to AutoCAD: Does the software require AutoCAD to operate, or does it run on its own? Does it read AutoCAD *.dwg* or *.dxf* files? Is the CAFM software company a Registered Autodesk Developer?
- upgrades: How frequent are they? What is the lag time behind AutoCAD upgrades? How much do they typically cost?
- relationship to Windows: Can you import/export data from or to other Windows applications? Does it have OLE capabilities? In short, can you reuse data you've already collected in another computer database?
- requirements for hardware, hard-drive capacity, and RAM
- training options: manual, on-line, provided by vendor
- documentation: manual, on-line, vendor-dependent?
- user base: What is the number of installed users? Is there access to user groups? Are users representative of the clients for which you are considering CAFM or of other design firms that are similar to yours in terms of project types and scope?

Most CAFM add-ons for AutoCAD are SQL databases. If you have or are willing to acquire in-house database programming skills yourself, you can develop your own CAFM package that you can then market as a proprietary product in conjunction with your other design services; a number of design firms have done this. With user-friendly databases such as Microsoft Access for Windows, you can produce unique variants on CAFM that meet the needs of your target markets. You can augment your products' capabilities by using other Windows features, such as OLE, to link text, numeric data, and graphics from other applications to your AutoCAD drawings.

If you go to the effort of acquiring a CAFM database or developing your own, the return on your CAFM investment will increase if you expand its uses beyond the conventional space management tasks. Remember that you've acquired a high-powered multiuse toolkit. As an example of an atypical but useful application, CAFM was used for "damage documentation" by a hospital following the 1994 Los Angeles earthquake.*

---

*"CAFM at Work in Earthquake Recovery." *A/E/C Systems Computer Solutions*, (May/June 1995), 36.

The top management of ARCHIBUS, one of the pioneers in CAFM software add-ons, has recently announced that CAFM is being superseded by CIFM—Computer-Integrated Facilities Management. This shift in terminology reflects organizations' needs to go beyond local CAFM functions and work with organization-wide systems as well as their desire to use CAFM data not only for typical real estate functions but also for general strategic planning, operations, and financial management tasks.

Color in AutoCAD

Traditionally, AutoCAD has been associated with black-and-white 2D drawing production. In the early versions of AutoCAD, only 16 colors were available, and their primary purpose was to allow drafters to separate entities by layer on-screen and by pen weights on plots. Recent AutoCAD releases, however, in combination with improvements in monitors and video cards, enable users to see and specify up to 255* colors at one time in AutoCAD. With the plethora of fast, reliable, and increasingly affordable color printers and plotters, along with the greater ease of 3D work and rendering, more and more design firms are starting to take advantage of AutoCAD's built-in color capabilities in their output.

Although AutoCAD does not offer the extensive color management features of desktop publishing or paint software, you nonetheless can plot with 255 colors in solid and halftones lines. Such a palette suffices for most situations. Color output can offer several advantages that hasten the return on your investment in color-producing devices:

- Although with color AutoCAD output you cannot replicate the appeal of beautifully hand-rendered watercolors or pencil sketches, you can nonetheless produce extremely sharp and compelling presentations.

- You have greater flexibility and options in your output than with only monochromatic printers and plotters and can make creative use of your existing AutoCAD files.

- You can eliminate hours of handwork with markers, pencils, transfer lettering, films, and patterned tapes, and when producing multiple color copies, you have greater control and consistency in output.

---

*On computer monitors that display 256 colors, AutoCAD will let you see up to 256 colors simultaneously, but one of those is assigned as the background color for the screen display. While this background color can be adjusted, it cannot be included in the colors available for output.

- You can quickly and easily test a wide range of color palettes and change color schemes.

- You can easily produce multiple copies of color output, and in a variety of sizes and scales, thereby eliminating the turnaround time and anxiety required in sending drawings to a reprographics service.

Uses for Color Output

You can use AutoCAD color output for a wide array of tasks. Typical design firm applications include the following:

- blocking and stacking diagrams
- presentation boards to show a new client how your firm's approach to design meshes with their philosophy
- composite plots of different building systems in a floor plan, presenting each major system—HVAC, electrical and data, lighting, and so on—in a different color to help highlight potential areas of conflict and the need for coordination
- full-scale mock-ups of signs and large scale graphics
- a collage of plans, building elevations, and street scenes for a real estate developer to present to prospective tenants
- titles and grid lines for a materials and finishes board

For routine drafting purposes, many CAD users find that 255 colors is overkill, so they resort to the original 16 AutoCAD colors. They also speed up their work during intense drafting sessions by adjusting their monitors to read only 8 or 16 colors; the screen regeneration time is then less than it would be for 255 colors.

The AutoCAD Color Palette

AutoCAD's unique palette of 255 colors is referred to as the AutoCAD Color Index (ACI). As Figure 7.3 shows, the ACI consists of 9 standard colors and six grays—the original AutoCAD colors—and 24 additional hues, which are each broken down into tints and tones.

The ACI is unique to AutoCAD and was designed initially with the intent of enabling drafters to visually separate on-screen entities they were working with as well as to create more variety in plotted line weights. The ACI does not directly map to the Pantone Matching System or other

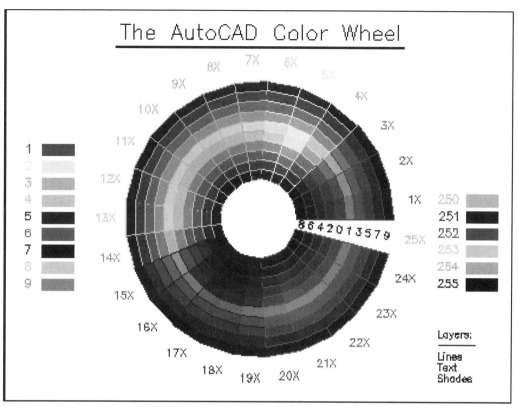

Figure 7.3 The AutoCAD Color Wheel can be found in the file Colorwh.dwg, which is included among the AutoCAD software files.

well-known international color reference systems used by printers but does apply indexes used to measure color on computer monitors; these can be accessed by selecting the **Color** option in the **Light** dialog box.

In addition to specifying colors by ACI numbers, you can specify ACI colors in terms of either the red/green/blue (RGB) or hue/lightness/saturation (HLS) color indexes. In Figure 7.4, for example, ACI color 11, a maroon tone, maps to the RGB triplet 1.00, 0.50, 0.50. Although the RGB and HLS indexes measure screen colors accurately, they cannot be used to specify process colors that print shops use. If you have a color reference book or a paint or desktop publishing software program that maps colors across all the major color indexes—such as cyan/magenta/yellow/black (CMYK), Commission Internationale de L'Eclairage (International Commission on Illumination or CIE), RGB, and HLS—you can attempt to map ACI colors to other color systems by extrapolation. According to the **Light** dialog box, AutoCAD Color 2 (yellow), for example, equals

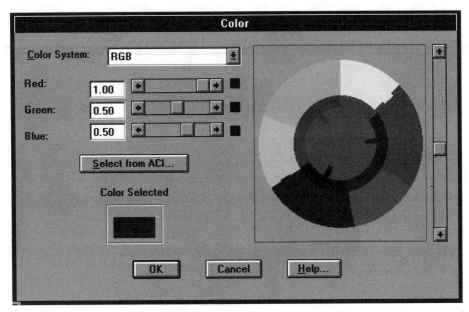

Figure 7.4 The AutoCAD **Color** option in the **Light** dialog box enables you to select colors according to both the ACI system and RGB (shown here) or HLS specifications.

RGB triplet 1.00, 1.00, 0.00. Color mapping programs would show this as HLS 16.67, 1.00, 1.00, and Pantone process color yellow 1-3. Such extrapolation is tedious, however, and doesn't work in reverse; that is, if you attempt to match a color from another color system with an ACI color. If you are skilled in AutoCAD Development System (ADS) programming, you can use an AutoCAD application, *Colext.c*, which maps the ACI index into seven color systems. You can also write a plot file containing algorithms that map ACI colors to Pantone, for example; this file can then be read by another graphics program. In most cases, however, if exact color specifying is important for your purposes, you'll often do best to export your AutoCAD images to a paint or desktop publishing program that offers more accessible color referencing and works with a range of color management utilities now available for Macintosh and Windows software platforms.

You cannot modify or expand upon the ACI colors for basic drafting purposes. However, when you use the **Render** command to work with color and light for models, you can go beyond the existing 255 colors by moving the scroll bars for red, green, and blue colors as shown in Figure 7.4. But, you can never apply a palette of more than 255 colors in any one drawing file.

Color On-Screen versus Color Plotting

With many paint and desktop publishing software programs, the correlation between color on the computer screen and on the resulting printout is high. Although a color monitor cannot show exactly what will print, you nonetheless can expect a teal rectangle on your screen to print (more or less) as a teal rectangle. Many desktop publishing programs now work with systemwide color calibration utilities that map colors consistently across different graphic software programs and ensure that printed colors are as close to screen colors as is technologically possible. Color calibration software is particularly useful if you are preparing materials to send to printing services; the calibration makes up for the fact that neither your monitor nor your own color devices offer accurate renditions of printers' inks.

In AutoCAD, color is not necessarily a "what you see is what you get" (WYSIWYG) feature either for on-screen colors or for output. First, although the ACI colors may map consistent RGB triplets, they may appear on different computer screens quite differently. The type of screen monitor and video card can make a major impact on how colors appear on-screen. Adjusting the monitor (if you have the choice) between 8, 16, 256, or 1 million colors will also significantly alter how AutoCAD screen colors appear. With a 16-color monitor, for example, ACI colors 17 through 255 will appear identical.

Second, on-screen colors assigned to layers and entities in AutoCAD files are not a direct indication of what will plot. As discussed in Chapters 4 and 8 and as shown in Figure 4.20, where output is concerned, each color assignment (depending on the output device) specifies a unique combination of colors, linetype, and width. Color numbers *may* be an indicator of colors that will plot, but only at the drafter's discretion and only if the output device can produce all 255 colors.

The way AutoCAD connects screen colors to output may seem unduly complex, but taking advantage of AutoCAD's approach increases your options for output. To do so efficiently, however, designers must take the time to plan and test AutoCAD color palettes and to document them thoroughly, as shown in Figure 8.4.

Even with the use of calibration software as described previously, color output varies among output devices; it can also vary within a single device depending on what print quality you specify (such as draft or final) or at what level of dpi (dots per inch) you print, such as 300 dpi or 600 dpi. The output media, whether drafting film, vellum, special or regular bond paper, can also affect the resulting colors. Thus, it is essential to test color output with *all* permutations of printers, plotters, output quality settings, and papers before developing office-standard palettes and before planning color-based projects. Offices that use color frequently can benefit from developing color palettes that will work successfully on *all* machines under *all* circumstances.

Office Color Palettes

Whether or not they formally calibrate colors on an officewide basis, many design firms find it useful to develop an office standard for computer-based color palettes just as they do with forms and typefaces. Many firms replicate existing palettes that have already been developed for color markers, pencils, and Pantone paper. As CAD-based color becomes more a part of your work, you should develop CAD-based versions of your existing standard color palettes.

Office-standard palettes offer you consistency in color printing and a uniform graphic image. As with office-standard typefaces, for example, color schemes often act as a statement of a designer's viewpoint and tastes. Many designers, such as Robert Adam, I. M. Pei, and Luis Barrigan, are distinguished as much for their use of color (even if the palette is monochromatic) as for other design elements. With regard to computers, standard color schemes also save time in two ways. First, if the palettes are fully tested as described above, you reduce the likelihood of encountering "surprises" in output, which might occur with an untested selection of colors. Second, an office-standard palette will reduce the need for designers to spend hours selecting from—as is feasible with some software—over 1 million colors. While color is a fun tool, it can distract you from accomplishing more urgent tasks if parameters aren't placed on its use.

In basic AutoCAD you cannot create and load custom color palettes at will, as you can, for example, with typefaces and linetypes. Nor can you easily build in defaults that would encourage designers to adhere to a particular color scheme. Therefore, to encourage use of a particular palette, you should design and print out reference charts such as that which is shown in Figure 7.5. These can be mounted on foam board and kept either in the design studio or next to color output devices. Today with the low costs of color printing and xeroxing, color copies of your palette can also be inserted into your office CAD standards manual. With these hard-copy references, designers who don't use CAD can nonetheless specify color schemes for CAD-based color presentations.

You can specify color palettes in various ways. One approach is by general color groups, such as "earth tones," "neutrals," "pastels," and "warm and cool grays." Another is by purpose, such as palettes for schematic and block diagrams, for renderings, and for different types of

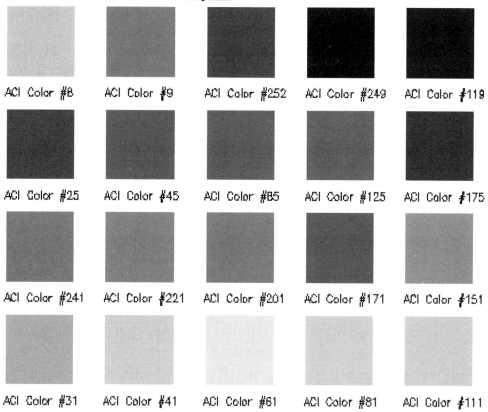

Figure 7.5 *Printing reference charts of office-standard color palettes, with the numbers of the ACI colors included, can help designers select suitable colors for CAD-based presentations and other output.*

presentation boards. For 3D renderings, color palettes can consist of not just hues per se but also of light, shade, materials, and textures.

 Many designers select on-screen colors by focusing on hues and thus select AutoCAD colors by using the ACI Color Wheel and RGB triplets. Note, however, that the HLS index can help build harmonious palettes by enabling you to select colors that are keyed to one another through shared properties. You can specify a wide range of hues that share a specific level of saturation or brightness. Alternatively, you can select colors that are close in hue but at extremes in intensity. You can also use the HLS index to plan illumination and colors for rendered 3D images.

Commercial Symbol Libraries

Many manufacturers of architectural and interior design products offer symbol libraries for use with AutoCAD and other drafting software. Andersen Windows, Herman Miller and Steelcase Office Furniture, and Kohler Plumbing are just a few of the companies that produce AutoCAD symbol library add-ons.

Most manufacturers' symbol libraries are essentially computerized versions of product catalogues. The symbols often include attributes, which contain fields for each product's name and number, finish options, prices, and other specification and ordering information; these attributed blocks can be linked to databases used for developing specifications, budgets, and product invoices. Indeed, many dealers and fabricators rely on these libraries to produce their bids and orders.

Current product offerings are listed in Autodesk's publication *The AutoCAD Resource Guide*. Other sources of information include manufacturers' representatives, Sweet's product catalogues, and the "New Product" listings in design publications such as *Architectural Record*, *Architecture*, *Interiors*, and *Progressive Architecture*. Many manufacturers also provide related products, such as electronic text versions of their specifications for importing into word processing software, or built-in cost estimation features. Some even provide CAD drafting services. Most symbol library add-ons come with a price, but some manufacturers provide complimentary copies for users with serious potential to specify their products.

The primary advantage in using these add-ons is that, provided the symbols are pictorially and dimensionally accurate, they represent actual products and building assemblies. Thus they may limit the possibility that a designer would draw a floor plan or detail that can't be built. As such, these libraries provide time savings in research and design and drafting time.

Manufacturers' symbols, however, should be used with extreme caution and never in lieu of generic symbol libraries that every design firm should develop for routine drafting. Indeed, manufacturers' symbol libraries raise some issues that designers should consider:

- Many of these symbols and details are obviously manufacturer specific. Their use could imply a bias toward a particular product, could limit consideration of alternative products, and may be potentially unethical in certain types of design projects, such as public work, that must be sent out for open bidding.

- Most products are manufactured according to specific modules and dimensions. Working with these standards can be particularly detrimental during the schematic design phase, when using specific products can inhibit creativity and preempt exploration of design alternatives.

- Commercial libraries can be unwieldy to use. Some provide a block for every single separate product in their catalogue. Instead of quickly inserting a generic block for an office task chair, a CAD drafter may be forced to select among dozens of blocks, which are often identified by cryptic product names and numbers. Unless these symbols are linked to a database where such product specificity is critical, using these blocks wastes time.

- Commercial block symbols consume precious computer memory. Because their ultimate purpose often is to generate specifications and shop drawings, many commercial blocks come laden with attributes. Just the presence of these attributes—even if designers don't use them—adds to the size of CAD files and reduces drafting speed.

- Commercial libraries come with their own graphic standards. Their block names, layer names and colors, scale factors, text styles, and symbol insertion points can violate your officewide standards. You may need to spend time addressing these issues, and you may also encounter problems when using the symbols with Scripts and AutoLISP routines and other aspects of your CAD system that you may have customized.

If you want to use the graphic aspect of a symbol or detail from a manufacturer's library but don't want the accompanying symbol-naming system, layering standards, or attributes, you can try exploding the symbol, removing the unwanted features, and reblocking it under another name. Make sure, however, that this procedure does not violate the manufacturer's copyright.

- Many designers use commercial library symbols purely for graphic representation; while others, such as specifiers and facilities managers, rely on them for the information that is attached to them. If CAD files containing these symbols are shared among parties who have different uses for commercial libraries, all parties should understand and agree on whether or not the attached information can serve as the basis for making decisions. Otherwise, major confusion can result.

A reminder: as with traditional hard-copy manufacturers' literature, you should always check to make sure that your electronic catalogues are current, so you won't mistakenly specify discontinued products. You also don't want to use databases containing product prices that are either obsolete or else inapplicable to your region of the country or type of project (for example, government projects often use a pricing structure different from that of commercial projects).

Desktop Publishing with AutoCAD

Desktop publishing (DTP) is a generic term for computer software that enables writers and designers to design and prepare publications for production, whether that production is done in-house or by an external printing service. Traditionally, desktop publishing referred to printed output, such as brochures, calendars, books, and other hard copy. Recently, it has begun to include slides, audiovisual presentations, Internet "pages," and other formats as well, and the distinction between DTP and multimedia is becoming increasingly vague.

Design firms find desktop publishing an indispensable tool in producing documents for marketing and programming, as well as for general document production. As CAD becomes more integrated with other software, designers are becoming aware that the production of construction documents constitutes a form of desktop publishing.

AutoCAD is not a desktop publishing program in the conventional sense. Despite ongoing improvements in text management, file exchange, printing, and other features, the software still lacks the basic features of high-powered DTP programs. It cannot handle complex text formatting and management, the use of color, or the integration of different file formats to the degree that programs such as QuarkXPress, Adobe PageMaker, or even high-powered word processors can. Nonetheless, AutoCAD's value in generating graphics for desktop publishing programs or in serving as a pseudo–desktop publisher upon occasion should not be overlooked; otherwise, you may duplicate effort and spend a great deal of time and expense going through many steps to replicate AutoCAD-generated images when you could work with them directly or quickly import them into other programs.

The trick to incorporating AutoCAD images into your desktop publishing operations is knowing when you should work within AutoCAD itself and when you should import AutoCAD files into other software programs. Here are some general guidelines; particulars depend on the desktop publishing software you use and its ability to import and export files.

Desktop Publishing within AutoCAD

Working *within* AutoCAD is preferable when the primary focus of your output is an AutoCAD-generated design rather than extensive text or a particular publishing format. If the intended output is a large-scale drawing or board with minimal text and no need for pagination, headers, and footers, working with an AutoCAD file makes sense.

Working directly in AutoCAD is also preferable if the image is an AutoCAD-based design that is still in development and you will need to produce an updated image on an ongoing basis. Although you can use the Windows OLE features to import or link AutoCAD files to other types of files, you may find in many situations that working directly with the original drawing file makes the most sense.

Another criterion is output. If you want to print oversize sheets and have automated and refined your AutoCAD plotting routines or just want to plot in black-and-white, you may find that plotting from AutoCAD is faster than plotting from other software. Many plotters require special drivers and plotting configurations to print files produced with desktop processing software; if you haven't invested in these but have configured AutoCAD, you'll find AutoCAD plotting more reliable. Plotting directly from AutoCAD files is also preferable if you want to retain a menu of line weights developed in AutoCAD.

AutoCAD within Desktop Publishing Software

Using desktop publishing software makes the most sense with AutoCAD files of designs that are complete and that are unlikely to need further revisions within AutoCAD itself. Typical applications include marketing brochures containing images of a finished project.

Desktop publishing is also preferable for projects involving large amounts of text in complicated formats or the blending of graphics, photographic images, text, and other elements from many different software applications. These packages automate many formatting and layout functions that AutoCAD would take more than twice as much time to perform (if it can at all).

For a piece that is going to be sent to a printer, you'll find desktop publishing gives you far more control over the output that your own computer produces. Desktop publishers provide more accurate color specifications, as described above, and color separation features for color printing. They also provide, as a matter of course, standard printing guides, such as cropping and registration marks.

Slide Presentations

A popular desktop publishing variant on hard-copy publications is the slide show. Most desktop publishing programs provide slide creation tools. Others, such as Microsoft Powerpoint and Harvard Graphics, specifically focus on the production of slides, overheads, and other presentation tools.

AutoCAD itself provides a built-in slide production function through its **Mslide** command. This command saves an AutoCAD screen image as a raster image file, which is identified by an *.sld* file extension. Such files can be viewed on-screen by using the **Vslide** command. Slides can then be stored together in a slide library if you use the AutoCAD utility *Slidelib.exe*. You can also produce automated slide shows by writing Script files that control the sequence and pacing of slides that you then present from within an AutoCAD file.

The AutoCAD slide-production capability provides a useful viewing device, but to use it, you have to have access to a computer that has AutoCAD installed on it. If you want to present slide shows to a large audience, you'll need to hook up your computer to a large television monitor or projection screen monitor. If portability is a concern, you can present the files on a laptop by using AutoCAD (ideally with a streamlined configuration dedicated just to this purpose), or else you can import your slides into a slide presentation package.

File Exchange with Desktop Publishers

As Chapter 10 describes, AutoCAD can exchange files with other software packages either through the Windows clipboard or through the use of file formats, such as *.dxf*, *.eps*, *.tiff*, and *.wmf*. The criteria for selecting among various file formats depends on three factors: The file format your desktop publishing program can read; the use to which the imported files will be put; and the amount of memory the file format typically demands.

Concerning the first factor, most DTPs can read files saved in universal file formats such as *.eps* and *.tiff*. Many can also import AutoCAD *.dxf*.

files. If, however, you are using AutoCAD on a Windows platform and you do your desktop publishing on a Macintosh, you cannot use platform-specific file formats, such as Windows *.bmp* and *.wmf* or Macintosh *.pict* without first performing interim file conversion (various software utilities are available for doing this). You need to review the documentation for both AutoCAD and your desktop publishing software to determine the optimal file-exchange process.

The second factor is the projected use of the imported file. If you merely want to import a graphic facsimile, such as a client logo or a simple building floor plate, a bitmapped file format usually suffices. If you need to be able to manipulate the image within your desktop publishing program, a *.dxf, .pict,* or *.wmf* file, which reads the image as a collection of separate entities, works better. Even within standard file import functions, both AutoCAD and other software programs provide a range of options that can affect the resulting files' functionality. When creating *.tiff* files from Auto-CAD, for example, you can specify various compression values, viewports, and windows. With the windows clipboard you can paste objects from another Windows application into an AutoCAD file as either a collection of polylines or merely as a simple metafile image.

The third factor is the likely size of the file you are importing. Because of the different ways they store information, file formats can vary considerably in the size of the files they generate. A moderately large AutoCAD *.dwg* file will produce *.dxf* and *.plt* files which are much larger, but a Windows *.bmp* file of the same image might be minuscule in comparison. Even within the same file format, file size can vary considerably depending on the various import specifications you select. An 8-bit, 72 dpi *.tiff* file of a photographic image could be as small as 25K. A 24-bit, 300 dpi *.tiff* file of the same image could expand to as much as 7,000K or 14,000K, depending on whether or not you elected to compress the *.tiff* file. While a robust DTP workstation can manage large files, you do want to be judicious when creating large files, as they will slow down your operating speed and also balloon the size of the DTP documents they are imported into. And many print shops and plotting services will charge extra for computer files which exceed a certain number of MBs in size.

Given the many options for doing file exchanges between AutoCAD and desktop publishing software, you'll need to do some testing before you can determine the best importing procedures. Once you've mastered the procedure, however, you'll save hours of work while achieving graphic consistency throughout your publications.

Field Conditions and Dimensioning

AutoCAD is notable for its powerful, flexible, and accurate dimensioning capabilities. The software not only lets you place dimensions in any format—linear, angular, radial, and so forth—according to your specifications but also gives dimensions "intelligence," meaning that you don't have to write out or calculate dimensional information yourself; AutoCAD can read dimensions from the entities being measured. In addition, AutoCAD dimensions can be associative, meaning that if a given dimension notation, including dimension texts, ticks, and extension lines, is read in AutoCAD as a single entity, you can resize the object which the dimension entity describes, and the dimension will automatically reflect the changes, as shown in Figure 7.6.

In addition to creating standard dimensions, AutoCAD provides strong tolerancing capabilities, including ANSI and ISO standard geometric symbols used in geometric tolerancing functions. Release 13's new Geometric Dimensioning and Tolerancing (GDT) feature offers many enhancements for engineering and mechanical drafting (see Figure 7.7). GDT, combined with AutoCAD's ability to adjust to any number of GDT international standards and units of measurement and its extreme accuracy, has contributed a great deal to the software's popularity both in the United States and abroad.

Although AutoCAD's dimensioning features are indisputably helpful when you are drafting and detailing new designs, they can be problematic if you attempt to use them on projects dealing with existing conditions. The majority of architectural design and space planning these days involves renovating existing structures rather than designing new ones from scratch; for such projects, AutoCAD's dimension tools should be used with care.

The primary difficulty of using AutoCAD's dimensioning tools with existing buildings is that the AutoCAD universe is based on a perfect, internally consistent X,Y,Z coordinate system. The program accounts for every

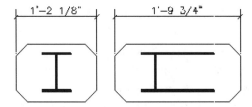

Figure 7.6 *When you use AutoCAD's Associative Dimensioning feature, changes in dimensioned entities are automatically reflected in the dimension text.*

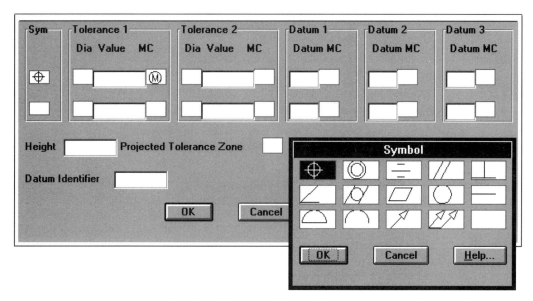

Figure 7.7 AutoCAD's GDT capabilities allow you to augment dimensional information with notations for tolerances, material conditions, and datum reverences.

unit of space in its universe, to eight decimal points. It permits no slack, no missing or unreachable pockets of space, no duplications or omissions.

The "real" world, on the other hand, is quite the opposite, especially when measured by ordinary humans. The real world is quirky, consisting of odd spaces and irregular shapes that cannot be documented by tape measures or standard rulers. Buildings settle, and many spaces, such as elevator and mechanical shafts, are inaccessible for taking measurements. Slack tape measures and hand-written field measurements that are vulnerable to human errors of transcription, illegibility, and omission only add to the imperfections inherent in actual conditions. Moreover, ordinary measuring tools cannot produce measurements to the level of precision a computer program can. As a consequence, relating imperfect reality to the perfect AutoCAD world, as suggested by Figure 7.8, is akin to stuffing a crooked grid into a pristine orthogonal box and expecting to be able to reconcile any differences.

Because the discrepancy between CAD and the real world, using AutoCAD for managing projects involving existing conditions requires special care. First, you must accept from the start of the project that with ordinary measurement tools you can only *approximate* existing conditions. Unless you apply professional surveying tools or high-tech and expensive applications such as photogrammetry, you cannot generate an internally consistent and complete AutoCAD facsimile of the building you are working on.

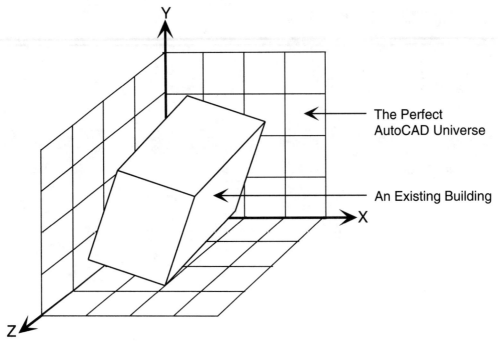

Figure 7.8 The real world cannot fit into the perfect coordinate system that AutoCAD provides.

It is important to discuss this limitation with clients at the outset of a project; if your clients expect a fully dimensionally accurate CAD file upon project completion, they should be prepared to pay you or a professional surveyor to perform what can be very extensive and expensive work.

Second, reconciling field documentation with CAD-based plans can be time-consuming and often frustrating for CAD users. In some situations, correcting a CAD plan simply requires moving or stretching a wall or other item. But in other cases, the work resembles a monkey puzzle, whereby you align two pieces properly, only to find all that the others are now out of whack. In such situations, it is often easier to draw a plan from scratch on CAD than to try to reconcile field conditions with AutoCAD's reliable but rather rigid coordinate framework. For this reason, project managers often need to specify to what degree a CAD drawing needs to reflect actual conditions and to what degree it should merely approximate them. Many managers opt to have CAD drawings modified only to the extent that at typical plotting scales, such as 1/8" or 1/4", the drawing *reads* accurately even if the various AutoCAD measurement commands would tell you otherwise.

The policy you implement for reconciling field dimensions with CAD has important implications for how you apply AutoCAD's dimensioning capabilities. If you have an internally "real" CAD base—as is feasible with new buildings or details—you can use AutoCAD's associative dimensioning features with ease. You will actually find associative dimensions helpful because they will be internally accurate, and their totals will be mathematically correct. For many existing structures, however, you may need to override the dimensions that AutoCAD automatically gives you and instead manually write dimensions based (ideally) on careful field measurements. You may also have to override default dimensions in order to specify distances that comply with building codes. For a major circulation path, for example, you would override the AutoCAD-generated dimension of 46 1/2" with the text "3'-10" Clear" in order to ensure that your design complies with local codes. In general, for dimensions that are not based on AutoCAD-provided information, you may want to "explode" dimension blocks so that they won't inadvertently change if the entities they measure do, or else you can disable the AutoCAD dimension variable (**Dimaso**) for associate dimensioning.

Your office policy on dimensioning is equally important where CAD file exchange is involved. Because the CAD interface is so precise, many CAD users have trouble remembering that the CAD world is hypothetical and that key conditions must be verified in the field. Each time you issue a CAD file to a client, consultant, or other party, include a combination disclaimer/reminder to this effect.

Although AutoCAD can automate and increase the accuracy of dimensions, it is not a substitute for the *craft* of dimensioning. Every experienced drafter knows, for example, that you do not dimension everything, but rather you need to show important and required dimensions, starting points, and alignments as a way to guide the contractor to what is essential; everything else is a resultant or falls into place. CAD can be dangerous because it can quickly produce complete strings of internally consistent dimensions throughout an entire drawing. If you issue drawings with such dimensions, however, you may be asking a contractor to build the unbuildable. For this reason, automated dimensioning should be used with care, and design teams must carefully guide junior drafters and review their work in this regard.

A related issue is dimensioning accuracy. Using the **Dimension Primary Units** box, as shown in Figure 7.9, you can specify that the precision of your dimension units be taken up to eight decimal points. A higher level of unit precision reduces the extent of rounding errors; on the other

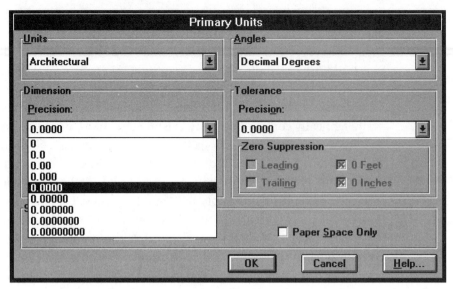

*Figure 7.9 The **Dimension Primary Units** dialog box allows you to specify the precision of both linear and angular dimensions according to your project's needs.*

hand, calling out room dimensions to 1/64 or 1/265 of a unit can be excessive, if not dangerous; construction trades cannot build to such precision. Many CAD users would find working with excessive precision distracting as well, since they rarely need to input measurements more precise than to 1/16 of an inch. Therefore, your working units and your dimensions rarely need to be more precise than 1/8" or 1/16", depending on the drawing's plotted scale. Where possible, avoid changing the unit precision factor in the middle of a drawing project; otherwise you may have to spend time reconciling rounding errors and inconsistencies. Ideally, the proper defaults for units should be imbedded in all your prototypical CAD files, as discussed in Chapter 4.

Government Work

Building projects for federal, state, or local government constitute a major form of work for architects, space planners, and members of related design disciplines. Even with political uncertainties and budget cutbacks, government clients can be the source of large, notable, and often technically challenging work.

As any veteran of government-sponsored design projects knows, government work requires dealing with bureaucracies, paperwork, regulations, and standards. AutoCAD can help you comply with government regulations not only because of its innate capabilities but also because it is heavily used in the federal government and, as a consequence, Autodesk has developed various products to help AutoCAD users who work on government projects.

Since federal and state governmental agencies are decentralized, naturally there is no single standard for CAD. Each agency may use its own preferred CAD platform. AutoCAD, however, is one of the predominant CAD packages used by government agencies. At present, it is the CAD standard of choice for the National Park Service, the U.S. Navy, the Bureau of Indian Affairs, and the Army Corps of Engineers, among others. Because these agencies are in essence building and managing huge facilities management databases, they run AutoCAD on very high powered workstations, often using far more powerful networks, stations, and peripherals than would be found in a typical design firm.

Many government agencies use AutoCAD as a tool not only to accomplish their individual departmental missions but also to create databases that are of use to the general public. Government agencies have been the major force behind the creation of Geographic Information Systems (GIS), which generate electronic maps used in many applications in both the private and public sectors. The National Park Service uses AutoCAD to document many historical monuments for its Historic American Buildings Survey (HABS) program.

If you wish to qualify as a contractor for many government projects, you should expect to be able to deliver your work in ways that conform to the needs of individual agencies. Each agency issues different requirements for design services; sometimes you simply need to produce a CAD file in a format that the particular agency's computer system can read; in other cases, you may have to provide CAD files not only using particular versions of AutoCAD but also using particular graphic standards or file formats. For an idea of how particular an agency is in this regard, refer to the listings for "Architect and Engineering Services—Construction" that appear in the *Commerce Business Daily*.

In addition to requiring that you have a compatible CAD system, government projects can affect your use of CAD in other ways:

Compatibility with other software applications. Government agencies increasingly rely not only on CAD but also on many other software applications. Consequently, they often require that eligible firms produce work with government-mandated computer applications. Sample applications include proprietary cost-estimation systems, word-processing programs, and specifications packages such as the Corps of Engineers Guide

Specifications (CEGS). Depending on the agency and the project, you may only have to provide files that are compatible with certain software; in other cases, you'll actually have to acquire the software and use it throughout the project.

Use of the metric system. In compliance with federal government requirements, a growing number of government branches use metric rather than, or in addition to, imperial units. Many government RFPs not only request the use of the metric system but demand that you demonstrate experience using it. With AutoCAD, you can work in metric units or in a combination of imperial and metric.

Standards for forms and graphics. Many government agencies are tougher than private corporate and institutional clients about using graphic standards and forms that reflect national or international standards. The federal government requires that many contractors comply with the ISO 9000 Series, a standard for document management that has as its goal the streamlining and standardizing of paperwork and implementing quality control procedures. Adherence to ANSI and ISO graphic standards may be expected. Since AutoCAD is used heavily in government work, the program offers built-in features, such as default GDT symbols, pen widths, and sheet sizes, that comply with several predominant national and international standards for graphic output; these features can facilitate compliance with various government standards. Note, however, that some government agencies, such as the Department of Defense, have their own specific drawing formats and standards. (Figure 7.12 reflects a state agency's standards for drawing borders and titles.)

Paperwork reduction. The 1995 Paperwork Reduction Act mandates procedures designed to reduce paperwork. One consequence is that electronic information transfer is encouraged, so you need to have access to a modem and the Internet. Another consequence is that with certain government departments, you have to submit electronic communications that comply with formatting standards such as ANSI X12.

Although you can use paper-based versions of federal and state publications, such as the **Federal Register and Commerce Business Daily**, these days you may find that the Internet provides more direct and timely access to the information you need.

Whether you already do government work or would like to, you may find it useful to contact Autodesk for information and publications. Since

AutoCAD and other Autodesk products are heavily used in government work, Autodesk provides special support for government applications. The company publishes a quarterly bulletin called the *Autodesk Federal Bulletin*, as well as a *Facilities Contract User's Manual* and prototypical bid forms for bidding on CAD systems. To encourage use of its products by government agencies, Autodesk offers substantial discounts (generally 20 percent to 45 percent) on its products through the General Services Agency (GSA).

Historic Preservation

CAD is most often associated with the design and drafting of *new* buildings, especially highly regular, orthogonal, and gridlike structures. Yet CAD in general, and AutoCAD in particular, is a logical tool for many tasks required in historic preservation. Indeed, the National Park Service uses AutoCAD to document historic monuments as part of its Historic American Buildings Survey (HABS); the Washington Monument and the Lincoln and Jefferson Memorials are important American sites that HABS has documented using AutoCAD.*

AutoCAD is a useful tool for all types of historic preservation work, including preservation, rehabilitation, restoration, and reconstruction. It proves useful not only because of its ability to render graphically complex items (see Figure 7.10) but also because of its capacity to serve as a giant database; much of historic preservation work involves meticulous documentation of existing conditions as well as compilation of pertinent historical information. For this, AutoCAD's ability to store information and separate it on an infinite array of layers can be invaluable. In addition, AutoCAD can help in analyzing conditions and determining the best among various options for treatment. In sum, AutoCAD is particularly useful for the following typical preservation tasks:

- documentation
- analysis
- construction documents and specifications

---

*For a fascinating description of the National Park Service's use of CAD, see Mark Schara, "Using AutoCAD and Photogrammetry to Preserve History," *AutoDesk Federal Bulletin,* Winter 1994, 5–7; and Mark Schara, "Measuring Buildings for CAD Measured Drawings—The Lincoln and Jefferson Memorials," *GRIST* (a publication of the U.S. Department of the Interior—National Park Service) (Winter 1994) 38, No. 1.

Documentation

Documentation is a major task in all aspects of preservation. It entails inventorying the physical aspects of a site and assessing its condition. Even if no form of restoration is planned, documentation is still essential for general inventorying and archival purposes. For sites that will undergo substantial changes, documentation helps you keep track of the original conditions and helps you build the rationale for proposed changes and modifications.

For older buildings, a complete set of original construction drawings or current as-builts is rarely available to serve as a base for documentation. In such instances, documentation often becomes a puzzle-solving process of collecting whatever pieces of original drawings remain and then relating them to the actual building site. Even when original drawings are available, they are not always reliable or easy to work with. They may be too fragile to handle, or so faded and torn that they are of little value; they may be stored in special collections of libraries and other sites, where access is limited. Even if drawings are available and legible, they are often from an era when there was little standardization of scales, sheet sizes, or other drafting conventions. Many architectural graphic standards, such as column grids, that we nowadays take for granted were not common a century ago.

Trying to build a complete picture of a site's original condition from fragments of drawings and an actual site that may be partially destroyed or altered beyond recognition is a form of archaeology. AutoCAD can serve on the archaeological team by acting as the central repository of whatever information exists. You can input whatever information you obtain and let AutoCAD reconcile it. AutoCAD will consolidate the building and site information regardless of the original drawings, scale, orientation, and graphic standards. You can add dimensions, notations, even attributed blocks of windows, bricks, cornices, and other architectural elements. You can track the source of your CAD data on a separate layer. Once you compile this information, you have a database that enables you to analyze the disparate elements in a cohesive format and drawings that you can plot at any scale, in clearly legible line weights.

CAD can also help you in the process of on-site documentation. One implementation, based on a technique used by the Canadian Inventory of Historic Buildings,* helps to speed up documentation by referring to a

---

*James Marston Fitch, *Historic Preservation: Curatorial Management of the Built World* (Charlottesville, VA: University Press of Virginia, 1990), chapter 15.

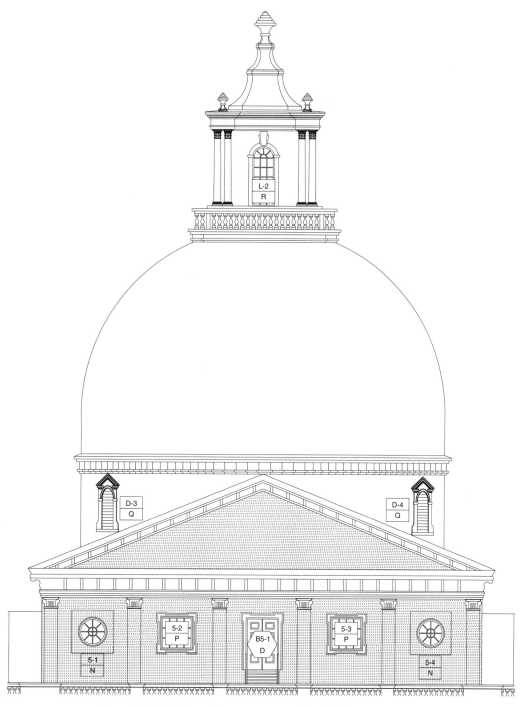

Figure 7.10 *CAD enables you to present detailed facades of historic buildings, as shown here in an elevation of the gilded dome of Bulfinch's Massachusetts State House in Boston, Massachusetts. AutoCAD drawing of the Massachusetts State House provided courtesy of Goody, Clancy & Associates, Inc., Boston, MA.*

HISTORIC PRESERVATION

graphic building inventory that contains pictures of common architectural elements, ranging from pedestals to columns to transoms. These elements can be represented in AutoCAD in the form of attributed blocks, which can then be inserted into a working electronic plan or elevation and modified to reflect actual conditions and dimensions. Rather than spending time and relying on the note-taking skills of a field verifier, you can simply select a symbol for the graphic equivalent and annotate the data fields as appropriate.

Inputting Historical Data into AutoCAD

The major challenges that using AutoCAD for historical documentation presents are the time and labor required to enter conditions into the computer and the difficulties of accurately representing often complex building conditions, particularly if you have no reasonably accurate architectural drawings to work with. Inputting data of this nature will always be time-consuming and labor-intensive, but several methods exist to help speed up the process and improve the accuracy of the data.

Scanning. If you have access to a large-format scanning machine, you can feed complete or partial sections of drawings or photographs into these machines and then convert the results into a vector-based computer file that AutoCAD can read. As described further in Chapter 10, the results of scanners depend on the quality of the software used and the quality of the original documents you scan. If the originals are clear and flat, they can provide a good CAD base for your work; they can be particularly helpful for replicating ornate architectural details. If the originals are faded, wrinkled, or torn, however, their value is limited (scanning from a black-and-white copy of the original may be more effective). Note that you may not be permitted to scan or digitize or even photocopy historically valuable drawings if they are considered too fragile; often you have to arrange for the drawings to be photographed and then have a full-size negative made as a basis for creating a Mylar or vellum reproducible. Alternatively, you can have the photographs developed directly on a CD-ROM, which can then serve as an ongoing reference to the images.

Photogrammetry. Photogrammetry (or stereophotogrammetry as it is sometimes called) is a photographic process that permits buildings and sites to be measured in two or three dimensions. Using a special camera, preservationists take several photographs of a building. The camera contains a grid that is superimposed on a negative as a frame of reference for measuring the building. The resulting pictures, or orthophotos, can then be calibrated and traced using special computer software, such as Photo-CAD (an AutoCAD add-on) by Desktop Photogrammetry, Inc., to produce a dimensionally accurate and consistent image.

Photogrammetry is one of the best ways to develop dimensionally accurate facsimiles of historical buildings in CAD. The major obstacles to using it are the time and expense involved; the software and camera costs alone are not inconsiderable, but what often costs more is the process of getting the pictures from the appropriate view. Rooftops, for example, can often only be documented by aerial views taken from helicopters or by standing on building scaffolding. For major historical sites, however, the resulting accuracy is worth the work. Photogrammetry has been used on sites as diverse as the Lincoln Memorial (see Figure 7.11), Mount Rushmore, the Great Sphinx, and the Acropolis in Athens.

In addition to improved accuracy, photogrammetry is often the best option in situations where conditions make it too difficult to document the site any other way. The National Park Service, for example, is using orthophotography to document a historical mill that was rendered inaccessible by fire and to record the fragile ruins at Colorado's Mesa Verde National Park.

Surveying Tools. In engineering and landscaping, a range of AutoCAD add-ons are used to survey conditions and sites. Using special surveying devices, you can collect information on the existing condition of buildings, roadways, and natural terrain into a digital format that can then be imported into AutoCAD files. Tools such as the GPS Pathfinder System devices by Trimble Navigation Ltd., for example, enable you to use signals from a system of international satellites to build relationships between a field location and a CAD-based map. A growing body of Geographic Information Systems (GIS) databases maintained by federal, state, and local government agencies is based on surveys of this nature. In fact, given that more and more of the world in being entered into a GIS database, one should always check with federal agencies to determine whether the information you need is already available in a CAD-compatible electronic format. Autodesk recently acquired a manufacturer of GIS software; expect to see Autodesk itself producing more GIS-related databases and tools in the future.

In comparing various forms of input, keep in mind that many of the tools discussed here produce enormous AutoCAD *.dwg* files. As discussed later in this section, you need a powerful workstation to effectively handle these files; you may instead want to split the information into smaller, linked **Xref** files.

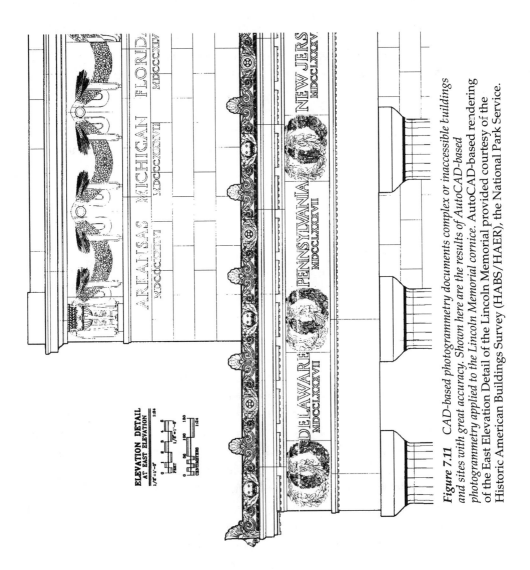

Figure 7.11 CAD-based photogrammetry documents complex or inaccessible buildings and sites with great accuracy. Shown here are the results of AutoCAD-based photogrammetry applied to the Lincoln Memorial cornice. AutoCAD-based rendering of the East Elevation Detail of the Lincoln Memorial provided courtesy of the Historic American Buildings Survey (HABS/HAER), the National Park Service.

Analysis

Once documentation is complete, architects and preservationists must then analyze the collected information to determine what work is appropriate and feasible. Having a compiled AutoCAD database for the project is a useful tool for doing analysis, as you can plot or view on-screen the data or subsets of data in a variety of scales and formats according to the issue you are analyzing.

The scope and purpose of preservation varies by site. It may require replicating the condition of a building at a certain time in the past or renovating it for new uses while preserving a certain historical character. If a building has a long history, preservationists often have to make judgment calls on which historical eras or combination of eras they should seek to replicate or reference. AutoCAD can help here by enabling you to overlay and compare in a single CAD file, or set of linked **Xref** files, the evolution of a historical site. By separating onto different layers the results of successive renovations, you can more easily understand what elements belong to a particular era and what is fundamental to the site itself.

As part of the analysis process, you can apply a growing number of 2D and 3D modeling or visualization techniques to review and present different options for preservation. You can merge your CAD-based plans and elevations with original photographs or paintings from the site's original era. You can generate animated walk-throughs or flyovers to review the site's appearance from all angles and to review options for design decisions such as historically appropriate color schemes and lighting.

Preservation is an umbrella term for different approaches to working with historical sites and monuments. Although many preservation clients may already have arrived at an approach before they hire a designer, they often may need assistance determining which treatment—preservation, rehabilitation, restoration, or reconstruction*—or combinations thereof, is appropriate. You can use AutoCAD to study the visual results (and, with links to cost estimation databases, the potential costs) of the various approaches.

---

*The National Park Service, Department of the Interior, has recently proposed revisions to its standards for historic preservation projects that would reduce the number of preservation approaches from seven (acquisition, protection, stabilization, preservation, rehabilitation, restoration, and reconstruction) to the four listed here.

Construction Documents and Specifications

AutoCAD is a powerful tool for producing working drawings and specifications for new construction in preservation projects. Although CAD is often considered only appropriate for repetitious designs, it actually is equally useful for accommodating the varied conditions and detailed notations that historical sites require.

Much historical architecture has a fundamental classicism, symmetry, and order, based on uniform building elements. During renovation, however, you often cannot prescribe work generically as you can with new buildings; for example, you cannot detail and annotate a typical column once and have the information be applicable to each incidence of that column. Rather, you often must notate and detail each architectural element, column by column, brick by brick, window by window. With AutoCAD, you can do this by developing blocks for typical items based on their *original* dimensions and features; then you can explode individual blocks as necessary and modify variances in actual conditions. This approach allows you to provide specific (and legible) instructions for each element while keeping in sight the visual uniformity that you may seek as a renovation goal.

If you use AutoCAD's SQL feature, you can link blocks to external databases to create the often complex schedules for doors, hardware, windows, and other building components that renovations entail. You can use these databases to do preliminary cost estimates and to compare different preservation options and strategies.

Another use of CAD is to provide full-scale details of historical elements for fabrication. If your CAD details are based on accurately scaled images created from scanning or photogrammetry, you can hand them over to millworkers, plasterers, and other fabricators to use as templates. If the fabricators use CAD themselves, they may be able to use your actual CAD file as the base of their design.

Preservation often entails coordinating large teams of consultants, including surveyors, art conservationists, engineers, professional preservationists, historians, and others. As with new building projects, preservation projects can rely on CAD as a common database and frame of reference.

One thing to be aware of when using AutoCAD on preservation projects is that, particularly if CAD files contain scanned images or highly detailed drawings, the files can become very large and unwieldy. Often, it is more productive to use AutoCAD's External References **(Xrefs)** to split the work into different files and then link them together to produce finished drawings. Using **Xrefs** also allows you to separate

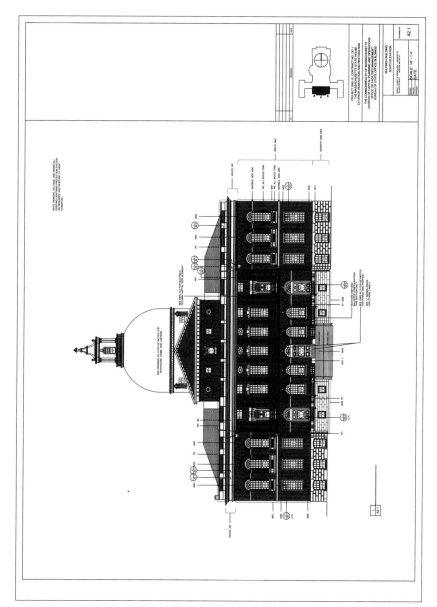

Figure 7.12 Design firms can use CAD to produce construction drawings not only for new buildings but also for renovations of centuries-old structures, such as the Massachusetts State House (1798), shown here. The building, designed by Charles Bulfinch, is undergoing comprehensive restoration by Goody, Clancy & Associates, whose CAD drawings reflect and annotate the complexities of the existing conditions and are based on extensive field surveys. AutoCAD drawing of the Massachusetts State House provided courtesy of Goody, Clancy & Associates, Inc., Boston, MA.

files containing original documentation of historical nature from those storing schematic ideas and new work; separating the files protects the integrity of the original data.

As with corporate and institutional work, historic preservation clients—especially if they are affiliated with government agencies—are likely these days to request CAD files of your work upon project completion. They will use these files for a range of preservation CAFM-oriented preservation applications, including in-house renovations, maintenance, and curatorial and archival functions. You can even expect that your CAD-based preservation work will someday end up in the U.S. Library of Congress.

International AutoCAD

A growing number of American design firms are seeking work abroad, either by setting up offices in other countries or by establishing partnerships with foreign design firms. As AutoCAD is the dominant CAD software not only in North America but worldwide, the odds are that any American firm working abroad will use a foreign-language version of AutoCAD and have to deal with different CAD customs.

AutoCAD is used in over 115 foreign countries and is available in 18 languages. AutoCAD will translate all the commands and routines so that the interface reads in the chosen language, except for terms that are too *computerese* to belong to any formal language. The basic organization of menus and functions remains the same, however, so that an American using a French or German version of AutoCAD could use his or her knowledge of the software to be reasonably productive without knowing what *"Keine Optimierung"* means. You do, however, need to have foreign-language skills to use the documentation that comes with other AutoCAD language versions, to read and write foreign-language AutoLISP routines and Scripts, and to access the wealth of third-party products, books, and journals that AutoCAD has spawned internationally as well as in the United States.

Although the AutoCAD interface is modified to meet language requirements in different countries, the underlying software remains the same just as it does for the different software platforms. Consequently, an AutoCAD file created in New York can be used in Singapore or Zurich. If you are using AutoCAD files on an international basis, however, you must be aware of the following aspects of international CAD relations.

Hardware locks. International versions of AutoCAD require the use of a hardware lock. This is a small electrical device that must be attached to any computer that has a licensed copy of AutoCAD installed on it (networked versions of American AutoCAD require this lock as well). Such devices typically are required in countries where protection for software copyrights is either not legislated or, if legislated, not aggressively implemented. Refer to the *AutoCAD Release 13 Installation Guide for Windows* for additional information.

If you work in other countries on a regular basis, you may find it worthwhile to create a variation of your AutoCAD customization—as described in Chapter 4—that reflects those countries' standards, such as metric units and ISO drawing sheet sizes. If you collaborate with firms abroad, you might borrow their standards or work together to build a joint customized setup.

Different characters. The American set of keyboard characters does not accommodate the letters and accents of many languages. AutoCAD, as have many other software packages, traditionally has relied on a set of 126 7-bit ASCII characters. Many foreign symbols—such as ä or £—rely on 8-bit characters, beginning with character 130. To work in Japanese, Czech, or Hebrew may require purchasing international font files as well as a differently configured keyboard. For some text, such as Eastern European languages, you can use the Unicode character encoding capability, which AutoCAD Release 13 supports; refer to *AutoCAD Release 13 User's Guide* for additional information. Sources of third-party international fonts for AutoCAD are listed in *The AutoCAD Resource Guide*. (The standard keyboard characters that AutoCAD supports are found in Appendix C of the *AutoCAD Release 13 User's Guide*.)

Big fonts. Some languages, such as Japanese or Korean, do not use Western-style alphanumeric characters and hence do not use ASCII characters. Rather, they use what are in essence pictograms. To handle Japanese Kanji or other multicharacter languages, AutoCAD provides a big-font file format. Refer to the *AutoCAD Release 13 User's Guide* for additional information.

Different text lengths. Compared to many other languages, English is fairly compact. If you attempt to translate aspects of your customized AutoCAD system, such as dialog boxes, AutoLISP routines, and title sheets, you should plan for text lines that are at least 15 to 20 percent longer.

Different preferences for drawing file exchange. In American design firms, *.dxf* is the most common file format used to exchange drawing files among different CAD software. You may find that, in other countries, file-exchange formats such as *.iges* prevail. AutoCAD can exchange files through many file formats, so a different format should not present insurmountable technical challenges, but you may have to become familiar with different file-exchange issues and routines.

AutoCAD provides a way to facilitate translation of AutoLISP routines from English to other languages: precede each command or key word in an AutoLISP routine file with an underscore (_). See the **AutoCAD Release 13 Customization Guide**, Chapter 8.

Different standards for drawing graphics. Other countries apply different standards for architectural graphics, drawing sizes, layers, and dimensioning techniques. Most of the rest of the world uses the metric system for dimensioning. In addition to international standards developed by the International Standards Organization (ISO), many countries have their own standards just as we often adhere to American standards such as the American National Standards Institute (ANSI) and the American Society of Mechanical Engineers (ASME). The Japanese, for example, have their own version of ANSI called the Japanese Industrial Standard (JIS). Australia, France, Germany, Great Britain, and Sweden have also issued their own national standards for CAD. In addition, CAD users in other countries have sought to develop standards for CAD layer names and other features; these differ from American standards as well.

AutoCAD not only gives you the flexibility to work with a range of standards; it also has imbedded in it the ability to set up metric-size sheets and units and offers ISO-standard GDT symbols, dimension styles, and linetypes. So it is not hard to adhere to international standards using AutoCAD; however, you probably will have to rework many of your own prototypical drawings and block libraries that are based on American standards.

Different building standards. Many American standards for building methods, components, and assemblies are unique to the United States. Other countries use different materials, dimensioning systems, and breakdowns of building trades. Your office standards for blocks, details, specifications, and notations will have to be reworked. If you use commercial symbol libraries, you'll have to acquire foreign versions (where available). In particular, you'll have to revise any standards based on Construction Specifications Institute (CSI) divisions, as the CSI categories are not commonly used outside of North America. You'll also have to comply with different building codes and regulations, or in some countries, with no requirements or guidelines whatsoever.

Different professional definitions. American architects and interior designers assume roles and responsibilities for which there is no exact counterpart in other countries. Many other countries split the roles of architect and engineer differently. In Great Britain, for example, the cost estimator is a well-defined and respected professional who assumes many responsibilities that in the United States are parceled out among architects, specification writers, contractors, and consultants. When working in other countries, you cannot expect to share CAD-based work with project team members in precisely the same manner that you do in the United States.

Computer regulations and copyrights. Much like the United States, other countries are finding that computer software brings along concerns about international copyright protection and general data transfer. Some countries are laissez-faire about computer usage, but others—particularly those in Western Europe—are often stricter than the United States about certain aspects of computer data exchange, especially with regards to protection of privacy. The European Economic Community (EEC) has issued special directives to protect database users.

When working in foreign countries, be sure to confer with local computer and legal experts before attempting to distribute AutoCAD files among colleagues, consultants, and clients. In many countries, copyrights to your work may, in concept, be protected by international treaties, such as the Berne Convention and the Universal Copyright Convention, but the actual implementation of such protection varies locally and may be either strengthened or weakened by local laws and practices.

AutoCAD sales and support. If you use AutoCAD in other countries, you are expected to work with local registered AutoCAD dealers. You may find that different pricing structures, license situations, and levels of support apply.

This chapter can only address general issues about international use of AutoCAD; for each country or region you work in, different regulations and procedures will apply. Often you'll want to contact local organizations, such as an AutoCAD dealer or a professional organization of AutoCAD users or designers, to gain an introduction to the local CAD scene. And although you may be hired abroad expressly to provide an American design approach, where local CAD practices are concerned, it is always wise to adhere to the old proverb When in Rome, do as the Romans do.

Layer-Naming Systems

In many ways, layers are the essence of transferring CAD information. They are the fundamental CAD feature that enables a single CAD file to produce multiple drawings, and they play an essential role in drawing productivity, coordination and quality control (see Chapter 4).

Because of the importance of layers, design firms, various A/E/C professional organizations, and developers of CAD-related software have sought to develop a standard layer-naming system that all design and construction professionals and CAD software packages can adhere to. As with other architectural graphic standards, these attempts at uniform layering have met with limited success in the A/D community, since it would be difficult to anticipate the varied range of project scopes and sizes that characterize design practice. Nonetheless, it is worthwhile to familiarize yourself with the various standard approaches to layering when you are developing your own layer system.

Despite efforts at industry uniformity, every design firm seems to use its own layering system, even if that system is based on one of the standard approaches. A shared goal of most layering approaches, however, is to find the optimal number of layers—a system that splits information into just enough layers to permit the creation of different drawing plots and pen weights, yet does not weigh down CAD files with superfluous layers; finding the right combination of layers is important because with Auto-CAD, the mere *definition* of a layer in a drawing file consumes memory.

AutoCAD Layer Characteristics

Basic AutoCAD installs with only one layer—Layer 0—and allows you to build a set of customized layers according to your needs. In selecting and developing an approach to layers, you should keep in mind several AutoCAD layer characteristics. First, you can specify layer names of up to thirty-one characters. However, very long names are tedious to type, and part of the name can be obscured in certain dialog boxes and in the status box of DOS-based AutoCAD. You can't use unusual ASCII characters or symbols such as dashes, dollar signs, and underscores for your layer names. If you do use characters other than standard letters and numbers, your layer names may cause problems when you try to import them into other CAD programs—and vice versa. You can type in names using lowercase characters, but AutoCAD will read and present layer names only in capitals and in alphabetical order.

A second consideration with AutoCAD layers is that, while you can use an unlimited number of layers and can have flexibility in designating layer names, you cannot create explicit layer *classes*. Layer classes, a feature found in other CAD programs as well as in various AutoCAD add-ons, help sort and select layers by their purpose or use. For example, layers for ceiling grids, soffits, lights, and HVAC devices might fall under the class RCP (Reflected Ceiling Plan). Demolition work layers might fall under the class Demo. To approximate layering by class in AutoCAD, you can purchase third-party add-ons that load layers and divide them into classes. You can also give related layers a common prefix or suffix. In Figure 7.13, all layers related to ceilings might have "RCP" at the start of their name. You can then use a "wildcard" to do a layer search in the **Set Layer Filters** dialog box by asking for all layers beginning with "RCP."

Finally, if you exchange AutoCAD files with other CAD systems, plan your layering system with a view to the layering characteristics of other major software. AutoCAD is relatively generous in permitting long layer names and an unlimited number of layers for each drawing. Many other CAD programs, however, restrict the number of layers to less than 10 (256 seems average), and some limit layer names to ten digits or less. Other CAD software may have trouble reading layer names containing digits other than standard letters or numbers and may have particular problems with blank spaces, underlines, and dashes. Unlike AutoCAD, certain other

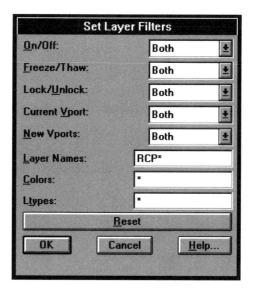

Figure 7.13
Using the *Set Layer Filters* dialog box enables you to screen for all layers with common names, colors, viewports, or other characteristics.

CAD programs may store layer names in combinations of upper- and lowercase characters. In AutoCAD, entities normally assume the color and other characteristics of the layers they're assigned to; you *can* override the layer defaults but, particularly for complex drawings, you will find that doing so slows down editing and analysis. With many other CAD systems, layers are used to determine whether an object is visible or editable, and they bear no relationship to an object's color or linetype. When you import a file from one of these systems into AutoCAD, you often must modify the imported entities so that their characteristics correlate to the layers they are placed on.

Typical Approaches to Layer Names

AIA-sponsored standard layers. A well-known approach to industry-wide layer-naming standards has been developed by the Task Force on CAD Layer Guidelines. Organized in 1988, this task force was made up of members of the American Institute of Architects (AIA), the International Facilities Management Association, the American Consulting Engineers Council, the American Society of Civil Engineers, and three federal government agencies. In its 1990 publication, *CAD Layer Guidelines: Recommended Designations for Architecture, Engineering, and Facility Management Computer-Aided Design*,* the task force presented a comprehensive alphanumeric layer-naming approach that organizes CAD layers by eight major groups, loosely based on A/E/C disciplines: architecture and interiors, civil engineering, electrical, fire protection, mechanical, plumbing, structural, and landscaping.

Within each of the eight major layer groups are layers for drawing information, elevations, sections, and building information details. Each group is identified by a single letter, such as *A* for architecture. (The guidelines present layer names in both short and long formats to allow for CAD software packages that limit the length of layer names.) In addition, you have the option of attaching modifiers that can help clarify whether that layer contains notes, dimensions, hatches, poché title blocks, or information on specific building products or assemblies. You can also indicate whether the building work is to be new, demolished, or "existing to remain," and whether the work is to be done on a particular floor. For

---

*Michael K. Schley, Ed., *CAD Layer Guidelines: Recommended Designations for Architecture, Engineering, and Facility Management Computer-Aided Design* (Washington, D.C.: The American Institute of Architects Press, 1990). Available from the AIA Bookstore, (202) 626-7475).

example, the layer on which you'd put an *existing* partial height wall that will remain during new construction could be named, according to your use of AIA-sponsored layers, any of the following:

- A-WALL (summary layer, long format)
- AWA (summary layer, short format)
- A_WALL_PRHT_EXST_05 (building information layer, long format)
- AWAPREX_05 (building information layer, short format)

The recommendations of the task force, although not universally agreed upon, are probably the most useful starting point for establishing a layer system for your design firm. This layering approach is the only one that fully reflects the way A/E/C professions set up drawings, and that addresses almost all typical A/D drawing applications, from 3D to poché. Consequently it has shaped the layer setups of many design and engineering firms. It is also reflected in various third-party add-ons for AutoCAD. For many firms, however, using the entire standards system is overkill; applying it selectively (as the task force expected it to be applied) is often more manageable.

Layers by number. Some firms use numbers for layer names. Layer 1 might be exterior walls; 2 is core walls, and so forth. Many firms group layers into multiples of ten or one hundred, so that layers to do with walls are numbered 1 to 10, layers pertaining to doors are numbered 20 through 29, and so forth. The primary advantage of using numbers for layer names is that you can type in the layer names quickly and with minimal risk of spelling errors. The primary disadvantage is that, since numbers are not intuitive guides to layer contents, users must either memorize the system or constantly refer to a written description of it. Note that AutoCAD will permit numbered layer names to start with leading zeros, such as 008, but other CAD packages may not.

One danger of using numbers for layer names is that if you merge your file with another CAD file that assigns the same numbers to different layer contents, you can end up with an unwieldy blend of two layer-naming systems. If Layer 15 represents stairs in one drawing file and lights in another, in a file that merges both drawing files you may end up with stairs and lights on the same layer. With text layer names, such merging is less likely to occur.

Layers by name. Many firms use simple names for layers. This approach is particularly common for projects that are simple or for which CAD is used only for basic drawing purposes. Layer names may be as basic and obvious as "Walls," "Doors," and "Windows." Names will become more complex when you need to distinguish, for example, between elements that are to remain, those that are to be demolished, or elements that are to be built new. Simple layer names often are the easiest to use, but can be prone to misspelling and must be well thought through if they are to accommodate the contents of complex drawings.

CSI Masterformat designations. Some CAD users use layer names based on the standard sixteen divisions of construction work developed by the Construction Specifications Institute (CSI). This approach is more likely to be used by contractors or facilities managers, since the CSI divisions don't directly relate to the way designers produce construction documents.

If you use CSI designations, you need to make sure that your CAD users understand the CSI system and have a detailed CSI index on hand. You also must specify in advance how finely you want to break down the CSI numbering system. While, on the one hand, limiting yourself to just sixteen divisions for naming layers may be too general for most design projects, you probably don't need to, for example, distinguish between division 09800 (Special Coatings) and 09900 (Painting). Therefore, CSI-based layers must be carefully thought through, and guidelines should be tested before they are put to use.

The CSI approach to layer naming goes hand in hand with efforts by proponents of the ConDoc system to apply CSI divisions to construction notations and specification references. If you subscribe to the ConDoc system (see Chapter 6), then CSI layer designations may be a logical layer system for you. Note that the CSI has been working with the National Institute of Building Sciences (NIBS) to develop a comprehensive format for the exchange of technology-based information: their effort includes consideration of CAD standards.

ISO layer standards. The International Standards Organizations (ISO) has developed a number of standards for electronic documents, with the goal of facilitating information exchange on a worldwide basis. The ISO has established standards for international drafting. In addition to issuing standards for graphic symbols, drawing size sheets, and other elements, it has made recommendations for international layer standards (ISO Standard TC10/SC8/WG13). The ISO's goal is to develop a layering system that can work not only with all CAD software but also across all different CAD-using disciplines and in all countries; ideally, such a standard, if implemented, would allow a CAD drawing's layers to be recognized *anywhere* in the world.

ISO layer standards are still evolving. Issues that have been addressed to date include an approved set of international letters, numbers, and other characters, and an approach to coding elements of layer names that resembles AIA-sponsored standard layers. Standards for colors, linetypes, and database organization are also being developed.

Since at this point ISO layer standards are still developing, at present they seem more applicable to industry than to the A/E/C disciplines. Even so, design firms, especially those doing international or government work, should monitor the evolution of these standards. A committee comprised of members of the AIA, CSI, and various government agencies are encouraging ISO to consider their recommendations for layer names.

Client names. A growing number of corporations, institutions, and government agencies that use CAD themselves now request that the designers they hire submit CAD files that not only can be read by their CAD system but also use the same nomenclature. Some clients merely expect layers to be renamed at the end of a project; others want the designer to use a particular layer-naming system from the start of the project to ensure compatible file exchange during the entire design and construction process.

Using a client's layering system throughout a project is only justifiable if the project scope, length, and fee warrant it. The effort involved in converting your system to a client's standards would not pay off for a small, short-term project. Indeed, tailoring your CAD system to a particular client's requirements requires doing serious cost-benefit analysis in the context of your overall marketing strategy.

If you do agree to use a client's layering system on a project, make sure to spell this out in your contract. If possible, append a *dated* copy of the client's layer standards to the contract. If the client changes its layering standards midway through the project, you should be entitled to ask for an additional fee or to delay your work while you adjust to the revised standards.

If you must convert your layer names to the client's upon completion of the project or even during the interim, you can write an AutoLISP routine or Script to automate the process. This is a great time-saver especially if you have to convert back and forth frequently.

Colors and pens. This layering approach maps layers to the line thickness and/or color they would plot at. With this approach, objects on a layer named Yellow_.35mm would plot in yellow with a .35mm line weight. This approach is little used in routine design work for two reasons: it bears no relation to design/construction work, and it only works if the layer names actually reflect the pen settings, which in AutoCAD is not necessarily a given (see Chapter 4). This approach would be of more value when you are plotting graphics where the focus is on color and line weight rather than on drawing content. The layer names would then serve as a guide for specifying colors based on their target output.

Whatever layer-naming approach or approaches you use, remember that you need to provide standard layer names for *all* your CAD drawings and all the major categories of information that the drawings contain. Although the focus of most standard layer-naming systems has been on construction plan layers, you also need to develop layer names for sections, elevations, details, 3D views, and rendered materials. You need layers for title borders and drawing information. Many firms also assign layers to work that is not always plotted but helps with general drafting; such layers include those used for drawing grids and guidelines, establishing guides for cutting oversize plots, and those "scratch" layers used for on-screen doodling, experimentation, notes, and reminders. While you cannot anticipate all the possible drawing layer names you might need, your general system should provide guides to logical names for additional layers that a project may require.

Implementing Standards for CAD Layer Names

Once a workable layer-naming system is developed in your office, it must then be implemented in all your AutoCAD drawings. Since layers are *drawing-specific* and not embedded in your overall system defaults, the easiest way to implement a layer system is to include it in all your AutoCAD prototypical drawings (see Step 6 as described in Chapter 4). In addition, you can automate layer setup by writing AutoLISP routines or Script files. Many third-party add-ons for AutoCAD automatically load layering systems as well.

In implementing standard CAD layers, design firms vary in what sets of layers they include in prototypical drawings. Since each defined layer uses up memory merely by being defined, many CAD users prefer to load only layers typical to all CAD projects, such as layers for titles, borders, and base plan information. They might use a Script or AutoLISP routine to load groups of other standard layers; this is preferable to manual layer setup, which is time consuming and often results in spelling errors and thus inconsistent layer names. Another approach is to set up prototypical drawings with a comprehensive set of layer names. When you clone the prototypical drawing file for a particular project, you can then use the AutoCAD **Purge** command to eliminate layers that are irrelevant.

CAD programs are not the only software programs that use layers. Many graphics, paint, and desktop publishing programs rely on layers to separate different graphic elements, or, for example, to facilitate color separations for files being sent out for printing. If you use such software extensively and create office-standard templates from them, you may want to include appropriate layer defaults for them as well. Note that if you import an AutoCAD .dxf file into software that uses layers, the AutoCAD layer names (and possibly the colors) may import as well.

Documenting Standards for CAD Layer Names

As mentioned in Chapter 4, an important step in the process of CAD customization is not merely the development and implementation of an office layering system but also its *documentation*. Many offices develop a matrix, such as the one shown in Figure 7.14, to document office-standard layers. This matrix should be annotated when the standard layers need to be

ABC Design Inc.

Master Layer/Drawing Matrix

| | | | | | Plan Number/Name | | | |
| | | | | | D-1 | A-1 | A-2 | A-3 |
Layer Name	Layer Contents	Layer Color	Line Type	Plotted Pen Weight	Demo Plan	Partition Plan	RCP Plan	Power & Signal Plan
0	Not used - base AutoCAD		Continuous					
Border	Drawing Border	1	Continuous	0.25	√	√	√	√
ASheet1	Drawing Title Information	7	Continuous	0.25	√	√	√	√
ASheet2	ABC Design Information	7	Continuous	0.25	√	√	√	√
ASheet3	Client Information	4	Continuous	0.25	√	√	√	√
Colgrid	Building Column Grid	1	Columngrd	0.25	√	√	√	√
Colbub	Column Bubbles & Numbe	9	Continuous	0.25	√	√	√	√
Limits	Limits of Work Designation	5	Limits	0.5	√			
Arcol	Base Building Columns	7	Continuous	0.25	√	√	√	√
Arwall	Exterior Walls	6	Continuous	0.35	√	√	√	√
Arglaz	Exterior Window Glazing	1	Continuous	0.25	√	√	√	√
ArCore	Base Core Plan	10	Continuous	0.35	√	√	√	√
Demo	Demolition Work-Floor	11	Demoline	0.5	√			

Figure 7.14 Documentation is key when you are developing and managing office-standard layers. Databases or spreadsheets prove helpful in this regard, as the information can be sorted by different fields. This diagram was produced using Microsoft Excel.

modified to address a particular project's needs. In addition to listing standard layer names, your documentation should spell out guidelines for naming new layers.

Layer documentation is an essential internal tool; it helps facilitate CAD communications with other engineers, clients, and external CAD users. Your office-standard layers and documentation should be distributed to a project team at the start of a project. As clients become more CAD-savvy, they too will appreciate, if not expect, layer documentation.

Modeling in AutoCAD

Modeling is one of the most important steps in the design process. Except for the most routine, mundane projects, modeling is the only way designers and their clients can fully visualize the results of an idea. Although traditional models have taken the form of small-scale physical structures, increasingly modeling takes electronic form. Thus, modeling today can be done using the standard 2D and 3D capabilities of AutoCAD or using related products by Autodesk and other manufacturers. You can use slide shows, animation, sounds, and other tools that are now grouped under the rubric *visualization*. Regardless of which approach to modeling you select, AutoCAD can prove a valuable model-making assistant.

Physical Models

If you are building a traditional physical scale model and have used AutoCAD in developing the initial design concepts that underlie the model, you can *re*use your AutoCAD information to make your model components. You can plot out your CAD files as scaled cutting templates for the model. Not only will you have accurately scaled models; you can reduce material wastage by positioning the CAD-based components so that you can fit as many pieces as possible on a standard-size plotted sheet.

You can use AutoCAD for templates and also for surfaces and backgrounds. Hatches combined with color can produce a range of patterns that resemble building materials, ground covers, and scenic backgrounds. If you use a 3D program such as Autodesk's 3D Studio or Autovision, or a high-quality paint or publishing program and have access to a color laser printer or copier, you can work from a plethora of materials and surface libraries to generate photorealistic images for everything from marble floors to Hawaiian sunsets. These can be pasted onto your built models as material finish surfaces or background scenes.

Paper is not the only material that AutoCAD can plot onto. Laser cutting machines, driven by AutoCAD plot files, can cut components in cardboard, cork, metal, plastic, Plexiglas, wood, and modeling materials (the process is called Computer-Aided Modeling and Manufacturing or CAM). After the pieces are cut, you can glue them together as you would hard-cut components. Many CAM machines are fairly expensive and generally still are the province of specialty model-making shops or service bureaus. Building models using this method is rarely cheap, but for very complex projects or multiple schemes, it can be more cost-effective than constructing in-house models by hand. CAM machines also cut materials with greater accuracy and speed than hand cutters can, and they cut arcs, circles, and ellipses more precisely and cleanly. In a few hours, you can laser cut the same number of pieces that, if cut by hand, would take days.

In addition to cutting actual model components, CAM lasers can etch patterns to represent stone, brickwork, and other building materials as well as design motifs and plan and elevation views of trees, shrubs, and windows. Lasers usually can cut materials up to 1/4" thick, so it is possible to produce textures and three-dimensional effects on the model pieces, thereby saving the time you would otherwise spend pasting materials on surfaces.

New building design is not the only purpose for assembling CAD-generated models. You can also use Geographic Information Survey (GIS) CAD files to generate dimensionally accurate models of natural terrain at different elevations. CAD files can direct cutting machines to plow through cork, foam board, and other materials at varying levels to produce a 3D scaled base for your site.

If you use external vendors to build models from your CAD files, make sure that your CAD files contain only information that should be plotted for models and not extraneous layers of information; otherwise, you run the risk that unintended design elements will appear in your model. Your CAD files should clearly delineate what the separate model components are. Model makers will find it helpful if you can provide small, annotated plots that show the intended model configuration.

CAD Models

The increased speed, affordability, and user-friendliness of computer-based modeling tools in recent years has made 3D modeling and related "visualization" presentations much more common means for design firms to use in developing their designs and conveying their ideas to clients. What just five years ago involved major effort and justified special fees is now becoming ubiquitous. Design firms that seek to remain competitive realize that clients increasingly expect to see computer models as a routine design tool.

Computer-generated models lack some of the qualities that make physical models so appealing: the tactile quality, the actual use of texture and shadow as opposed to the rendering of it, and the ability of clients themselves to hold the model and move it around. Nevertheless, for certain projects and purposes, computer-generated 3D models are more powerful tools for the following reasons:

- Changes and alternatives can be produced more quickly.
- More combinations and permutations of variables, such as scale, color, texture, light, and shadow, can be studied.
- Scale becomes irrelevant; you are not limited by the size of modeling materials, display surfaces, or transportation logistics. Models in 3D are produced in "true" scale and then displayed on-screen or plotted at the scale of your choice.
- The viewer can participate in the model. Unlike small-scale physical models, where the viewer is a giant peering down onto a Lilliputian world, 3D models can be viewed from a person's actual eye level (see Figure 7.15). Viewer participation is particularly useful when you need to convey the feeling of tight and complex interior spaces. With animation, viewers not only can "see" the space from an ordinary viewpoint but also move through that space.
- You can simultaneously see multiple views of the same model, and in a single CAD file (see Figure 7.16). With a physical model, you'd have to use mirrors to see more than one view at a time.

Despite the advantages of modeling in 3D CAD, AutoCAD's 3D modeling capabilities when compared to its 2D drafting features are underused. One reason is that constructing 3D models requires more care and skill in setting up, as well as more powerful computers, yet being able to create 3D models using AutoCAD is not usually considered as essential to a project as producing CAD-based construction documents. In addition, while most young designers have been trained on 2D CAD at school or on the job, 3D is still a more unique specialty that drafters learn by choice or at a design school that stressed CAD-based design work. Finally, "plain vanilla" 3D AutoCAD (like basic 2D AutoCAD) is generic. Its modeling functions have not been customized specifically for architects and interior designers; users have to customize their own system or purchase an add-on

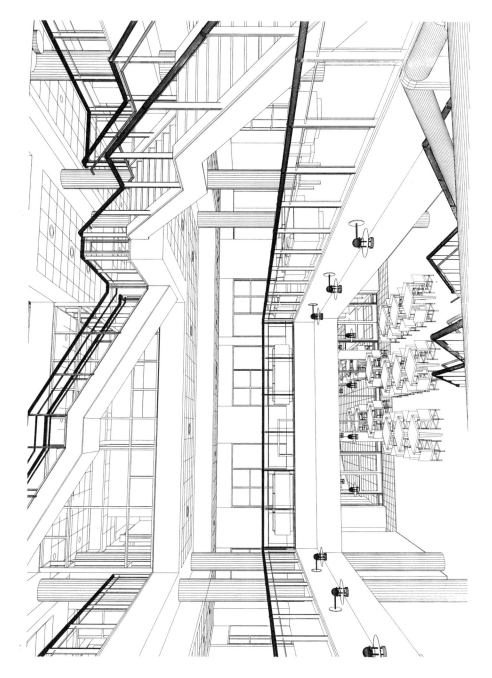

Figure 7.15 This AutoCAD-generated interior perspective of a new college building allows viewers to see a realistic "viewpoint" of a space from the angle and height of their choice. AutoCAD drawing of the Babson College F. W. Olin Hall provided courtesy of Goody, Clancy & Associates, Inc., Boston, MA.

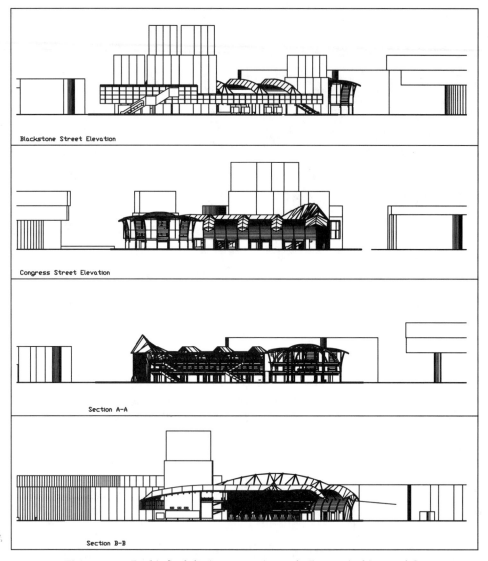

Figure 7.16 *For his final thesis presentation at the Boston Architectural Center, architectural student Olaf Vollertsen replicated in AutoCAD a physical model that he had already built to develop his thesis project: a design for an urban marketplace, the Blackstone Market, in downtown Boston. Working with a single AutoCAD file, he then plotted for his final presentation a complete set of plans, elevations, sections, and interior and exterior perspectives. AutoCAD drawing courtesy of Olaf M. Vollertsen, Architectural Intern.*

tailored to their professional needs. For these reasons, 3D skills remain underdeveloped in many design firms—although the trend is changing—and are optional enough that many firms subcontract 3D work to external consultants who specialize in 3D and animation.

Conceptually, 3D AutoCAD is not difficult. If you recall your high school geometry, you can understand that using 2D AutoCAD entails working in an X,Y coordinate system. AutoCAD in 3D simply adds the Z coordinate, so that you move through an X,Y,Z coordinate system. While a three-coordinate world is not difficult to envision conceptually, the addition of the Z coordinate greatly complicates matters.

To understand exactly why this is so, imagine playing a computer-based version of tic-tac-toe in both 2D and 3D, as shown in Figure 7.17. In 2D, you simply have to deal with a flat grid composed of nine squares. Your flat 2D disks can move in only three directions: horizontally, vertically, and diagonally. You play in only one view—a flat X,Y plane. In 3D you have many more options to contend with. First, you have to consider twenty-seven spaces—now cubes instead of squares—through which you can move your tic-tac-toe pieces. You can also move your pieces in more than twice as many directions. In addition, you can move through a range of environments: see-through wire-frame, surface-covered, or solid cubes. And you can view the game from any possible surface and angle outside—or even within—the cube. Moreover, your disks no longer need to be flat circles;

Figure 7.17
Playing tic-tac-toe—or doing anything else—becomes geometrically more complex in 3D compared to 2D.

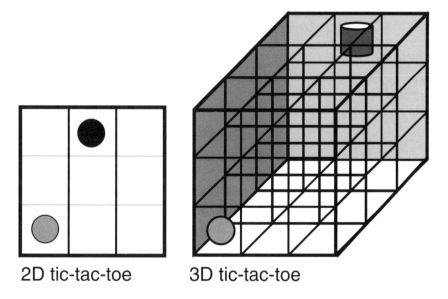

2D tic-tac-toe   3D tic-tac-toe

MODELING IN AUTOCAD

they can be spheres or extruded cylinders. The entire game can be rendered in an infinite combination of colors, textures, patterns, light sources, and shadow types.

Because of all these elements, 3D models take longer to construct than 2D drawings. First, 3D model makers must make more decisions about every model component. For a 2D line, you only have to specify it in terms of two coordinates, or a length and angle, thickness, color, and linetype. Once that 2D line becomes part of a 3D plane, you have to specify more complex coordinates and angles in the X,Y,Z plane, and the characteristics of all the plane's surfaces, which may vary. Second, all these features demand more of AutoCAD in terms of compiling and rendering. Regeneration of 3D models, therefore, generally takes longer, although in recent years 3D software programs, in combination with faster computers, have made great improvements in speeding up the regeneration process.

Given that 3D is more time-consuming and complex, the most critical aspect of creating 3D models may not be the actual construction process as much as planning, scheduling, and budgeting for the work. Although 3D model makers are responsible for the actual creation of the model, project managers must take responsibility for the overall production. Otherwise, the model either may not be created within a specified budget or time frame, or it will not be useful.

Many instructional manuals on 3D encourage model makers to view themselves as film directors. This approach is suitable, whether or not animation is involved, because modeling requires blending the skills of directors, cinematographers, set designers, lighting specialists, costume designers, script writers, and even sound editors and composers.

If, then, model makers function essentially as film *directors*, then project managers serve as film *producers*. Project managers may not know the exact particulars of 3D model construction, but they need to understand all the options. Their responsibilities include scheduling, casting, marketing, and financing. And just as standard film productions can lose money if the producers or directors make production changes midway through the project, so design firms can lose time and money with 3D models if they attempt to overhaul the model concept after production starts. Although some aspects of 3D design can be modified very quickly, you may have to start others from scratch. Therefore, it is vitally important to plan the general features of 3D models in advance and to address the following issues:

- What is the purpose of the model?
  - What issues should the model address? What are the key variables under consideration? What can be ignored at this point?

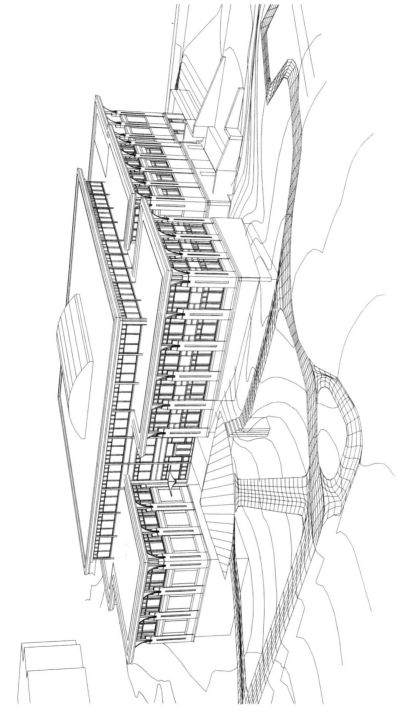

Figure 7.18 This 3D model for a new college building combines carefully constructed 3D components with topological information in order to show the building in the context of its site. AutoCAD drawing of the Babson College F. W. Olin Hall provided courtesy of Goody, Clancy & Associates, Inc., Boston, MA.

MODELING IN AUTOCAD

- Will one model suffice, or do you need to compare different options?
- Do you anticipate reusing the model in other forms and for other purposes? Or will it be used as a quick one-time study?
- Do you want to "flatten" the model into 2D for additional studies or for use in construction drawings?
- How important is absolute dimensional accuracy and internal consistency? Will an approximation suffice?
- Do you need to see the model from many angles and views?
- Is rendering of colors, surfaces, and materials required? If so, how complex or "photorealistic" should the model be?
- Are you prepared to do some aspects of work by hand?
- What is your time frame?
- What is your budget? Can you charge a special fee for the model?
- What staff—whether in-house or external—is available to work on the model?

Once the model is under construction, there are still some design elements that will need your attention, including

- viewing angles
- lens angles (which can mimic actual camera lenses)
- lighting type and intensity
- surface materials, colors, textures, and degree of shine and roughness
- shadows
- pace of views if slide shows or animation is used
- sound effects if any

Since each decision requires selecting among infinite options, it is often best to sketch a mock-up or storyboard of your intended model, making as many decisions as possible on paper before you have them translated into 3D.

When planning 3D models, you might take inspiration from the great British film director Alfred Hitchcock, who mapped out all his films to the minutest detail before the cameras ever started rolling. His approach did not allow for much flexibility or modifications during the filming process, but the resulting films met his goals to great acclaim, and his films generally did not exceed budget.

258   SPECIAL AUTOCAD APPLICATIONS

AutoCAD Modeling Options

Basic AutoCAD offers four built-in methods for creating 3D images.

1. Isometric models. This option is actually a 2D drawing technique that allows you to simulate 3D views by altering your drawing coordinate system so that the Z coordinate axis is vertical and the X and Y axes lie 120° rather than 90° apart from each other. Isometric models can't be viewed from multiple angles and measurements, except along isometric axes, are distorted. You also cannot apply any rendering tools to them. Such models, however, can be useful for quick, one-time studies of 3D interactions of planes and surfaces and to give some sense of how an object fits in 3D space. Apart from the **Isoplane** command, you can construct isometric models by using standard 2D AutoCAD drawing commands, and, therefore, you do not need to have extensive knowledge of 3D AutoCAD. Isometrics are also easy to plot.

2. Wire-frame models. This option enables you to build wire-frame skeletons composed of lines and curves representing the edges of objects. Unlike surface or solid models, AutoCAD does not provide any time-saving tools for creating standard geometric shapes working with the wire-frame option. Each model component must be separately constructed, just like intricate physical models. Moreover, you can't apply surfaces, light, and shading to them. Many designers find wire-frame models most useful as underlays for hand-rendered perspectives, although when time permits, these models can also be useful for working out highly complex construction assemblies that can't be as easily replicated using surface and solid tools (see Figure 7.18).

3. Surface models or meshes. With this option you can create models with faceted surfaces, or meshes, that conceal objects behind them. The difficulty of constructing meshes depends on whether your object is a dimensionally and geometrically simple shape, such as a box, sphere, or torus; AutoCAD provides Surfaces tools to automate the creation of basic mesh shapes. More complex shapes, or complicated interactions of meshes, require more modeling time as you must specify coordinates for the vertices of each surface face. You can apply AutoCAD's limited shading, lighting, and material rendering tools to surface models. Surface models are often used for presentation renderings or topographical models. As mentioned above, surface models have faceted surfaces, so you can only approximate curves; if you need true, smooth curves, you must use other software, such as the AutoCAD add-on, AutoSurf.

4. Solid models. This option allows you to create solid models by making solid extrusions of existing 2D objects; by using AutoCAD's 3D **Shape** tools (see Figure 6.6) for boxes, spheres, cones, and other basic shapes; or by building complex composite solids. Solid models are useful for studies of interactions of shapes and building masses. Depending on the model's purpose and complexity, solid shapes can be the easiest 3D shapes to construct. Like surface models, solids can be rendered with light, shade, and material surfaces. They can also be assigned physical properties, such as mass, weight, and centers of gravity.

As AutoCAD software has evolved, 3D modeling has become more integral, accurate, and easier to use. Nonetheless, project managers must understand the limitations of 3D AutoCAD and investigate alternative modeling tools when project needs exceed AutoCAD's capabilities. Some of the particularly noteworthy limitations of 3D AutoCAD are noted here.

- Although 3D AutoCAD does offer rendering capabilities, they are limited. You can show colors and materials on surfaces and solids, and you can modify the reflectivity or roughness of surfaces. As following sections describe, you can generate some useful rendering effects if you take the time to customize AutoCAD's rendering tools. But you can't apply true texture mapping or "photorealistic" effects or backgrounds. Your options for lighting and shading are also limited and basic.

- You can create some sense of movement through your model by creating drawing sheets whith multiple views or by designing carefully paced "slide" shows. You cannot, however, do actual 3D walk-throughs or any other forms of animation. You can incorporate an occasional noise, but you can't attach a soundtrack.

- Although you can create dimensionally sound models, you do confront limits, particularly with curved surfaces, in applying precise dimensions and making calculations.

For any project that warrants highly rendered presentations, animation, or a high degree of accuracy, you need to use a third-party add-on, import an existing AutoCAD file into another software program, or use a specialized program to draw a new model from scratch.

In selecting and using other modeling and animation software with AutoCAD, you have many options, and your criteria should be based first of all on your goals for the model, as previously discussed. In addition, you should bear in mind that 3D AutoCAD as well as many add-ons are, just like 2D AutoCAD, *generic* modeling and rendering tools. They do not address the particular modeling needs of architects and interior designers; indeed, many seem most suited for industrial design. So while you certainly want to consider generic software, such as Autodesk's own AutoVision (a rendering add-on) and 3D Studio (a separate, high-powered animation program), you should also investigate a number of third-party products created specifically for use on building construction projects. Many such products are listed in *The AutoCAD Resource Guide*; Autodesk dealers who specifically tailor their services and product lines to the A/D community can direct you to pertinent products as well.

As with all third-party products, some run *inside* AutoCAD; others are completely separate packages that usually can work with drawing files imported from AutoCAD, but they do not necessarily need an AutoCAD file to work. Some packages offer complete 3D modeling and rendering functions; others apply selected tools to enhance existing AutoCAD models. Robert McNeel & Associates offers two AutoCAD add-ons that help you enhance existing AutoCAD models: AccuRender for sophisticated, building-oriented rendering and WalkAbout for walk-through and flyover presentations. Facade, by Eclipse, is an AutoCAD add-on for creating architectural models. Many comprehensive add-ons for designers, such as Softdesk's Auto-Architect and KETIV's ARCHT (see Chapter 4), include basic 3D tools in addition to their 2D capabilities. You can also purchase a number of block libraries for AutoCAD and other software; such libraries offer 3D building products, furniture, plumbing fixtures, landscaping materials, vehicles, exotic finish materials, and "mesh" people.

In deciding how to work with other 3D software, one key issue is whether you keep your model in AutoCAD—enhanced by an add-on—or whether you import it into another package. With completely separate software, you can use tools that AutoCAD doesn't support. However, you lose the ability to integrate information with your original AutoCAD file, making it difficult to use a single model for all stages of design, including construction drawings. With most software packages, you can import the model back into the original AutoCAD file, but you may lose some elements of the model, and you have to be prepared to spend time checking for dimensional accuracy and modifying the model for regular 2D use. Modeling outside of AutoCAD, therefore, is often most useful for major, highly rendered modeling projects where you are allowed sufficient lead time or for when you are prepared to lose connection to the original 2D drawing files.

General 3D Modeling Guidelines

Experienced users of physical models know that effective models have certain characteristics and conform to certain guidelines. Most of these suggestions apply to 2D models as well, but because of the additional complexities involved in 3D models, they are reiterated below with special attention to 3D contexts.

Focus on a just few important variables. Most models, particularly those done for the purpose of study rather than presentation, are built with the view of addressing certain important issues. These issues may concern scale, circulation, massing, facade design, or site orientation. Whatever the issue, the model should emphasize what is key and downplay or disregard considerations that are not of immediate relevance.

Focus on *content*. Especially with some of the high-powered rendering tools that are now available, you can achieve some remarkable visual effects. It is easy to get sidetracked by exploring all the options for controlling the *appearance* of your model and consequently to lose sight of the model's intended *content*. Therefore, focus on the story the model needs to tell; the visual effects should evolve naturally from that.

Use variable elements only for options you want to study. If the purpose of your model is to compare different options for a particular design issue, then vary those particular elements while keeping other elements uniform. If the issue at hand is general building circulation, then don't present circulation options in radically different color schemes.

Apply a scale and level of detail appropriate for the phase of the project. If you are in the early stages of design where decisions about broad issues such as site planning, general circulation, or massing are still evolving, you normally don't want to present close-up views of ornately detailed window or floor patterns. Be particularly careful about the level of detail you present when using AutoCAD 3D. Since you can so easily add detail prematurely, you can easily make what should be rough 3D models seem polished. Because you construct CAD models in "true" scale, and can present them from any vantage point, you risk losing sight of the scale at which you present your work.

Limit the level of detail and realism. You can render a 3D model to a high level of photo realism if you have the time and computer memory. Even so, be cautious of using highly precise details even at the final stages of design development. Designers generally find that the more a model approximates reality, the more likely it is that viewers will find flaws in it. Perhaps this is because the viewers begin to lose sight of the fact that they are working with conceptual models. Regardless, your models are more likely to be a success if they do not pretend to be anything other than models. With fewer finishing touches you are also less likely to be held responsible for any discrepancies between a visualized design and the actual implementation.

Beware of unintended meanings. Many model viewers, particularly if they are unfamiliar with the design process, may read meanings into your model that you did not intend. For example, you may select a palette of colors and patterns merely to differentiate between different surfaces or spaces; your client, however, may read these as proposals for paint colors, finishes, and materials. Similarly, a palette you consider to be neutral may have all sorts of symbolic references for others. Despite the plethora of rendering tools that all 3D software offer, use them sparingly if they are not actually needed for your model.

Test preview your model. As any Hollywood film producer would affirm, no production should be released until a test audience has previewed it. This holds true for models as for movies. Make sure to schedule time to have your 3D model viewed by colleagues who can bring a fresh and critical eye. You want to note, and address, errors and omissions before you present the model to your client. As with 2D drawings, 3D also requires quality control.

Modeling with Light

The design element that can make or break a design is lighting. Without lighting, nothing is rendered. With lighting, color, scale, and proportion are enhanced. Using lights in physical models, although important, is often difficult since so many variables are involved and light fixtures can rarely be mimicked at small scales.

Although you can find model-sized lights for physical models, you have many more options and much more flexibility if you can apply lights through 2D and 3D programs.

If you want to accurately replicate the effects of actual lamps and fixtures, you need to purchase software designed particularly for this purpose. In basic AutoCAD, however, you have a range of light tools to test broad lighting options.

The AutoCAD **Light** command or toolbar icon is accessed through the **Render** toolbar or command. As the **Lights** dialog box shows in Figure 7.19, you can create a library of predefined lights and then select from among them to light your model.

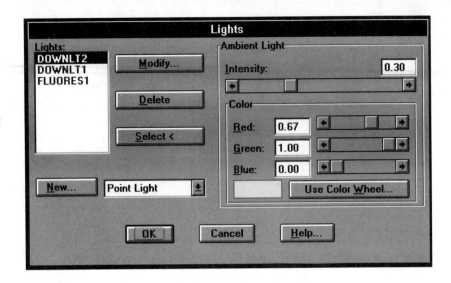

Figure 7.19
The AutoCAD ***Lights*** *dialog box enables you to create, select, and modify lighting options from a collection of defined light sources.*

The **Light** command gives you many different options for creating your light sources. You have essentially four light types: *ambient light*, which is always present; *distant lights*, which project light uniformly from no obvious source and are a good source of daylight; *point lights*, which radiate light in all directions from one source and are used for general illumination and light bulb effects; and *spotlights*, which tend to function most like light fixtures in that they direct cones of light in defined beams and angles. Each light type has different variables that you can modify, such as light falloff and attenuation rates for point and spotlights.

For all types of lights, you can specify the light color in terms of red/green/blue (RGB) values on the color wheel or in terms of hue/light/saturation (HLS) values.

In selecting lights for use in rendering models, you can specify one or more lights that you can store along with a view of your model using the **Scenes** dialog box. In creating scenes, you must consider the interaction of light with materials, surfaces (in terms of their roughness and smoothness), and color. You must also determine how light affects your palette of shades. With AutoCAD's **Shade** command, you can select one of four methods for rendering edges of shaded surfaces as well as the degree to which light diffuses as it reflects off each surface. Each shade specification will interact with light in different ways, so you need to spend some time testing each option.

If you render models frequently in AutoCAD, you can save time by building up a library of standard light types and scenes. Since you have complete flexibility in naming lights and scenes, as you do with all customized applications, your names should make sense in the context of your work. For light names, for example, instead of using cryptic names such as Light 1A, you might specify whether the lights represent incandescent or fluorescent lamps or replicate an office-standard light fixture. You can also link defined lights to an office-standard pro forma light fixture schedule.

Although AutoCAD's light feature produces some powerful renderings, it should not be used as a reference for specifications of actual lamps and light fixtures. If you want to specify particular types of lights, one option is to import your file to software designed specifically for lighting design, such as Lightscape by Lightscape Technologies. Another option is to work with a lighting consultant or engineer to approximate a fixture's luminance within the constraints of the AutoCAD **Light** command.

In selecting and using electronic lighting models, you will find that there are two basic methods of simulating light. The first is *ray tracing*, which calculates the path of a light ray as it travels between a viewer's eye and a light source. Ray tracing works best if you are working with exterior light sources. The AutoCAD add-ons Autovision and 3D Studio offer ray-tracing tools. The second technique is *radiosity*, which calculates light as variable based on its interaction with 3D surfaces. Unlike ray tracing, radiosity works independent of any particular viewer standpoint. With this technique you can more easily move around models to see them from various angles. Since radiosity is a newer technique, it is less likely to be incorporated in standard 3D packages, but its applicability is growing. In some software, radiosity can be used in conjunction with ray tracing.

Basic categories of light, such as daylight, tungsten, and sunlight, can be measured on the Kelvin scale according to their color temperature. It is possible, using somewhat complex mathematical algorithms, to convert Kelvin into RGB or HLS triplets, which you can then replicate in the AutoCAD **Color** dialog box. You can use AutoLISP and ADS routines to build relationships between light characteristics and RGB and HLS color triplets.

Modeling with Materials

Along with light and shade, materials are another important component that can make models more convincing to viewers. As usual, AutoCAD comes with a library of standard materials that you can modify through the **Materials Library** dialog box, which is accessed with the **Matlib** icon in the **Render** toolbar (Figure 7.20).

Materials are saved in **Materials Library** files with an *.mli* extension. You can import *.mli* files from Autovision or other programs. In addition,

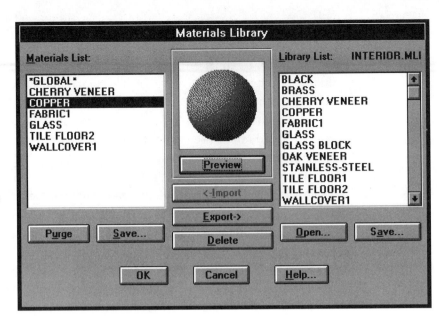

Figure 7.20
*You can use the AutoCAD **Materials Library** dialog box to specify electronic versions of your material samples library. This example shows a materials palette taken from a custom library of interior finishes and materials.*

you can modify materials using the **Rmat** command (see Figure 7.21). With **Rmat,** you can apply new colors, whether from the ACI, RGB, or HLS color systems; you can specify the degree of shade and reflective values; and you can adjust the level of smoothness or brightness. To adjust other characteristics, however, such as a surface transparency or specific textures, you have to edit the file in AutoVision or other more advanced rendering software.

If you frequently use the rendering capabilities of AutoCAD or other software, you may want to develop a sample materials library, containing materials you frequently use for modeling or specifying. As with standard color palettes, you should plot out and paste up samples of the materials as reference boards, analogous to manufacturers' sample boards, so that designers can specify 3D materials with greater confidence and spend less time on unnecessary experimentation. You can save various materials collections as *.mli* files with meaningful names, such as *Intfin.mli* for interior finishes and *Extfin.mli* for exterior materials.

AutoCAD's **Materials** commands allow you to link materials with specific layers. This feature can be useful for global editing. If you assign materials to layers, however, you should incorporate layer names—or at least a layer-naming policy—for materials into your office CAD layer-naming standards.

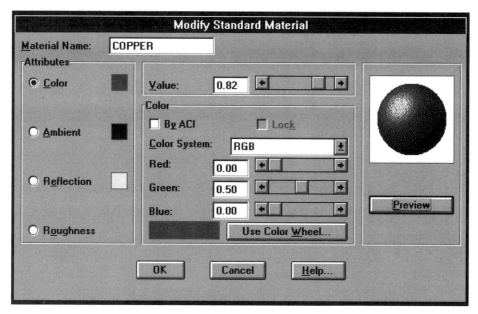

Figure 7.21 Through the AutoCAD **Modify Standard Material** dialog box, you can alter the characteristics of defined materials. In the example shown here, the material is changed from a bright, gleaming copper into a rough, matte verdigris.

As the use of AutoCAD's 3D capabilities becomes more widespread, more and more materials libraries—from both Autodesk and other vendors—are becoming available. Apart from AutoCAD's base materials file, *Render.mli*, you can purchase third-party add-ons that provide materials and textures for everything from marbles and bubbles to earth, wood, and water. For use with AutoCAD files used in AutoVision or 3D Studio, you can purchase Autodesk's Texture Universe, which is a CD-ROM containing digitized textures and backgrounds. For a useful summary of such add-ons, ask Autodesk to send you an independent periodical called *The Autodesk Multimedia Partner Catalog*.

Presentation Options

One of the most important steps in planning 3D models is to consider presentation options other than paper plots. You can present slide shows, video animations, walk-throughs, and flyovers. Most models can be printed out in static black-and-white plots, but some highly rendered models need to be shown on a video monitor or a slide screen or be output as a high-quality color laser plot to convey the model's full effect. You'll need computers with sound boards or VCRs if you want to feature sounds in conjunction with images.

Given the different output considerations, it is important to determine from the onset of the project how the model will be presented. If the presentation exceeds your present in-house output resources, you may have to acquire, rent, or borrow a color printer, slide printer, photo-CD, or video screen to complete your presentation. In some situations, you may end up demonstrating your model at your clients' office, using their equipment. Regardless, make sure that you schedule, budget, and test all output devices needed for presenting your model effectively.

Internal versus External 3D

Given the expense and time involved in mastering 3D, many design firms with limited in-house capabilities and/or expertise go outside to create 3D models. They often contract with an individual consultant who builds the model on his or her own machine. Consultants working in 3D offer the same benefits and carry the same risks as do consultants for other aspects of CAD (see Chapter 4). Just as many design firms have not been able to retain full-time the services of a renderer, perspectivist, or model maker, so, also, they often can't justify hiring a skilled 3D/animation modeler. It can be cheaper to maintain an ongoing relationship with an external consultant they know and trust.

A variation on this option is to use a CAD service center, which provides not only model makers but also a full range of equipment and reprographics services. CAD centers may provide models in all forms of 3D as well as animated models; set up presentations in media, such as slides, transparencies, or videotape; and make plots in black-and-white or color on bond, vellum, or drafting film. As with most reprographic services, the costs of CAD centers are often passed along to clients as direct reimbursables.

The services offered by CAD centers are not cheap, but they can bear the acquisition costs and overhead involved in acquiring, configuring, and maintaining equipment that the average design firm couldn't afford. If you use a CAD center, you don't have to hire and train staff; paying for a couple of hours of the center's staff time, while expensive, can cost less than carrying an expensive full-time 3D modeler in-house. CAD centers also offer you the opportunity to become familiar with different equipment and output options; you can become a well-educated consumer should the time come that you need to invest in such hardware.

As with any plotting or reprographics service, you must investigate CAD centers carefully and evaluate them not only for price but also for reliability, turnaround time, and range of services offered. As anyone who has incurred problems with orders for simple blacklines or photocopies knows, you need to be able to provide instructions for your models in such a way that your orders won't be open to misinterpretation.

Some Additional 3D Tips

- Designers usually learn 3D AutoCAD most easily when they can implement it in an actual project in which they can address particular issues of concern; otherwise, they can become distracted by encountering too many options for too many variables. This problem is often compounded by the fact that many generic 3D tutorials use sample files for mechanical engineering or industrial applications—fields that apply 3D differently from the way designers do. If you don't currently have a billable client project to convert to 3D, then do a test project using your office or some other project that addresses issues similar to what your projected 3D needs are likely to be.

- 3D model making actually requires more meticulous drafting skills than 2D does. In ordinary 2D construction drafting, for example, you can get away with sloppily drawn lines and polylines that don't connect or entities that appear orthogonal but actually aren't. In 3D, however, you come upon the moment of truth: a nonclosed polyline cannot be extruded into a 3D shape, and an "orthogonal" line drawn at 89° rather than 90° can cause problems in the X,Y,Z coordinate system. Any form of 3D modeling, such as wire-frame construction, that requires precise alignment of separate components requires the same care and accuracy that physical model making does.

- Although you have several model-making options to choose from, you should avoid mixing model types or converting from one model type to another in the midst of a project. Note that you actually face limited conversion options in any case. You can convert solid model types to surfaces and surface models to wire frames, but you can't convert wire frames to the other model types. If you try to combine different types of models, your rendering results may be unpredictable or inconsistent.

Multiple Drawings, Floors, and Project Phases

Much routine CAD drafting is done for fast renovations of small single-building or single-floor projects. For such projects, the building floor plate can easily fit at 1/8" scale on a single standard-size drawing sheet, and the resulting drawing set is quite compact.

Large projects, however, involve more complex drawing sets. You may have multiple floors, building floor plates that stretch across two or more standard-size drawing sheets, or construction work that is to be completed in different phases. AutoCAD can handle these conditions, but planning is required on your part to realize the benefits of CAD for coordination, efficiency, and speed.

Multiple Drawings

Drawings for buildings or sites that are extremely long or wide often must be spread over two or more drawing sheets, depending on the drawing scale you're using (see Figure 7.23). Small-scale key plans and matchlines are then used to help contractors and other parties relate each individual portion of the work to the others.

With AutoCAD, you have several options for working with multiple sheets:

Use a single CAD file. This is a standard approach where the floor plate of the entire building(s) is kept in a single CAD file. Separate title sheets are overlaid onto different parts of the floor plate, as shown in Figure 7.22. Each title sheet is stored on separate sets of layers. When the east side of the site is plotted, for example, all title sheet layers except for the east side titles are turned off; when the west side is plotted, only the west side title layers are visible, and so forth.

With this approach, your floor plate is likely to "bleed" to the drawing edges at those points where the different sections of the building overlap. As shown in Figure 7.22, for example, the floor plan's central portion will cross through the A-1 title area. You can address this overlap either by setting up the drawing titles and borders in paper space or by applying special "masking" techniques.

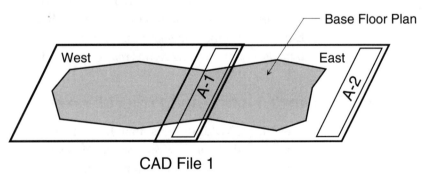

Figure 7.22 *The single CAD file approach to multiple drawing sheets*

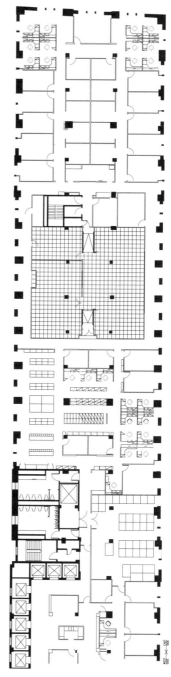

Figure 7.23 When an office renovation site involves multiple floors and exceptionally wide or long floor plates (as shown here), the use of standard-size sheets and drawing scales may require that each floor plan be split onto multiple sheets and, often, multiple CAD files. Understanding when, and how, to split the CAD-based work is an essential part of project planning. AutoCAD file of a space plan for Warren, Gorham, and Lamont offices provided courtesy of Griswold, Heckel & Kelly, Inc., Chicago, IL.

Single files work best when the project scope and drawing file size is small; a single file requires less file management and less effort to track and coordinate floor plate revisions, blocks, and graphical standards. You simply have to remember which drawing elements can be used for all plotted sheets and which appear on selected ones. You can use tools such as Script files to set up the correct layers and plotting windows for plotting each floor plan correctly.

Use multiple files. Many CAD users split drawing information into separate files; the information that each contains relates directly to the plotted output. Separate files are easier to manage in terms of speed and regeneration and managing title sheets in model space. However, discipline and coordination is required insofar as using and updating blocks for elements common to all buildings.

With split plans, the graphic convention is to show a matchline with a portion of the other plan alongside for orientation. With multiple files, the matchline should be a Wblock that is updated and reinserted as required.

Use External References (Xrefs). The AutoCAD External References (**Xrefs**) feature enables you to link or embed one drawing or another. When the first drawing is changed, that change will be reflected in the second one (see Figure 7.24). **Xrefs** are particularly valuable when more than one drafter is working on a project. Multiple drafters can simultaneously "share" drawing information by viewing the same files, although they will not be able to alter them.

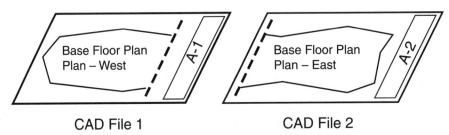

Figure 7.24 The separate CAD file approach to multiple drawing sheets

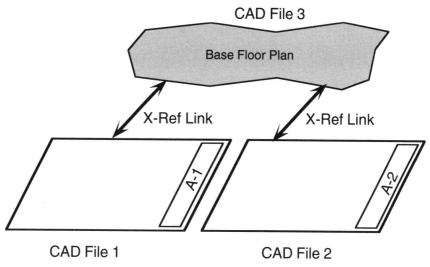

Figure 7.25 The **Xrefs** approach to multiple drawing sheets

As Figure 7.25 shows, drafters working separately on different aspects of a design can, at the same time, reference the same bare floor plan information. **Xrefs** can be used for drawing an element that appears in more than one drawing and is subject to revision. You might use **Xrefs** for the title sheets, for the base building perimeter and column grids, and for shared construction notes and legends.

Xrefs must be filed in computer directories in an organized manner; otherwise the paths linking them will be lost. They do require team management and careful division of work (see Chapter 5), but they are invaluable for linking together large sets of information, for tracking and coordinating changes, and for keeping drawing files "lean and mean." Sharing **Xrefs** among different computers, however, requires access to a centralized network file server. All Xref files that comprise a drawing must be kept together if they are to be used effectively by consultants, clients, and other external parties.

Multiple Floors

The options for working with multiple floors parallel those for working with multiple buildings or sections of buildings. The difference is that you are working vertically rather than horizontally, so common vertical elements such as building facades, cores, stairways, and shafts apply to multiple floors and therefore to multiple drawings. Following are some guidelines for working with multiple floors.

Use a single file. For small or simple buildings, some CAD users combine floor plates into one file and turn different combinations of layers on and off to produce plots for each floor plate. The advantage of this approach is that shared elements, such as a building core, can be created once, and you don't have to worry about updating multiple files should a shared element change. In addition, you can work with multiple layers simultaneously and can quickly analyze the relationship of one floor to another.

As with multiple buildings and sections, the single file approach can result in files that are quite large and unwieldy, requiring that you create title block layers for each drawing that the file produces. Many designers, however, find this a convenient approach during the early stages of a project when the building basics are still being developed. As each floor takes on a more individual character, it is then split off into separate files.

Use separate files. With this method, each floor plate is stored in a separate file. Shared elements, such as column grids or building cores, must be reblocked into each file every time the elements change. Separate files take up less memory, and in some respects they are easier to manage, especially for multiple drafters. However, you risk losing coordination.

Use External References. The AutoCAD External Reference feature works similarly for floors as for plans, offering the same pros and cons when you are working with multiple drawings. External References are helpful particularly when you have to convert information on standard scale plans to enlarged scale plans.

Multiple Phases

Particularly for large and complex projects, the design may be completed all at once, but the construction and installation may be split into different phases. This requires either splitting the bid documents into different drawing sets or creating a single drawing set that contains separate drawings for each phase.

With AutoCAD, you have the same options for multiple phases as you do for multiple floors and buildings. Your choice, however, should be dictated by an additional concern: accounting for every building component or piece of furniture or equipment during each phase. The major challenge of phased work is ensuring that nothing is inadvertently eliminated or duplicated during the various phases of work. Thus you want to try to build into your CAD system a way to coordinate work between phases.

One way to do this is to separate out every element that changes or moves during each phase as an External Reference or external block. These items would be inserted into base drawings on different layers that correspond to each phase. You can then track the files, using various query commands to ensure that each item appears as appropriate on each phase layer.

Signage

Signage and other architectural graphics are important finishing touches for building exteriors and interiors. They are at once necessities, as mandated by building codes and the need for people to find their way, as well as artistic design features. On very inexpensive projects and certain preservation work, they often are the sole element where attention to design elements and creativity is allowed and affordable.

In terms of computer applications, signage design is often considered the province of desktop publishing programs, reflecting the need to handle the complexities of text and to ensure that typefaces specified on-screen resemble what is actually produced. In addition, these programs have access to the abundance of printer fonts available for both Windows and Macintosh platforms. Although AutoCAD has expanded its typeface repertoire to include PostScript and TrueType typefaces, it is by comparison a limited type foundry.

Despite these limitations, AutoCAD should not be overlooked as a useful signage design tool. There are several reasons you should consider using it, particularly in conjunction with architectural drawings. With AutoCAD Release 13's enhanced file-sharing capabilities, you can import AutoCAD-generated designs into other programs as required.

Some reasons to using AutoCAD for signage applications include the following:

To see signage in context. While some signs are simple plaques that are not architecturally significant, others are architectural graphics or dimensional objects that not only affect the overall architectural appearance of a building but also the building's structure, electrical system, and major systems. Often signage is considered after the fact, postconstruction, when it is more expensive to install. Using AutoCAD to design signage in conjunction with other design work produces a more integrated package and at less cost to the client. You can view the same design elements both at full scale and on 1/4" or 1/2" elevations and sections, and in 3D as well as 2D.

To produce full-scale mock-ups. With AutoCAD, you can plot full-scale, dimensional plots of letters and logos. These can be applied to walls, soffits, doors, and other surfaces as full-scale mock-ups. While small-scale prints enable you to view test prints of small-scale text with ease, seeing full-scale text is enormously helpful when you are designing large signs which contain text and logos of 1'-0" or higher. The sign will "read" differently in full-scale than it will at a fraction of the projected size.

To refine layout and spacing. AutoCAD does not provide some of the text management tools, such as kerning and leading, that desktop publishers do. But it does enable you to modify individual letters and numbers if you create them using one of AutoCAD's outline fonts. You can then modify their height, width, and spacing. AutoCAD treats individual characters as entites that can be scaled, mirrored, and rotated. When designing large signs, billboards, and dimensional letters, such precision is important.

To create production templates. Your AutoCAD files can be used by signage manufacturers to create templates for cutting, engraving, and printing. Many fabricators use computer-driven cutting machines that are guided by AutoCAD-driven software add-ons. These capabilities are useful if you use AutoCAD to create and modify letters that do not match known printer typefaces or to replicate custom logos. AutoCAD signage templates also space text and graphics precisely and clarify edge details of dimensional signs.

To link signage drawings to specifications. As with other architectural products, signage is bid through a combination of drawings, written specifications, and signage schedules. You can use AutoCAD's SQL capabilities to link signage drawings to schedules in database programs. You can also imbed signage schedules created in Microsoft Excel or Lotus 1-2-3 in the drawing files.

Working with the Metric System

For centuries American architects and interior designers have been working happily in their insular way with imperial units—feet and inches—while most of the rest of the world has converted to the metric system. In the past decade, however, American designers have had to mute their isolationist measurement tendencies. More and more Americans seek design projects abroad, and oftentimes they need to go metric. Many federal government projects already require use of the metric system as well.

AutoCAD can very easily accommodate the use of metric units on design projects; this is one reason the software has achieved worldwide popularity. There are three approaches to using metric units in your projects.

Creating a new drawing in metric units. As discussed in Chapter 4, when you create a drawing, you specify your preferred units. You do this first by specifying metric in the International Settings field (Figure 7.26) of the **Preferences** dialog box. Then, using the **Ddunits** command, you select your preferred unit format from decimal, fractional, or scientific formats. (Architectural and Engineering units refer to inches.) Once you have specified "Metric," you are ready to draw using metric units.

Using dual dimensioning units. If appropriate, you can work in feet, inches, or millimeters, but show your dimensions in both *metric* and *imperial* units. To do this, you need to activate the AutoCAD **Dimension Dimalt** (dimalternate) variable. In the **Alternate Units** dialog box, you need to specify in the Linear Scale field, as shown in Figure 7.27, a scale (or conversion) factor so that the alternate dimension will appear in metric units, in brackets, next to their imperial counterpart. This is shown in Figure 7.28. AutoCAD's default scaling value is 25.4, which represents the number of millimeters per inch.

One consequence of working with dual dimensions is that the dimension text becomes much longer and thus harder to place on a drawing, especially when tolerances are called out. You may need to adjust the

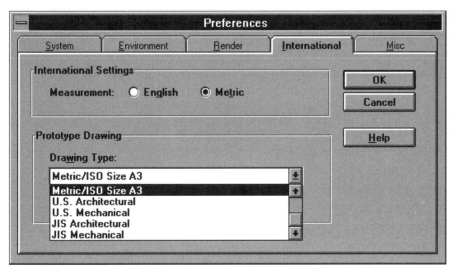

Figure 7.26 *The International Settings field of the* **Preferences** *dialog box*

WORKING WITH THE METRIC SYSTEM

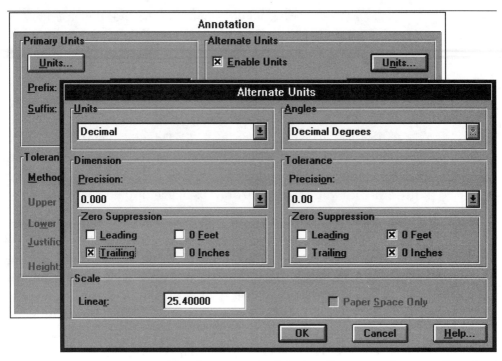

Figure 7.27 The **Alternate Units** dialog box allows you to show your dimensions in both metric and imperial units.

Figure 7.28
The results of applying alternate dimension units

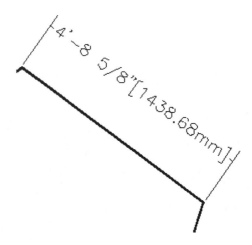

width or height of the dimension type style, reformat text lines, or, if appropriate, suppress leading and trailing zeros. AutoCAD lets you specify a degree of precision for alternate units that can differ from that specified for the base drawing units. Since inches are larger units of measurements than millimeters, always specify a level of precision for imperial units that is at least one decimal point greater than for metric.

Converting imperial drawings to metric. If you have an existing drawing that is scaled in feet and inches but that you need to convert to metric, you need to use a utility AutoLISP routine called *Cvunit*, which specifies a scale factor for converting from one set of dimension units to another. Refer to the *AutoCAD Release 13 Customization Guide* for further instructions. Alternatively, you can try switching the drawing units to a metric-compatible set of units and scaling the entire drawing. Your scaling factor must be very accurate, however.

Although AutoCAD itself handles metric units easily, you should be aware that switching between unit systems during a project can have an impact on the graphic appearance of your output and may render aspects of your customization meaningless. Specifically, converting to metric midstream may require modification of the following elements:

- base title sheets sizes and drawing limits
- dimension styles
- linetype scales
- block insertion scales
- hatch pattern scales
- plot settings
- industry-standard block symbols, details, and other drawn components
- viewports

8 Managing AutoCAD Output

Inputting and manipulating data in AutoCAD files is only half the CAD equation. The other half is getting printed output from your CAD files.

With the early versions of AutoCAD, plotting was generally a mysterious and perplexing process, prone to error. AutoCAD Release 12 eased the configuration of plotting devices, and its new **Plot Preview** feature provided CAD users with more control over their plots. Plotting has been further aided by ongoing improvements in plotting technology. The result: machines that are simultaneously quieter, faster, more manageable, and relatively cheaper.

Although plotting with AutoCAD has become much easier, large output devices still constitute major investments for many small to medium-size design offices. Only the largest design firms with many workstations and extensive networks can instantly justify the costs of the more expensive printers and plotters. For smaller firms, a careful analysis must be made of the pros and cons of using internal devices versus external services. This analysis should be made in the context of a design firm's general CAD strategy as discussed in Chapter 2.

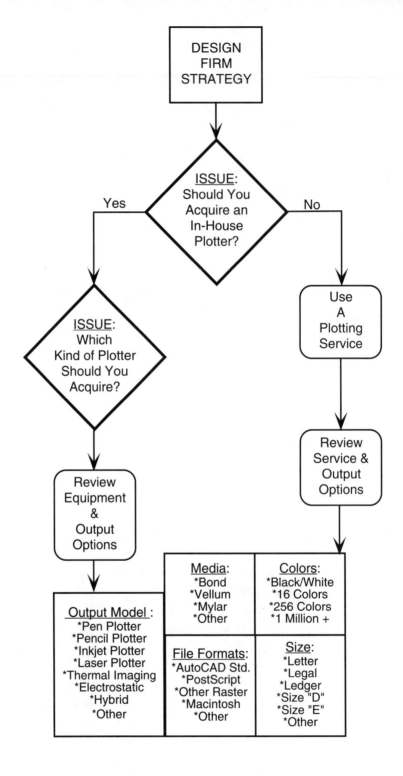

Figure 8.1
Your firm's decisions on output reflect your overall CAD strategy.

282   MANAGING AUTOCAD OUTPUT

Production: Choosing between Internal and External Options

All design offices need at least one output device, whether a printer or a plotter, that enables them to print their AutoCAD files. At the very minimum, they need access to a laser or inkjet printer that can print on standard letter (8 1/2" x 11") or legal (8 1/2" x 14") sheets; most offices already have such printers in use for doing word processing and project management. To print AutoCAD files from these printers, however, you need to install a compatible AutoCAD Autodesk Device Interface (ADI) driver. With many printer drivers, you can easily print an AutoCAD-generated image, but the print may not show any variety on line weights, colors, or tones.

Standard printers, while useful for outputting "miniature" drawing snapshots and close-ups of portions of drawings, do not accommodate the standard architectural drawing sizes, ranging from 36" x 24" to 48" x 36" or larger. For this reason, you must have access to a plotter that can produce drawing plots on large sheets.

A key management decision, therefore, is whether to have an on-site plotter or to use external plotter services; if you do decide to acquire a plotter, you then must choose among various types and models of plotters. Your decision cannot be based solely on initial acquisition costs, which nowadays can be relatively minimal. You also need to consider the relative costs of maintaining the machine, the costs and types of supplies you'll need to stock, and the nature and quality of output that you seek. Decisions about buying plotters versus using plotting services need not be mutually exclusive: many design firms with their own plotting devices also use plotting services as backup when their own machines are either down or overloaded or for generating types of output they can't produce themselves. Many design firms find that, as CAD permeates their practice, they must repeat the decision-making process shown in Figure 8.1 in order to determine if they should replace their existing plotter with a more powerful machine or purchase an additional one.

Although plotters are a major acquisition, they are now far more affordable, faster, and easier to maintain than they were a decade ago. For any design firm seriously committed to CAD, an in-house plotter rapidly becomes essential. Many firms find that even when their average plotting volume is moderate, having an in-house plotter prevents production bottlenecks, reduces production turnaround time, and encourages more creative forms of output, not only with CAD but also with any software that can send output to plotters. Access to a good in-house plotter can increase your sense of control over the entire CAD input/output cycle, which in turn helps build confidence in the CAD process and reduces the level of doubt and anxiety that any CAD skeptics in your office may feel when CAD production seems beyond their control. That, in turn, will encourage your firm to use CAD, thereby increasing the return on your CAD investment.

Reasons to Purchase a Plotter

Apart from the need to reduce production bottlenecks and to build confidence in CAD production, there are a number of other reasons why a design firm should consider purchasing a plotter.

The plotter will have a high usage rate. If you will use the plotter on a daily basis, that is, have a high plotter *usage factor*, you can justify the use of floor space and staff time. If most of your plotting is applied directly to billable projects, the machine can pay for itself within six months to a year.

You are not getting your current work done. If lack of a plotter causes you to miss project deadlines, creates stress because external plotting services prove unreliable, or creates production bottlenecks, you need to get a plotter. Don't let the cost of a plotter deter you from gaining the ability to use CAD to service your clients properly and make your staff more productive.

You don't want to rely on plotting services. If you are not located near a plotting service, feel uneasy about depending on messengers and delivery services, or can't find a service with a turnaround time that meets your demands, then you need an in-house plotter. These days, lack of physical proximity to a service can be irrelevant, thanks to modems and overnight delivery services, but to use plotting services effectively, you still have to be able to work with their production turnaround time and delivery schedule.

You can make the plotter a profit center. Many design firms can recover at least some of their plotter investment by providing plotting services to other firms. If your own plotter usage rate is low and you are located near other design or engineering firms that need access to a plotter, you can either sell access to your machine or run plots for them yourself. If your price per sheet and reliability is more competitive than that of a plotting service, you have the opportunity to make a profit on your investment. Just don't let your plotting operations disrupt your own operations or distract your staff from their other tasks. Determine in advance what a fair price per plot should be; the price should cover the costs of plotting media, ink, machine wear and tear, and staff support. Many plotter manufacturers can give you estimates for the cost of plotting a standard-size architectural sheet. You can use these figures as a base from which to calculate your own rates, although bear in mind that manufacturers' estimates tend to be based on assumptions about paper quality and ink usage that may not be applicable to your clientele's.

Reasons Not *to Purchase a Plotter*

Although many firms find in-house plotters necessary and useful, others—particularly small to medium-size firms—find reasons to avoid, or at least delay, purchasing a first plotter or, in many cases nowadays, an additional device. Listed below are some reasons why not purchasing a plotter may make sense for the time being.

The plotter will have a low usage rate. If you run only a few plots per week or month on average, that is, you would have a low usage factor, don't invest in a machine that will frequently sit idle. You will only be tying up floor space and funds in a nonproductive asset when you could invest in something of more immediate value to your office.

Your office space is tight. Plotters themselves have a large footprint, usually 3 to 5 feet long by 2 feet deep, and they generally weigh well over 100 pounds. They need space for user circulation, storage of plotting supplies, as well as adequate electrical power and ventilation. Some models, particularly pen plotters, can also be very noisy. If your office has never found the space for a diazo machine or a large floor-mounted copying machine, you will probably have difficulty accommodating a plotter.

You have access to reliable plotting services. If you have found a plotting service that is reliable, has a delivery schedule that meets your needs, and offers competitive prices, you may find using the service preferable to having your staff spend time setting up and watching plots. Your staff will still need to spend time writing plot files, arranging for them to be picked up by—or modemed to—the plotting service, and reviewing the completed drawing plots when they are delivered back to you.

You can't afford a plotter *now.* Despite ongoing cost decreases, acquiring a plotter ties up funds that could be used for paying salaries, financing office improvements, marketing, purchasing another computer, or other expenditures that seem more urgent. Moreover, especially since they are always being superseded by newer and better models, plotters never increase in value—they become technologically obsolete rapidly. So although you may realize tax benefits by depreciating your investment, you will be lucky if you can eventually sell your plotter for more than its official book value. Under some circumstances, therefore, purchasing a plotter may not make immediate financial sense, although this course must be weighed against the long-term costs—financial and otherwise—of not having one. Delaying a purchase may also make sense if you have reason to expect a model you'd consider purchasing to drop radically in price within the next six to twelve months.

Determining the Break-Even Point for Plotter Purchases

As with other major investments in CAD, plotters justify doing a break-even analysis to determine if and when purchasing a first (or additional plotter) is justifiable. As Figure 8.2 shows, the break-even point for buying a plotter occurs when the total cost of acquiring and maintaining a plotter equals the revenue that can be generated by the plotter within a given amount of time.

As with all break-even analyses, assumptions about different variables affect the break-even point. For plotters and other output devices, the key variables are the initial outlay for equipment, the per plot costs, the assumed time frame for seeing a return on investment, and the anticipated plotting volume. Suppose, for example, that a firm purchases an inkjet plotter for $7,500 and determines that each 42" × 30" vellum plot

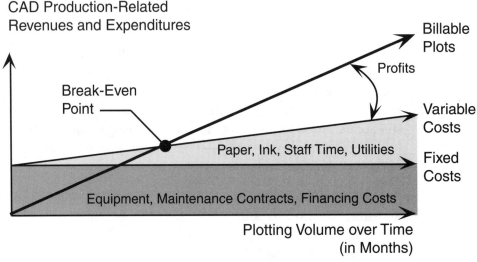

Figure 8.2 *Determining a break-even point for plotter usage helps justify the purchase of a plotter.*

run on the plotter has a unit cost of $3. External plotting services charge $15 for the same plot size and media. The break-even point, therefore, occurs when the in-house plotting volume reaches a point where it absorbs both the acquisition costs of the plotter and the unit costs of making each additional plot:

$7,500 + $3X = $15X, or $7,500 = $12X,
where X equals the number of plots run

If the firm runs only 10 plots on average per week, it would take 62.5 weeks or over a year

($7,500 ÷ 12) / 10

to cross the break-even point. If the average weekly plotting volume increases to 25, then the break-even point drops to 25 weeks

($7,000 ÷ 12) / 25

or just under six months.

To calculate the break-even point, you need to consider not only the costs of acquiring and operating a machine but also the explicit and implicit costs of *not* having a plotter. Tracking invoices from plotting services can give you a good indication of how much in-house plotting volume you must generate to achieve a profitable utilization rate with your own machine. You also should note whether plotting services are charging you for frequent rush orders which signals a need for immediate access to a machine. Also consider less quantifiable factors, such as the comfort level that an in-house plotter can provide or the ability to meet project plotting deadlines with less stress.

You'll cross the break-even point for a plotter purchase very early on if you make sure its use extends well beyond the routine plotting of black-and-white AutoCAD-drawn construction drawings. Plotters can be used with other software for a wide range of applications, ranging from desktop publishing, signage, presentation charts, and project schedules to important office announcements, such as staff vacation schedules.

Plotter Options

If you choose to purchase a plotter, you confront a wide range of choices in terms of plotter types, speed, features, and output, as Figure 8.3 suggests. There being no free lunch, you can usually expect to face trade-offs between cost, speed, plot quality, and other features.

Figure 8.3
Selecting a plotter requires weighing different options for a number of product characteristics and accepting trade-offs.

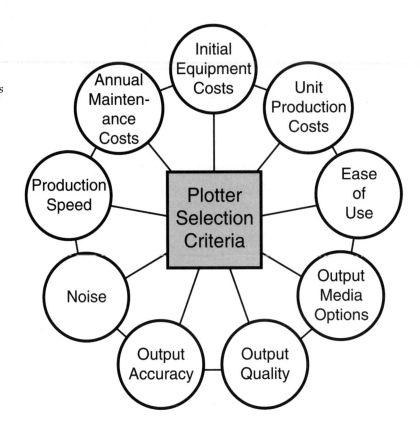

Over the past decade, several categories of plotters have been available to the A/D community. They are discussed below. Note that over the past several years, however, two plotting technologies in particular have come to dominate the marketplace for plotters: inkjet and laser/LED.

Pen Plotters

Until recently, pen plotters were the standard large-size output devices found in many design offices. Pen plotters use carousels of pens to plot drawings, and designers value the resulting plots for their quality line weight and clarity. Because they read vector files, pen plotters plot graphic entities as continuous lines or arcs, thus producing clean and precise output. Pen plotters generally work well with traditional architectural media, including archival-quality drafting film.

Despite their quality output, pen plotters have lost market share to other plotter species because of features that have less to do with output quality and more to do with speed and maintenance factors. Because they are vector-driven, pen plotters draw each line separately, and thus they bounce around the plotting area; this causes them to be both slow and noisy. Pen plotters usually require monitoring because their pens can clog at any time without any warning, and because the media is prone to slip. Pen plotters are also limited in the graphic effects they can produce. Although AutoCAD now lets you can specify 255 separate "pen" colors for your plots, most pen plotters can only handle 8 or fewer pens at any one time (usually in a limited palette of black, red, green, and blue) and you can achieve gray or halftones only by using red or green pens. Plotter pens come in a finite set of standard pen widths, such as .25, .35, .5, and .7 millimeters; you have to use certain AutoCAD commands such as **Pline** to produce other line weights, and even then you need to consider the interaction of the line weight specified in the drawing file with the assigned plotter pen thickness.

With list prices averaging between $1,500 and $5,000, pen plotters are among the most affordable plotter devices now sold. Newer models offer improvements in speed and reliability. Nonetheless, pen plotters continue to lose out to other types of plotters, notably inkjet plotters. Recently Hewlett-Packard, one of the major manufacturers of pen plotters, announced that it would discontinue manufacturing its pen plotter models.

Pencil Plotters

Designers who have had the opportunity to use them very much like pencil plotters. These plotters operate much like pen plotters except that they use colored leads instead of colored ink pens. Pencil plotters can hold many more leads (several hundred) than ordinary technical pencils can, and you can use a range of different colors. Pencil plots are not only good quality but erasable. Because the pencils don't skip, pencil plotters can draw faster than pen plotters can. Some models, however, do accept pen carousels, so they can double as pen plotters.

Despite their general quality and affordability, pencil plotters have a limited market in the United States, partly because only one manufacturer, which is Japanese, is well represented in the United States. Expect to find that pencil plotters are more prevalent abroad, especially in Japan, where most of the pencil plotter manufacturers are located.

Inkjet Plotters

In recent years, inkjet plotters have become extremely popular with design firms, and now they are the preferred choice for many. Instead of pen carousels, inkjet plotters rely on ink cartridges. Black-and-white plotters use a single black ink cartridge; color plotters mix colors from black, magenta, cyan, and yellow ink cartridges. Most color plotters let you plot from a palette of from at least 256 colors to over 1 million, and many can map colors to the Pantone Matching System and other color reference systems. Most inkjet plotters can produce plots in a range of output quality—typically draft, final, and enhanced—enabling you to focus on speed *or* quality, depending on your immediate plotting needs.

Inkjet plotters produce raster-driven plots. The plots come out of the machine in smooth, continuous rows of dots making the plot process at once faster and quieter. Especially when run in draft mode, inkjet plotters can plot as much as ten times faster than pen plotters. A draft-quality D-size (33" × 21") drawing can plot at 2 to 3 minutes; a good or final-quality plot takes between 5 and 8 minutes. The raster files plot entities not as continuous graphics but as a series of dots. The current resolution standards range from 300 to 700 dots per inch (dpi). Although a quality inkjet plotter can generate sharp-looking plots when run at high resolution, many designers find that the quality up close seems jagged and less sharp.

Although inkjet plotters can handle a wide range of media in terms of size and output, they generally require that you use special inkjet–suitable vellum or drafting film so that the ink will take, and the specialized media are often more expensive than standard plotting media. Because the inks used are water-soluble, the plots must be allowed to dry before they can be handled. Inkjet plotters do not produce archival-quality plots, but they can print on a wide range of other media, from photographic films to specialty papers.

Despite the limits in output quality, many design firms favor inkjet plotters for cost and maintenance reasons. Good-quality inkjet plotters with color capabilities now cost under $8,000; some black-and-white inkjet plotters list at around $2,400. There is also the advantage that inkjet plotters plot continuously rather than moving around the page; you usually have advance warning if the plotter's ink or paper supply is running low or if your plot contains an error in output and therefore should be canceled. Inkjet plotters are easy to configure and maintain, although for very high plotting volume, they require more attendance than thermal plotters.

Many design firms use inkjet plotters for routine draft plots and color presentations, and then rely on other types of plotters, often through a plotting service, to produce archival-quality final plots. As the market for inkjet plotters seems strong, however, expect to see ongoing improvements in output quality, inks, and media, as well as in speed.

Laser /LED Plotters

Laser/LED (electrophotographic) plotters are overgrown versions of office laser printers. They offer sharp quality and a range of line effects; color is available. Like inkjet plotters, these plotters are raster devices, so up close the line quality is not perfect. They also have problems printing on film, as the laser printing process is heat driven and causes the medium to "wave," although new products are addressing this shortcoming. Laser/LED plotters are fast, quiet, and require little ongoing supervision; the primary maintenance task, as with small laser printers, is replacing toner. Because they are so fast—they are capable of producing a D-size sheet in seconds—lasers/LED plotters are often used to make multiple copies of the same plots, and in some offices have replaced diazo machines.

For smaller design firms, the major obstacle to purchasing laser plotters is their cost. Initial equipment costs start at well over $10,000 and can exceed $35,000. Some cost over $100,000. Consequently, laser plotters tend to be found in larger architecture/engineering (A/E) firms or at plotting services, where the plotting volume justifies the investment. Some models can handle large format scanners, and these may be required by offices that need to manage a substantial number of large format documents.

Thermal Imaging Plotters

Thermal (or direct) imaging plotters use a heat-based process to produce high-quality, accurate images on paper, vellum, or film. Output colors are limited to black and red, and because printouts will react to heat, the output is not truly permanent; indeed, you can use special erasers to make alterations.

List prices for thermal plotters range from approximately $10,000 for a D-size low-resolution black-and-white plotter to over $25,000 for full-featured models. Standard media costs are moderate, and maintenance is extremely low. Thermal plotters are dependable, quiet, and especially appropriate for higher levels of plotting volume.

Electrostatic Plotters

Electrostatic plotters are fast, high-powered, and expensive output devices. They produce high-quality raster or vector-based plots in black-and-white or color. These devices are very fast, producing a standard-size drawing in as little as under a minute. Although they are quiet and dependable to operate, if you want to do your own maintenance, you'll need special training.

For most design firms, the major obstacles to purchasing an electrostatic plotter are cost and environmental concerns. Most models cost well over $25,000, and the media is not cheap. Moreover, the inks used are chemical-based and cause fumes and wastes that require special ventilation and disposal. Although some manufacturers are working hard to make electrostatic plotting a more environmentally sound process, others, such as CalComp, are dropping electrostatic products altogether.

Electrostatic plotters only make sense for firms that have extremely high plotting volume levels and can afford to deal with the accompanying environmental requirements. Although such plotters are beyond the pocketbooks of most design firms, many designers will commission electrostatic plots through plotting services when plots of exceptional quality and volume are warranted.

Hybrid Printer/Plotters

When purchasing a plotter is not immediately an option and your regular office printer generates output in a format that is too small, consider one among the growing number of *hybrid* printer/plotters that print paper, vellum, and film in paper sizes ranging from A (10 1/2" x 8") to C (21" × 16"). Many models use inkjet–based technology, AutoCAD-compatible printer drivers, and are reasonably fast and quiet. Depending on the output technology, color capabilities, output resolutions, and paper sizes they accommodate, these devices typically have list prices ranging from $1,350 to $9,000. Because they can sit on a desk or table, they do not have the space requirements that full-size plotters do.

Many design firms have found that printer/plotters suffice for all but final full-size plots and have learned to tailor their check sets and other drawing formats to work within the constraints of these machines. Even if you have a full-fledged plotter readily available, you may find a printer/plotter useful for routine plots as well as desktop publishing and marketing projects.

Internal Production: Equipment Selection Guidelines

AutoCAD Release 13 includes plotting drivers and plotter configuration files for standard plotters made by CalComp, Canon, Hewlett-Packard, and Houston Instruments.* The plotter drivers are listed in the *AutoCAD Release 13 Installation Guide*. If you purchase a model other than the ones listed, you may need to acquire plotter drivers and configuration information directly from the plotter's manufacturer; many plotters, however, will emulate Hewlett-Packard plotter configurations, since most plotter manufacturers conform to the standards that Hewlett-Packard has set with its Hewlett-Packard Graphics Language (HPGL) and HPGL/2 plotting languages.

Most of the major manufacturers of plotters strive to have competitive prices as well as speed, good-quality output, and special features and accessories. Therefore, when selecting a plotter, the features and price of the machine should not be your sole selection criteria. You also need to consider product warranties, support, availability of supplies, and the manufacturer's financial stability. You don't want to be stuck with a plotter that you can't maintain because the manufacturer went out of business.

When evaluating plotters and printers in terms of speed output, don't make decisions solely on statistics provided by the manufacturer. The speed of output can vary enormously according to a plot file's size and contents and the media and quality of the plot. A plot file containing elements such as large blocks of solid colors or half tones can slow down plotting speed (and use up much more ink and thereby cost more than the "average" plot). When seriously considering the purchase of a plotter, ask to run test plots from one of your own CAD files, preferably one which is illustrative of the file size and graphic complexity of your typical projects. Track the plotting time for the various levels of print quality, such as "draft" or "enhanced," that you will frequently require.

The technological development in plotting devices is constant and rapid. The machine you buy is already technologically—although not functionally—obsolete the minute it is installed in your office. Therefore, it can be worthwhile to stretch yourself financially to purchase additional memory capacity as well as certain high-powered "new" and "extra" features that are likely to become standard in future models. This way, you can stave off product obsolescence. Avoid paying for features that the

---

*Note that many other CAD software packages do not include plotter drivers in their base price; instead they sell plotter drivers at an additional cost.

nature of your practice doesn't warrant but stretch for capabilities that will enhance your ordinary routines and ease of use. Among the features you should consider are the following:

Color plotting capabilities. If you want to use CAD for color presentations and diagrams (see Chapter 7), consider paying an extra $1,000 or so to acquire a color inkjet plotter. Nowadays, the premium you pay for color is relatively small; moreover, the costs of color ink cartridges may be offset not only by reduced labor costs but also by a reduced need for color markers, pencils, and overlay sheets.

The ability to accommodate large-size drawing sheets. Many plotters come in different models for handling D- and E-size sheets; the cost differential between such models ranges from $500 to $1,000. Even if your typical drawing sheet size is D-size or smaller, purchasing a plotter that can handle larger paper will increase your plotting options.

The ability to "mirror" or create reverse-read plots. This feature is especially useful if you routinely plot title sheets and base plans on the *back* of film or vellum.

The ability to program a range of pen settings, line weights, and plot routines within the plotter itself. Programmability can expand your plotting options.

The ability to support a high degree of accuracy. Although the average person may not register it, plotter output does vary in its dimensional accuracy or the degree to which it is truly to scale. Extreme accuracy is essential in CAD applications such as computer-aided manufacturing, but it is useful in scaled architectural drawings as well. For most plotters, a maximum error tolerance of ±0.015" to ±0.20" is normal; some models, such as thermal plotters, purport to offer a ±.0005" tolerance factor.

The ability to plot from rolls of paper rather than just from individual sheets and to cut each plot off the roll automatically when the plotting is finished (some models handle two or more rolls of paper). These features will save you time feeding individual sheets of paper into the plotter for each plot, and they let you batch a group of CAD plot files to the plotter and go out to lunch with a clear conscience.

A carriage for collecting and holding completed plots as they are cut off the paper roll. This device collects sheets in an orderly pile and prevents plots from falling on the floor and getting stepped on, wrinkled, or smudged.

The ability to support multiple plotting file formats. Any plotter for which AutoCAD provides drivers can read AutoCAD plot file formats, such as the HPGL plotting languages. If you also use other software, however, your plotter may need to read plot file formats that are specific to that given software program, or plot file formats for Macintosh or Windows platforms. If you use software with PostScript capabilities, you will want a special driver for that. Some plotters provide a range of drivers automatically; others sell them as options.

The ability to support software for plot spooling. This enables you to run plots in the "background" while you continue working in AutoCAD or write other plot files.

The ability to support software utilities for plot queuing and management. Both third-party software ("plot managers") for AutoCAD as well as Windows and Macintosh network and print managers and other software utilities enable you to monitor the status of plots and to determine who sent them and how large they are. Many enable you to reorder the sequence of plots in order to accommodate deadlines; the program may also alert you when a plot file that requires that you load special paper, pens, or ink is ready to plot. Such software is particularly essential on networks where multiple users share output devices. The more intensively the plotter is used and the more complex the range of output formats, the more plot management capabilities you need.

Expandable memory. Like computers, plotters require RAM in order to work faster or more efficiently. Even if you don't need more than the base RAM at the time of the initial purchase, make sure you have the ability to add more, especially if you anticipate that your plotting level is going to increase. For many plotters, 4 MB of RAM is standard; the ability to expand to 20 or 32 MB is useful.

Flexibility in output size. Some software programs and plotters only read drivers for specific sizes of media. If your plotter, for example, only reads 30-inch- or 42-inch-wide rolls, and your standard size drawing is 36", you will waste considerable paper, not to mention time trimming 42-inch-wide sheets down to the proper size.

Appropriate options for extended warranties and service contracts. In addition to standard product warranties, most plotter manufacturers offer extended warranties and service contracts. The costs of these packages must be evaluated against the likelihood that the machines will break down and disrupt your work. In reviewing these services, look at several items in particular. First, see if the contracts include a provision for you to have the use of a substitute machine if your plotter is going to be out of commission for more than a day. Second, determine whether the company's service contracts are based on a flat annual fee or on plotting volume; depending on your expected plotting volume, one option will be more cost-effective than the other.

Being a savvy consumer of plotting devices can be a challenge since the products available change constantly, and the technical features often are presented in terms that are meaningless to most designers. Remember that the bottom line for you is ease of use and quality of output, and, of course, compatibility with AutoCAD and other software.

There are various ways to research and locate the plotter device that is right for you.

CAD publications. AutoCAD-oriented publications such as *CADalyst* and *CADENCE*, as well as general computer magazines, contain reviews on, and advertisements for, plotters. They often include Reader Service Cards, which enable you to easily obtain product literature directly from the manufacturer. Magazines may publish the results of formal test comparisons, listing key features of plotters, such as speed, plot quality, media options, and reliability.

Product fairs. Every year, major cities throughout the United States and abroad host product fairs or conventions during which manufacturers have a chance to present their latest products. These events, which may be produced by members of the design/build community or by computer companies, are an excellent opportunity to see plotters in action, to view their output, and to collect technical specifications and dealer names. Look for notices of such events in local newspapers or in the local chapter newsletters of the American Institute of Architects, the International Facilities Management Association, and other professional organizations.

Local dealers. Every major producer of plotters has dealers in the major cities. These dealers are always happy to show you what they have in their showroom; often they will offer to loan you a machine on a trial basis. Many dealers are also aligned with plotting bureaus and therefore can offer valuable technical expertise.

Professional peers. Your professional colleagues can be a useful source of advice, and they may offer the most honest and relevant guidance. Apart from looking through your Rolodex, you could look up a local professional group, which may have a computer users committee that provides useful contacts. As Chapter 10 discusses, much of this information can now be gleaned through the Internet.

External Production: Vendor Selection Guidelines

As CAD has permeated the design community, it has spawned plotting bureaus—firms that will take your CAD drawing or plot files and generate plots for you, in the scale and format and on the media of your choice.

Apart from the aforementioned reasons for using a plotting service in lieu of purchasing your own plotter, there are valid reasons for establishing a working relationship with a plotting service *even* when you have in-house capabilities.

Plotting services are faster. Most bureaus use the fastest, most high-powered machines available, and they keep their machines primed for plotting. They can produce ten plots in the time it takes you to set up one.

They provide a backup option. If you are experiencing a bottleneck at your plotter because of multiple deadlines, plotting services can pick up the slack. They also prove essential when your machine is malfunctioning or being serviced.

They absorb the costs of failure. When you plot in-house, you (not the client) usually absorb the costs of plots that failed because the ink smudged or ran out, or the sheet slipped or got chewed up by the rollers. When a plotting bureau has a plotting failure (assuming that the plot or drawing files you provided have no flaws), they must eat the costs. They can only charge you for successfully completed plots.

They help track reimbursables. With most plotting bureaus, you include a client or project number with each plotting order you issue. This number usually appears on the invoice that the plotting service includes with the completed plots. This invoice can easily be presented to the client as a reimbursable. With in-house plotting, you have to develop a fair and reliable methodology for tracking plots (computer software exists to facilitate this) if you want to charge plots as reimbursables.

They promote plotting discipline. As anyone familiar with word processing will confirm, if you have an easy way to generate output, you have the temptation to generate an endless but often unnecessary series of drafts of your work. This process not only wastes time and paper but can produce a potentially confusing pile of paper. With plotting bureaus, the turnaround time involved in getting plot files to and from a service, as well as explicit plotting charges, deters casual plotting. (Your plotting service may also offer discounts for extra large plotting orders, which further compels judicious use of plotting bureaus.)

They offer the benefits of ongoing research and development. Most plotting firms, especially those that purport to be on the cutting edge of technology, have a professional incentive to invest in, and experiment with, the latest output devices and media. In the ordinary course of work, they can test and learn about devices that a typical design firm does not have access to. You can nonetheless benefit from this R&D in terms of their services and their applying its results to your work, and you can use what you learn in any future purchases of equipment.

In developing a relationship with a plotting service, make sure to clarify the following parameters:

The cost per plot. Most plotting services distribute price schedules that show the unit (usually by square foot) costs for different media—bond, vellum, film, and so on—and for black-and-white, color, and other output formats. Find out if you can get discounts for larger volumes in plotting. Will there be discounts if you order a minimum number of plots per month? Will there be setup fees for each order—which would encourage plotting in bulk—or for cutting and trimming sheets? Some services run all plots using maximum-width rolls of media, and they charge you for each linear foot used rather than for the actual square footage of *your* plots.

The quality of the output. Ask for sample plots of all media—vellum, bond, and film—that the service uses, since the quality of media can vary considerably from service to service, and the media quality can affect the quality of plot output. Special attention should be given to film plots, as some devices have problems with this media, and not all films are truly of archival quality. If the service offers an array of output devices—thermal, laser, inkjet—make sure to get samples of output from each type of machine you wish to use.

The turnaround time. Most plotting services provide standard schedules for turnaround time and delivery. Will these work for you if you're in a pinch? Is delivery feasible during rush-hour traffic? Will you be charged extra for rush orders? Can you make an "appointment" to have your work done within a particular time frame?

The plotter configurations. To write AutoCAD plot files that a plotting service can use, you must use the AutoCAD **Configure** menu to specify and set up the plotting service's equipment. Ask the plotting service to provide not only the manufacturer's name and model number but also a detailed rundown of its configuration. This configuration should be provided in written form or as a *.pcp* text file, which you can then access through the

AutoCAD **Plot Device and Default** dialog box (see Figure 4.19). If you use multiple services, the same device can have many configurations, and you don't want to spend time adjusting a single configuration for multiple users. Even if the services use the same machines, specify separate plotting configurations in AutoCAD for each service. AutoCAD allows you to name these devices in a user-friendly and recognizable fashion, for example "ACME Plotting Service," as distinguished from your in-house plotters.

Subtle differences in plotter configurations and plot file specifications, such as the plot origin, can affect your plot output. To test the settings, have the plotting service run a complimentary test plot of one of your plot files, preferably one based on your prototype drawing files.

CAD file requirements. Most plotting services accept both AutoCAD drawing files and plot files. They're more likely to charge you, however, for the time they spend writing plot files from your drawing files, and usually they won't accept responsibility for any mistakes they make in writing the files. Therefore, most design firms distribute plot files. When doing so, find out if plotting services have preferences for how plot files are written. For example, some services prefer that the AutoCAD plotting area be specified by the drawing extents rather than by a windowed area of the drawing screen. They may also charge you for plot files that are larger than 3 MB or so.

One reason many designers prefer to send AutoCAD plot files to plotting services rather than drawing files is that plot files contain all the information needed to plot the drawing. Drawing files write plot files based on the overall AutoCAD configuration. If your drawings depend on special customized features (such as third-party typefaces or overlays) or external references, then your plot files must be written on a computer that has those features. Plotting services normally install only "plain vanilla" AutoCAD on their computers; consequently, their system may not recognize third-party add-ons or customized features of your AutoCAD configuration. Unless you provide supplementary configuration files and utilities, your plotting service won't be able to write a plot file that meets your expectations. If you do provide your plotting service with third-party software, you must be careful not to violate any software copyrights in the process, and you may have to pay extra to have the additional software installed on the plotting service's system.

CAD file transfer capabilities. Make sure that your plotting service can read your plot files. Most services accept 5 1/4-inch and 3 1/2-inch diskettes as well as SyQuest or Bernoulli cartridges. Very large drawing or plot files won't fit onto standard diskettes, so you may have to compress them; make sure your plotting service will accept compressed files and that it has the means to decompress them.

Compatible versions of software. Most plotting services are actually more aggressive than many design firms about upgrading to the newest releases of AutoCAD and other software. Even so, you do need to make sure that they have a version of AutoCAD and AutoCAD Device Interface (ADI) drivers that can support your plot files.

Modem options. A growing number of plotting services now accept—and often encourage—clients to send plot files via modem. Transmitting files over a telephone line saves enormously in turnaround time as well as in time spent copying plot files to disks. Unlike disks, modems don't place limits on the size of files that can be transferred. This is very useful, since plot files are larger than the original drawing files. Your service may provide a software equivalent of its hard-copy order forms, which helps enormously in clarifying orders, record keeping, and billing. In order to accommodate all their customers, most plotting services use modems with high transmission rates; make sure you get a modem that uses the proper protocol. While using a modem definitely reduces turnaround time and investment in disks and other storage media, you may have to resend plot files if they don't transmit properly over the phone lines on the first try.

Technical expertise. Plotting services vary enormously in terms of their technical expertise. Try to use services with staff who are experienced not only in running plot machines but also in working with AutoCAD. And when you call their references, try to gauge whether the staff are proactive—that is, whether they will call you immediately if they suspect a problem with your plot files or whether you will have to wait until you receive the plots to see whether they were successful.

Options for Plot Media

Most plotters can handle the major types of paper that designers use to make reproducibles and other kinds of plots. In addition, specialty opaque papers are now starting to become available. Following are some things to note about standard plotter paper types.

Vellum. Most high-quality vellums have one side that is preferable for plotting because of its "tooth," which absorbs the ink better and limits bleeding and fuzziness. The best plotting vellums are 20 lb.

Film. Film is often packed with a powder that prevents the sheets from sticking together. Wipe off both sides of each sheet with a wet towel before plotting, or the powder will corrupt the ink. With laser plotters, ink adheres to film by a heat transformation process. The heat generated by this process may cause the film to "wave." If the plotter temperature is reduced, the film may be smoother, but the ink may flake off. Fortunately, new developments in film are starting to address these results.

Bond Paper. Many firms that have pen or pencil plotters save on plot media by using the reverse side of previously-used sheets of bond paper for test plots. You can also run plots on print paper that has been run through a diazo machine. Most plotters can also plot on letter- or legal-size bond paper or odd-sized paper scraps, so you can save your paper stock—and a bit of the earth—by "recycling" paper for test plots.

Bond paper for plotters comes in a growing number of varieties, depending on the type of plotter you use. Bond paper comes in a range of weights, with 20 lb. and 24 lb. being the most typical; it can have a flat, satin, or glossy surface. Special presentation bond is required for inkjet-based color and best-quality plots.

Other Media. Although most design firms run CAD plots on the standard media—bond, vellum, or film—a growing number of output media are emerging. Depending on the kind of plotting device you have, you can plot on photographic film, clear overheads, specialty papers such as 90 lb. Arches watercolor paper, and coated canvas. An inkjet plotter will handle almost any media that can absorb water-based ink and that is of the appropriate weight, thickness, and flexibility. For more routine office production, smaller laser and inkjet printers can print text and images on overhead transparencies, card-stock report covers, and specialty papers with preprinted designs. Major manufacturers of labels, such as Avery International, now sell printer-ready labels for almost any labeling function imaginable, as well as computer-based templates for formatting the labels.

Configuring Plotters for AutoCAD

As Chapter 4 discusses, Step 7 of the CAD customization process involves configuring your output devices. This means configuring *all* the output devices you'd normally use. In addition to each device, you can write configuration setup files for *all* the permutations of plotters, printers, pen menus, and paper sizes that you might require.

AutoCAD Release 13 provides ADI drivers for the major models of plotters. When you specify an output device in your plotter configuration, you select an ADI device, either accept or modify its defaults, and save this configuration under the name of your choice. This name need not be the plotter model; indeed, a single machine may warrant several configurations, so it is often better to have a generic and meaningful configuration name, such as "In-House Color Plotter" or "In-House Laser Printer #1."

Information about configuring the major models of plotters is provided in Appendix C of the *AutoCAD Release 13 Installation Guide*. For most routine plotting setups, you won't need to bother with many of the configuration options described in this appendix, but you should be aware of the following points:

- With AutoCAD Release 13 for Windows, you use Release 13 ADI drivers—whether provided by AutoCAD or by third-party vendors—that are designed to work with Release 13's 32-bit system. You may be able to use some Release 12 ADI drivers with Release 13, but when you upgrade to Release 13, you may need to revise any plotter configurations built on Release 12. If you upgrade to Microsoft Windows 95, you may encounter similar issues for any Windows system printer drivers.

- You should specify a "time-out" value for plotting. This value specifies how long AutoCAD will hold additional information from a plotting file while the plotter plots the information it already has collected. You can specify from 0 to 500 minutes; 30 seconds is the default, and many software manufacturers recommend more. The slower your plotter or the more complex the plots, the higher the time-out value should be. The optimal value often is determined after several weeks or months of experience and may change as your plotting needs evolve.

- Every plotter and printer that offers color output mixes color inks somewhat differently. You will want to test your output devices by printing the AutoCAD *Chroma.dwg* file that comes with your AutoCAD software. With ASCII or binary files, you also specify whether you want to plot in 16 or 256 colors.

- As Figure 4.20 shows, the colors and line thicknesses you see on-screen are *not* necessarily what you get on plotted output; this can cause surprising results in output. Pen settings therefore should be documented so as to minimize confusion and so that you can replicate them in the future. Tracking plotter pen settings on a chart such as the one shown in Figure 8.4 helps drafters understand the effect on their plots of specifying particular colors and line types for layers and entities.

Figure 8.4
A chart of pen settings is essential for relating AutoCAD screen colors to output.

ABC Design Inc.

Plotter Pen Settings Menu #3
Plot Output: Standard Black and White 1/8" Scale Plots
Output Device: Office Inkjet Plotter #1

Color Number	Pen Color	Pen Width	Comments
1	7	0.10	
2	7	0.25	
3	7	0.35	
4	7	0.50	
5	7	0.25	
6	7	0.25	
7	7	0.70	
8	3	0.10	Green pen for half tones
9	7	0.10	
10	7	0.25	
11	7	0.25	
12	7	0.35	
13	3	0.35	Green pen for half tones
14	3	0.25	Green pen for half tones
15	7	0.35	
16-256	7	0.10	

Particularly if you run AutoCAD on a network, you should bear in mind that the way you configure your output devices is not determined solely by AutoCAD. You also need to consider various options provided by Windows or Macintosh system software, by individual software applications, by your network software, and by the printers or plotters themselves. Determining how all your software and hardware work best together takes some technical knowledge and experimentation, but once you have set up the proper working relationship between the two, the result is a very efficient, powerful, automated output system.

Some Plotting Tips

Adjust your files to the limits of plotting devices. Although your plotter may nominally plot a 42" × 30" sheet, for example, it usually has built-in margins in order to accommodate plotter rollers. Design your sheets, particularly the titles and borders, with these margins in mind. If you want the drawing to "bleed" to the edges, you usually must run your plot on a larger sheet and trim it down.

Maintain your plotter regularly. Although some of the more "intelligent" plotters may have an interface that warns you in advance if the ink cartridge is drying out or if the sheet is improperly loaded, plot failure usually comes as a nasty surprise (and always in compliance with Murphy's Law). The best way to protect against plot failures is to follow the manufacturer's instructions on maintaining the machine and to have it professionally serviced on a regular basis.

Minimize the use of full-size plots. With AutoCAD, you can plot easily at any scale (including some not typically used by architects and interior designers). Knowing that, you should feel comfortable curtailing the temptation to constantly plot full-size sheets for routine checking of minor items. Increasingly, designers work with check sets at 50 percent the size of full-size sheets and laser print "snapshots." As mentioned above, you may use small plotters or printers for all your test and interim plots, and save the full-size plots for the important coordination sets and the formal permit, bid, and construction sets. By using small plotters, you will save time and paper, and you will also keep your flat files less cluttered and reduce the risk of distributing the wrong versions of your work.

Develop and implement a policy for naming plot files. The names of plot files can become critical, especially when they are shared on a network. As discussed in Chapter 10, plot file names should be planned with two goals in mind. The first is to prevent one plot file from inadvertently overriding another; this can easily occur if two files with the identical plot file name are directed to the same plot "holding" file directory. The second goal is to enable users to understand what a plot file contains. The plot file name *Plot1.plt* is meaningless to all except the person who wrote the file. The name *9677-A1.plt* at least suggests that the plot file pertains to project number 9677 and probably plots Construction Drawing A-1.

Plotting Reimbursables

Although most design contracts today classify routine CAD operations under ordinary overhead expenses, they usually consider CAD plots, whether 42" × 30" full-size drawings or 8 1/2" x 11" laser prints, as reimbursables. Having an efficient way to track plots is critical for billing purposes.

If you use a plotting service, you will normally receive with each completed plot a transmittal with the client's account number on it. These orders are then summarized in monthly invoices. For most purposes, these transmittals and invoices can be used to bill reimbursables to the client. If your plotting service can't handle multiple account numbers in one order, then place orders *separately* for each account.

If you do plots in-house, you need to develop a method for tracking plots accurately and easily. Many firms clone the log form that they use for in-house diazo prints to create a plot log as shown in Figure 8.5. This log is normally kept in a binder next to the plotter, and the totals are collected by the accounting department at the end of each month. You can also purchase software that prompts computer users to enter the project account number, plot purpose, and other relevant information before they can run a plot. Such software can track plot quantities, quality level, and failures as well. The software then generates a monthly log that can then be imported to accounting software for generating invoices.

Regardless of your tracking approach, your plot log should be specific enough to serve multiple purposes. The paramount one is that you bill accurately for reimbursables; the information that the log provides should reflect plot media, size, and any other factor that may affect the reimbursable amounts. In addition, the log can be used to track inventories of supplies as well as the plotter utilization rate; with a concrete picture of plotter usage, you can determine whether the purchase of a second plotter is justified, for example.

Every firm that charges plots as reimbursables should have a policy for ensuring that plots are recorded fairly. Clients usually should not be billed for plots that failed because the plotter ran out of media or the drawing sheet slipped or got "eaten" by the plotter. It is debatable whether clients should be charged for drafts or test plots.

Plotting Schedules

Regardless of your firm's plotting capabilities, budget time not only for running plots, but also for plotter downtime, maintenance, and running the occasional test plot. The more *un*reliable an in-house plotter or an external plotting service is, the more you need to build in time as a safety net to ensure that your plotting gets done on schedule.

As Chapter 5 discusses, ultimately project managers are responsible for building plotting time into project schedules. They should rely on

ABC Design Inc.

PLOT LOG FORM

Client Name: Acme Corporation

Month: March, 1996

Client Project Number: 96077.23

Plot Date	Plot Size	Quantity	Plot Media	Output Quality	Black/White	Color
3/2/96	42" x 30"	10	Vellum	Draft	X	
3/2/96	48" x 36"	2	Special Bond	Enhanced		X
3/10/96	42" x 30"	2	Mylar	Final	X	
3/12/96	42" x 30"	2	Mylar	Final	X	
3/12/96	42" x 30"	10	Vellum	Draft	X	
3/15/96	42" x 30"	16	Mylar	Final	X	
3/22/96	36" x 24"	4	Vellum	Draft	X	

Figure 8.5 A sample plot log. This form can be created with any spreadsheet or word-processing software package.

CAD managers and drafters, however, for guidelines on how to calculate plotting time. Figure 8.6 shows one approach to collecting information for plotting time. Such a chart must be adjusted to reflect the capabilities of the actual machines or plotting services you use.

ABC Design Inc.

STANDARD PLOT SCHEDULE INFORMATION

(Estimated Times shown for plotting a standard 42" x 30" sheet)

Media Output	Turn-Around Time		
	In-House Plotters		Plotting Service
	Pen Plotter	InkJet Plotter	Laser Plotter
Draft	N/A - Only Final Quality	4 Min. per plot	N/A - Only Final Quality Output
Final Quality - Vellum B/W	45 Min. per plot	10 Min. per plot	Typical turn-around time is one working day regardless of plot quantity
Final Quality - Mylar B/W	One Hour	15 Min. per plot	Typical turn-around time is one working day regardless of plot quantity
Enhanced Quality - Bond Color	N/A - Same as regular plots but with multicolored pens	16 Min. per plot	N/A
Comments	Must budget staff time to monitor plot and feed sheets; limited color options	Assumes media is automatically fed and cut; can plot over 256 colors	Call one day in advance to schedule large plots or special orders; special charges apply for trimming sheets and files larger than 3 MB

Figure 8.6 An example of a chart used to collect information for estimating plotting turnaround times and for planning production schedules. Such information should be distributed to all staff who are dependent on CAD plots to complete their own work.

As a general guideline, you should schedule plots to be completed at least one day prior to the deadline for issuing a drawing. You need to budget time both for errors and last-minute changes. For large projects, scheduling one or two weeks for final plots is not uncommon. In addition, time should be scheduled for reviewing complete plots with a fresh and objective eye. Having plots reviewed by an experienced designer/drafter who is not on the project team but who has a sharp eye for inconsistencies and errors can be the best way to ensure quality control of plotted drawings.

Other Output Devices

While plotters and printers are the primary output devices with which CAD users are concerned, a growing number of other machines are now available and are increasingly important for use with AutoCAD. While the selection criteria and technicalities for each device differ from plotters, the general approach to knowing when to acquire them versus paying for an outside service remains the same. Listed below are some of the output devices that you might consider using with AutoCAD and other software.

Color laser copiers. You can hook up your computer or computer network to color laser copiers and use your computer to specify multiple color copies in letter, legal, and ledger sizes. Although laser copiers are not inexpensive, you may be able to justify the investment if you rely on color laser media not only for design output but also for graphics, marketing publications, and other forms of desktop publishing. When high volume is a concern, having in-house color laser capabilities can also be more cost-effective than rendering multiple color presentations by hand.

Specialty color output devices. Although not yet as prevalent or affordable as their inkjet or laser printer and plotter cousins, color printers are now available for printing color images on materials ranging from fabric to cork, plywood, and rubber. You can print your firm's logo and company information on mugs, mouse pads, and T-shirts. Specialty printers are available for every sort of output, ranging from photo identification cards to banners. Since the technology that drives these machines is fairly new and expensive—depending on what they do, they range from $10,000 to over $500,000—and require special output media, such machines generally are the province of print bureaus and other production houses. But you can reduce costs and increase control over your output by providing the images you want printed in a computer file format. Most devices typically read PostScript or *.tiff* files, which AutoCAD and most other graphic-oriented software can produce.

Cutting machines. AutoCAD files can be read by a range of machines that cut metal, plastics, cork, foam boards, and other materials. These machines are used for a range of Computer-Aided Modeling and Manufacturing (CAM) applications, from making architectural models to cutting dimensional letters for signage and clothing patterns. CAM can be particularly useful for complicated topological models or intricately textured building facades. Most design firms do not have the level of work of this nature to justify purchasing CAM cutting machines. If you work consistently with a CAD-based model maker, however, you may want to specify a custom CAD configuration to write plot files for such machines.

Video projectors and large screen monitors. For AutoCAD slide shows as well as for 3D renderings and animation, a standard plotter or printer often cannot properly replicate your work. The standard 14-inch or 17-inch computer screen is too small for doing presentations to large audiences. Having a large screen monitor (over 21 inches on the diagonal) or a video projector is essential for demonstrations that do not involve paper-based plans. These devices are also generally useful for viewing instructional videos, presentations provided by product manufacturers, and other presentations. Sound capability is of increasing importance with these machines.

Large-format scanners. Large-format scanning machines aren't actually output devices but rather are the input equivalent of large plotters. Often similar in size and form to plotters, they can scan drawings up to sizes D (33" × 21") and E (43" × 33") as raster-based computer files that can then be converted to vector-based files to be used in AutoCAD and other CAD programs (see Chapter 10). In the not too distant future, expect to see a growing number of these scanners merged with plotters into one input-output device. Large-format scanners range in list price from just under $10,000 to over $20,000 depending on the size of sheets they can handle and the dpi resolutions they offer. Most design firms do not have a large enough scanning volume to justify purchasing their own large-format scanner, but firms can now farm out large-format scanning to a reprographics/plotting service. As with inkjet plotters, scanners have been dropping in price while improving in quality and speed. Many management departments of corporations and other institutions find them essential in managing their architectural drawings.

AutoCAD's basic documentation does not provide information on output devices other than plotters and printers, but you can obtain names of manufacturers from *The AutoCAD Resource Guide* and from local vendors and CAD-oriented magazines.

9 Administration of an AutoCAD System

> *"Well, in our country,"* said Alice, still panting a little, *"you generally get to somewhere else—if you ran very fast for a long time as we've been doing."*
>
> *"A slow sort of country!"* said the Queen. *"Now, here, you see, it takes all the running you can do, to keep in the same place. If you want to get somewhere else, you must run at least twice as fast as that."*
>
> —Lewis Carroll, *Through the Looking Glass*

You know that AutoCAD and other computer software have begun to have a full impact on your office when you find that traditional ways of doing things have to change. Computers affect all aspects of a firm's operations, from the ways fees and budgets are developed to staffing to filing systems.

Although the methods for administering an electronic office differ from traditional ones, in general the goals of efficiency, profitability, and greater enjoyment of work remain. Used effectively, computer tools can help you achieve these goals more easily, with less angst and less effort. Ideally, computers permit you to automate the boring, routine work and let you focus on the fun and creative stuff.

A Traditional Breakdown of Project Phases

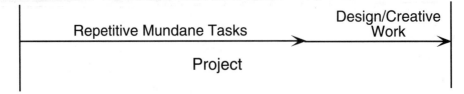

B Ideal Revised Breakdown of Project Phases, using CAD

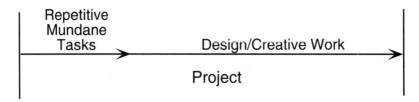

C Unfavorable Revised Breakdown of Project Phases, using CAD

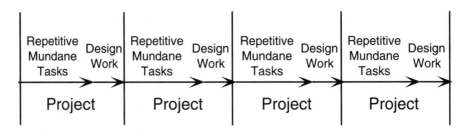

Figure 9.1 While the ideal use of computers should permit us to evolve from the professional paradigm as shown in A to that in B, many designers feel that their practice resembles what is shown in C.

Some argue that computers do not save time; they merely allow us to cram the same amount of work into shorter time frames, as suggested in C of Figure 9.1. Whatever proves true for you, most of your competition uses computers, and if you want to stay competitive, you must either run with computers or invent an effective "mousetrap" that enables you to profit without them.

Financial Administration

As your office acquires more and more computers, your expenditures for all things computer-related—from machines to employee training—will increase correspondingly. This is perfectly acceptable, provided that your computer-related capacities parallel two other developments:

- A reduction in non-CAD design and production expenses. Effective CAD use *shifts* resources from traditional design and drafting methods to computer-based methods. Therefore, some of your increased CAD expenses should be offset by reductions in non-CAD expenses. For example, you would expect to spend more on computer diskettes and less on pencil leads. If you do not experience this reduction, then CAD is not making inroads into your office organization, and you may be essentially duplicating efforts and expenses.

- An ongoing increase in CAD-related projects, revenues, and profits. Not only should CAD enable you to handle your existing workload more efficiently; it should also increase your possibilities for attracting new clients and new types of projects. You should be able to fit more projects within a given time and staff level or else be capable of taking on larger or more complex projects. Over time, you should expect to see the project margins on typical CAD projects widen.

In order to monitor and meet these developments, you can rely on a range of tools to track your CAD-related income and expenses and to ensure that your computer expenditures do not become a bottomless money pit but rather that they are an essential element in your firm's strategy. Two of the most important tools for assessing computer usage are budgets and calculations on returns.

Budgeting for AutoCAD

Anyone who has ever purchased a computer knows about the sometimes hefty hardware costs and the often equally hefty software costs. To acquire of a CAD-capable workstation costs at least $4,000, even if you're not using AutoCAD. Then you must factor in many other related costs, some of which are not explicit. Calculations become more complex when you purchase multiple workstations, as at some point you can realize certain economies of scale. Moreover, as CAD becomes more integral to your operations, you will find it harder to determine which sources of income and expense are truly CAD-specific.

As Chapter 2 discusses, your CAD-related decisions should be considered in the context of your overall firm strategy and financial projections. You need to view CAD expenditures in the context of all your other projected outflows, revenues, and debt/equity situation. A key issue in your overall budgeting is your *time frame* for earning a return on your investment (see the section on "Calculating Returns on AutoCAD" later in this chapter). While every firm wants to earn a high return on any investment in CAD instantly, the reality is that you may not see a return for several years. Pressures to earn an immediate return may tempt you to consider a CAD package that *appears* to demand the lowest upfront outlay. While you certainly want to avoid extraneous costs, you should not skimp on the quality of software and hardware or in your investment in training and customization. In the long run, a CAD investment that is shaped primarily by what initially seems cheapest may ultimately prove unsatisfactory and do more damage than good—in ways not only financial. Consider making an investment that takes longer to pay off but is effective and productive.

As important as your time frame for paying off your investment is the way you categorize costs, typically fixed, variable, and semivariable costs. As Figure 9.2 shows, the higher your fixed costs, the longer it will take for you to reach a break-even point on your investment in CAD; therefore, you usually have an incentive to make as many costs as possible be variable and applicable to billable projects. As firms increasingly incorporate CAD into the routine costs of doing business, they are more likely to apply CAD-related expenses to general overhead rather than to projects, a process that can reduce what can be classified as billable costs. Your

Figure 9.2
You need to categorize and track revenues and expenses carefully in order to determine when your CAD investment has become profitable.

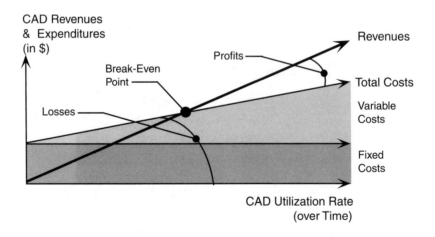

accounting procedures usually determine how you classify most costs, yet in some situations, depending on certain strategic decisions, you can have a choice. Some decisions result in a capitalized asset; others in an operating expense. If, for example, you pay for computers with a single cash payment, you reduce your cash reserves and now own a depreciable asset, but you do not have the monthly interest expenses that you'd assume if you leased the machines or used extended credit to acquire them. Your potential returns, therefore, depend on decisions other than merely what CAD system you invest in.

CAD-related Income

The first step in budgeting for CAD requires you to project target revenues for the coming three to five years. Investments in either individual workstations or entire CAD systems, including networks, warrant making projections for at least this much time. You cannot realistically expect an immediate return on a new system unless you are adding an incremental station to an existing system and have a project in mind to earn back the investment.

As described later in this chapter, it is increasingly less likely that you will have opportunities to earn income by billing out special fees for *routine* CAD-based production work. Most clients would balk at paying extra for what has become a normal mode of production. So for typical CAD activities, you often simply multiply the number of hours for a project by the billable rates of the CAD users who will be assigned to the project. Even if you do not charge explicit fees for CAD use on such projects, you may choose to track CAD usage internally by requesting that drafters indicate on their time sheets a separate line item for hours spent using CAD; such documentation helps you determine your CAD *usage factor* (discussed further on) and offers especially valuable management information for design forms that are just starting to shift to CAD.

Although ordinary CAD production may not justify charging special fees, you can generate specifically CAD-derived revenue if you offer certain special services that rely on computers and CAD. Sources of special CAD-based services include making 3D models and animated walk-through studies, doing area calculations, and providing external facilities management services. Ideally, you should be able to relate fees for special services directly to special expenses, such as the acquisition of software or staff training for a particular project.

In addition to explicit forms of CAD-generated revenue, you should look for *implicit* forms of revenue and related benefits that accrue from using CAD. Without explicitly generating fees, CAD can help you get placed on short lists and win projects for which you would be otherwise ineligible if you had not acquired CAD—or agreed to do so.

CAD-related Expenses

Estimating CAD-related expenses often seems easier than estimating CAD-derived income. You can easily pull out invoices for machines, software, service contracts, and supplies.

Hardware

Hardware is a major part of your CAD-related expenses. This remains so even though hardware prices have fallen over the past decade. These price decreases are even more impressive when measured in terms of cost per byte; the relative costs of hardware devices have dropped as power and capabilities have improved.

The most expensive aspect of hardware is its rapid rate of obsolescence. Major improvements in hardware appear every three years, and manufacturers often discontinue making a particular model after only one year. Although you may get good use out of a computer for five years, its life expectancy as a CAD station rarely goes beyond three. So you have to factor into your budget the impact of high rates of depreciation, low salvage values, and ongoing investments in upgrades.

In budgeting for hardware, don't include the purchase price of only a CPU, monitor, keyboard, and mouse. You likely will be spending, albeit in smaller increments, hundreds of dollars for cables, drivers, and other devices for making your system work better as well as for items that you didn't know you needed until after the initial purchase. If you purchase hardware from many different manufacturers, you may incur additional costs trying to ensure compatibility across your various machines.* Careful research and planning of hardware purchases can reduce these costs somewhat.

---

*An important new feature of Microsoft Windows 95 is "plug and play" technology, which is intended to ensure compatibility between Windows 95 and those peripherals that subscribe to Microsoft standards for "plug and play." To take advantage of this technology, however, you have to purchase hardware that is "plug and play"compatible. Expect a lag time for this standard to become the norm and to be priced comparably with other equipment currently available.

Software

The major component of software-related costs is the initial purchase price of the software, whether it is AutoCAD or another program. When purchasing AutoCAD, you have to consider whether you are purchasing an individual stand-alone license or a networked version with multiple licenses. In addition to the initial purchase cost, you may need to factor in software add-ons from either Autodesk or third-party developers. When purchasing software for more than one workstation, always inquire as to whether you can obtain a "bulk" discount for ordering multiple licenses. Many software packages offer networked or multiuser versions that can be more economical than separate licenses even if you end up with more licenses than you currently need. A growing number of software producers are offering financial incentives to purchase their program on CD-ROM rather than on diskettes.

When budgeting for software, note that in addition to paying for the initial license, you must plan on keeping up with periodic upgrades. With AutoCAD, historically, upgrades have been every two years or so, and many other software vendors have similar upgrade schedules. You must budget not only for the cost of upgrades themselves, but also for the time spent on installation, testing, and updating your overall system (see Chapter 10) as well as training your staff to use the upgrades.

Maintenance Contracts

Hardware manufacturers and dealers generally offer maintenance contracts for equipment you purchase from them. These extend or augment the hardware's basic warranty. In general, these contracts are money losers for owners of machines that are rarely used. For heavily used machines, for which failure would cause serious production problems, however, maintenance contracts may be worthwhile if they include a provision for having the use of a substitute ("loaner") machine when your machine is under repair. Some contracts also include routine maintenance, which is worth considering if you do not have in-house staff capable of performing such maintenance.

Customization

As discussed in Chapter 4, with newly installed CAD programs, and AutoCAD in particular, you need to plan for staff time spent customizing the software for your particular office. Regardless of what approach to customization you choose, expect to spend money on some combination of billable and nonbillable staff time to achieve this important task. The nonbillable portion will pay for itself with future CAD projects, since routine tasks can be automated. Although the initial, major customization expenses occur when your CAD system is implemented or comprehensively overhauled, you might budget for partial revisions every eighteen to twenty-four months or so.

Staffing

Firms that use CAD effectively find that they can endure short-term fluctuations in project work without expanding or contracting their staff. Such firms see substantial savings in the cost of hiring (posting job notices, interviewing time, training, and so on) and firing (paying unemployment taxes, coping with a leaner staff), and they also maintain high employee morale. However, as you add more computers, you may need to hire a CAD manager and CAD-literate staff, which can add to overhead expenses.

Training

Even with staff who are already CAD-literate, you must budget time to train them on *your* system. New employees must be given at least two days to become familiar with your setup. And you must provide training with each acquisition of new software or upgrade or when major changes to the firm's configuration are made. Some firms apply a budgeting rule of thumb for training costs: they set them at the equivalent of 60 percent of hardware and software acquisition costs. You can lower this percentage if you provide a user-friendly CAD environment, effective training programs, and strong professional incentives for staff to master the software.

Service Bureaus

If you use external service bureaus for CAD input and production, you may be spared some or all of the costs of in-house acquisitions, configuration, and training. However, you still have to budget time for staff to establish a relationship with the bureau, to prepare CAD files and instructions, and to perform quality control on the completed work. By not purchasing your own equipment, you pay opportunity costs in terms of control and turnaround time that can offset the savings that service bureaus may provide.

Financing

The way you finance your CAD system also affects your cash flow and returns. If you *pay cash* for equipment, you've put money down on products that immediately start to lose value; however, you can usually realize a tax shield by depreciating the equipment down to its salvage value if you hold it for a minimum number of years. If you *lease* equipment, you can write off interest charges and avoid being stuck with obsolete machines, but you have to abide by the terms of the lease, and you may fall victim to unfavorable interest rates. If you *rent* equipment, you can easily add more machines when you're faced with a short-term CAD production crunch, and then unload unused machines once the need for them disappears. But unless you have a rental contract that gives you credit toward purchase for each week or month you pay of rent, renting is a money-losing option if the rental period goes beyond several months. Consulting a tax professional prior to making a major investment in equipment is always recommended, especially if major changes in the tax code are impending or have just been passed by federal or state governments.

Office Planning

The traditional drafting stations with parallel rules and rear storage surfaces don't work with CAD. You need to provide ergonomically sound and functional workstations for all your employees who use computers (see Figure 9.4). The ideal CAD station provides space for a CPU, monitor, keyboard, mouse and mouse pad, related computer components, drawings, and lighting. Depending on what workstations currently exist in your office, prepare to shell out some money to modify or replace existing workstation pods. You may also have to rearrange office space to accommodate plotters, printers, scanners, and other shared peripherals.

Supplies

With a CAD system you need to maintain an inventory of plotting paper, pens and inks, diskettes and diskette storage devices, backup devices and tapes, training and instructional manuals, and so forth. While individually some computer supplies are small-ticket items, collectively they can add up to considerable annual expenditures. Yet if you use CAD aggressively, your outlays for traditional drafting supplies, such as pencils, erasers, films, and so forth, fall inversely.

Utilities and HVAC

If using CAD brings more machines into the office instead of merely causing you to replace existing ones, then the demands on your electrical capacity will rise. The machines themselves will consume more energy, especially those that operate twenty-four hours per day, and they will also place demands on the office air supply, creating the need to increase air circulation. Many offices have to spend money to upgrade their electrical service and, if modems and networks are installed, to add telephone lines, and to have the office HVAC rebalanced. Although an officewide energy reduction program will help you control utility costs, expect these costs to rise.

Opportunity Costs

CAD expenditures include implicit opportunity costs. If you spend money on a CAD system, you are spending funds that could go to staff raises, training and professional development, marketing programs, general office renovations, debt reduction, and other equally valid expenses. By *not* spending money on CAD, however, your potential production levels and efficiency will be reduced, you may limit your eligibility to qualify for new projects, and you run the risk of losing existing clients. Considering the opportunity costs of CAD is essential in evaluating projected expenditures; going through the CAD decision-making process as described in Chapters 2 and 3 helps identify these opportunity costs.

CAD Economies of Scale

In calculating the income and expenses relating to CAD, bear in mind that in a CAD-efficient office the more workstations you acquire, the less the incremental *cost per workstation* should be. As your CAD system expands, the initial costs and learning curve incurred in setting up one workstation will save you time, and thus costs, when the next one arrives. The unit costs of customization will drop when the benefits of customization are applied over more and more workstations. Your workstation unit costs will be reduced because not every CAD station will require a separate plotter,

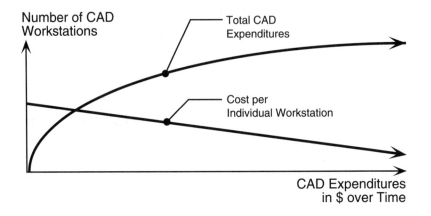

Figure 9.3 In CAD-efficient offices, the acquisition cost of each additional workstation should decline.

printer, or other peripheral; the costs of each peripheral can be shared by a growing number of workstations. Similar economies of scale exist in training. So when you plan to acquire multiple workstations, look not only at the totals but also at the effective cost per workstation (see Figure 9.3). Even when you must upgrade to a higher level of hardware, you should see your unit cost drop.

Sources of Budgeting Information

A budget is only as good as the accuracy and timeliness of its components. You'll have to spend time collecting information from a variety of sources; fortunately, most of these sources are easy to locate.

Historical Information

If you have already invested in CAD or in major computer equipment of any sort, you have a base from which to start collecting information. Moreover, your accounting department probably has set up the accounts necessary to track CAD-related expenses and revenues. Historical information, of course, is only useful as a starting point. You must factor in the costs of both inflation and deflation as prices for hardware and software will decline over time.

Vendors

Vendors, such as AutoCAD dealers, can provide current price information for specific products. You can also acquire useful price information from mail-order catalogues (not for AutoCAD, however). Make sure that your requests for information are as detailed as possible and include drivers, cords, service contracts, shipping, and other seemingly minor items that can affect overall costs considerably.

Trade Shows

Trade shows are an excellent source of information about products and services. Within a few hours, you can collect a lot of general information about CAD stations, plotters, and other items you may plan to invest in. You can then follow up with calls to specific product sales representatives.

Periodicals

Professional periodicals directed to architects and interior designers often publish periodic surveys on CAD usage that list current costs for workstations, CAD drafters, and so forth. Although the figures are often simplified and do not reflect local markets, they can provide helpful insight into trends and pricing practices.

Colleagues and Professional Organizations

Peers and various professional organizations can offer guidelines on estimating costs and fees. In several major U.S. cities, local chapters of major professional organizations, such as the AIA, have roundtables for managers and financial officers. As Chapter 10 discusses, nowadays you can often "download" product information from electronic bulletin boards via the Internet.

Calculating Returns on AutoCAD

To evaluate and compare budgeted and actual income and expenses for AutoCAD, you need a methodology with which to calculate returns on your AutoCAD investment. Several standard ways to do so are described in the following paragraphs.

Bear in mind, however, that all pro forma return on investment (ROI) calculations are simplistic. You must refine and expand upon each line item to reflect your actual operations. In addition, you should experiment with different scenarios. Even if the current market for design services seems strong, for example, you need to provide a safety cushion in the event of an economic downturn. Using a spreadsheet program such as Lotus 1-2-3 or Microsoft Excel is an excellent way to track and analyze this information. Indeed, several such programs come with built-in ROI formulas.

In applying any measurement of returns, augment hard financial figures with more qualitative factors such as the probable impact of a particular scenario on the marketability and quality of your design services, on employee interaction and morale, and on general office efficiency.

The AutoCAD Return on Investment (ROI) Calculator

For many of its products, Autodesk itself has published ROIs based on general assumptions. A number of years ago, the company introduced the AutoCAD Return on Investment Calculator, a simple computer program based on Lotus 1-2-3. Although Autodesk no longer distributes this program, the example presented here, The Calculator, is a useful starting point from which to develop your own analysis, as it specifically addresses the costs of investing in AutoCAD. The Calculator allows you to specify values for seven variables, from which it generates five-year projections for net savings/costs and productivity gains, as well as annual and cumulative ROIs, and an internal rate of return. Shown in the accompanying chart is an example of output from The Calculator, using the program's default variables except with regards to the hardware and software costs, which are adjusted to reflect recent prices.

Table 9.1 AutoCAD Return on Investment (ROI) Calculator: A Sample Analysis

$3,000	Cost of Computer Hardware/Operating System
$3,500	Cost of AutoCAD Software
$35	Cost Per Labor Hour (including overhead and benefits)
180	Labor Hours per Person per Month (Assumes 22.5 days per month, 8 hours per day)
4	Training Time Required (in months) (Enter value between 0 and 12)
40%	Productivity LOSS During Training Time (Enter value between 0 and 1; Example: 0.50 = 50% Loss)
50%	Productivity GAIN After Training is Completed (Enter value between 0 and 2; Example: 0.75 = 75% Gain)

All Other Annual Savings (Costs):

($1,000)	Year 1	Includes Maintenance Contracts,
($500)	Year 2	Service Bureau Charges, Upgrade
($1,000)	Year 3	Costs, Additional Training, etc.
$6,000	Year 4	
$6,000	Year 5	

	Year 1	Year 2	Year 3	Year 4	Year 5
Hardware System Cost	($3,000)	0	0	0	0
Software Cost	($3,500)	0	0	0	0
Productivity Loss	($10,080)	0	0	0	0
Annual Productivity Gain	$16,800	$16,800	$25,200	$25,200	$25,200
All Other Savings(Costs)	($1,000)	($500)	($1,000)	$6,000	$6,000
Net Annual Gain(Loss)	($780)	$24,700	$24,200	$31,200	$31,200
Cumulative Gain(Loss)	($780)	$23,920	$48,120	$79,320	$110,520
Cumulative ROI	95%	244%	390%	578%	767%
Average Annual ROI	95%	122%	130%	145%	153%

Net Present Value @ 10%	$78,569
Internal Rate of Return (IRR)	3166%
Payback Period	Year 2

Example shown here based on materials provided by Autodesk, Inc.

The Calculator, while a useful starting point, is somewhat simplistic in breaking out costs. You can produce a more accurate analysis, especially when budgeting for multiple workstations and shared plotters, printers, and other peripherals, if you use a spreadsheet or database software. The Calculator's assumptions, in particular, should be reviewed carefully:

- The costs of acquisition, training, and so forth, are shown up front. You must factor into the "Other costs" category the ongoing costs of upgrades for RAM, new releases of AutoCAD, and additional training and customization.

- Financing costs are not explicitly spelled out, nor is the anticipated rate of inflation. Both have a major impact on your expected internal rate of return (IRR) and payback period.

- The Calculator calculates returns on productivity loss and gains, using a percentage factor. You can perform an analysis based on dollar figures, using figures for revenues and training expenses. You must determine which additional revenues and profits result from greater productivity and which from special billing rates for CAD.

- While it does consider training time, The Calculator does not explicitly address the costs of customization. This is an essential cost that is tied directly into productivity gains.

- The Calculator assumes training is required, which is not always the case, since you may already have CAD-proficient users on board. The program also assumes that the training "takes," (that is, that the staff whom you trained absorb what they need to know); this is not a given and depends on the quality of training and on your staff's responsiveness and interest.

- The Calculator assumes there is downtime for training and learning CAD but doesn't explicitly state whether or not it occurs during ordinary office hours. In some offices, training might be done on evenings and weekends. You might need to pay staff overtime, but that way you avoid lost productivity and missed deadlines during regular work hours.

- Apart from productivity losses, The Calculator does not include *opportunity costs*—the costs of *not* having CAD and thereby of possibly losing work, of gaining work because you *do* use CAD, or of shifting funds from other resources to CAD.

- The Calculator has a five-year time frame. Some computer gurus consider three years to be the only realistic time frame for CAD systems;

others would say longer. Your assumptions about the lasting power of your purchases is critical, as is the depreciation rate and salvage value you assume for obsolete equipment.

Revenues and Expenses per Employee

Financial analysts often judge a firm's productivity by its revenues and net income per employee. You can apply this approach to testing the cost-effectiveness of using CAD in your firm by calculating overall changes in revenues and expenses per individual employee and for each member of your staff before and after acquiring CAD. If your employees explicitly list on their time sheets whether they use CAD on a project, you can track whether individual CAD users are more productive than their counterparts, whether CAD users or not.

In calculating this rate, you need to think of ways that CAD may play a role in producing additional revenues. Some revenues might result from increased productivity on the part of an existing employee pool; others may reflect the results of hiring additional employees whose productivity remains level but who appear more productive because the unit costs of additional CAD workstations have fallen while the number of CAD-based projects in your office has increased. It is also essential to look at qualitative factors; talented and hardworking employees should not be penalized because they do not explicitly show productivity gains when CAD is implemented in your office.

CAD Utilization Ratio

Some firms justify purchases of additional CAD stations by applying a *CAD usage factor* or *utilization ratio*. This ratio calculates the number of hours per week or month during which each CAD station is in use. You can use this ratio to estimate how quickly you can recover the costs of your investment in CAD. If, for example, a CAD station is used for 30 hours per week during a 40-hour workweek, the utilization ratio is

$30 \div 40 = 75\%$

A key factor in the resulting ratio is your assumption about the number of hours per week a CAD station will be used. If the maximum average number of usable hours assumes a 40-hour week, the resulting ratio will be different than if a typical workweek is assumed to be 60 hours or more. Since computers don't officially need sleep, you can calculate utilization ratios based on a 168-hour workweek, but then you must factor in the operating costs of keeping your office open twenty-four hours per day, seven days a week (not to mention the possibility that your firm may acquire a reputation for being a sweatshop). Your usability rate will also be affected by whether or not your CAD stations are shared; if stations are shared, the number of hours for which a CAD station is unoccupied may be limited, but you can encounter bottlenecks if more than one user needs to work at the station at one time. Note that firms that insist that their CAD users be fully rounded design practitioners are content with a CAD utilization ratio of *less than* fifty percent, providing they are realizing the goals they had established for CAD.

Utilization ratios can be applied not only to CAD stations but also to peripherals such as plotters, printers, and modems. You can apply the ratio both to equipment you already own and to outside vendors if you use them in lieu of in-house resources. If, for example, you send drawing files to a plotting service several times per day, and if you equate each order with two hours of in-house plotting time, you may be able to justify purchasing an in-house plotter (see Chapter 8).

Developing a CAD utilization ratio assumes that you have the ability to track CAD use. Firms that have designated computers and people for CAD can track CAD use fairly easily through time sheets and computer network monitors. As CAD becomes more integral to your firm, however, determining this ratio becomes more difficult, and the process of tracking time will be more tedious for CAD users. If you take advantage of the growing file-exchange capabilities between AutoCAD and Windows and get away from having workstations designated purely for CAD, you may lose altogether the ability to track CAD utilization and instead be able to follow only general computer usage. If the computer becomes the preferred tool in your office, calculating such ratios is no longer necessary or worthwhile.

Setting Fees for AutoCAD Services

All design firms want and need, of course, to make a profit on their projects in order to survive financially. As part of this, firms need to find ways to generate a return on their investment in CAD and to avoid giving away their work for free. Understanding how to set fees for CAD-based projects will help ensure better project profit margins and help you compete in the design marketplace.

Scope of Fees for CAD Work

In the early days of CAD, firms charged separate and special fees for all CAD-generated work. The extra charge was justifiable because only a small number of firms used CAD and did so for projects where the time frame and technical requirements demanded faster and more precise output. In those days, clients were willing to pay extra—often $5 to $15 more per hour—for CAD. Now CAD is so prevalent that to charge a premium for it is akin to charging clients more for producing design specifications on a word processor rather than on a typewriter. Clients will not pay extra for a process that is part of the industry mainstream.

Given the ubiquitous nature of CAD these days, a general pricing guideline is that you should not charge special fees for routine production work that you could produce *either* by hand or by CAD. Clients should not be expected to pay a premium for *your* choice of production methods. And if you are not making a profit on routine CAD-based production, then you need to review your internal production processes; with proper training and customization, you should be able to generate a return on your CAD investment.

While you ordinarily would not charge special fees for routine production, you often can charge additional fees for particular CAD-generated services. Some of these services, described below, are considered supplemental whether or not they involve using CAD. You should also review the schedule of supplemental services in pro forma design contracts, such as the Schedule of Designated Services in AIA Document B163.

Use of special CAD standards and graphics. Some clients—particularly major corporations and certain government agencies—require that drawings be produced according to particular graphic standards, for example, using sheets of unusual dimensions or particular text styles, layer names, or architectural graphics. Such standards help the clients realize benefits of consistency but do require that their design firms spend extra time tailoring their own CAD setup to these standards. Time spent accommodating these requirements may be billed separately unless the proposed fee will cover it.

Use of supplementary software. As more clients develop their own in-house computer databases, they may require that the design firms they hire use some of the same software. If the client does not provide you with a copy of this software at no cost to you, you should negotiate about who will pay for the software or else absorb the cost in your design fee.

Modifying your in-house standards or acquiring a special piece of computer hardware or software in order to obtain or keep a client ultimately can be a money loser unless your client is long-term, loyal, and generous in granting work and paying good fees. Do not purchase such items unless and until you have won the project. Moreover, don't forget that the actual costs involve not only the price of the software but also installation, training, and coordination.

Special studies. Particularly during the programming or schematic design phases, clients may wish to perform special studies, such as reviews of sites they are considering purchasing or leasing. They may want you to analyze certain design features, such as building facades, employee workstations, or lighting layouts. Special studies generally fall under the category of special services and often involve customized use of AutoCAD or other software. Such studies normally justify fees for additional services.

Visualization. Visualization studies include model construction, whether of traditional physical models or of computer-generated models. Examples of computer-based modeling are studies of elevations comparing different materials and lighting designs, shadow studies, and animated walk-throughs.

CAD as-builts. Depending on the nature of your contracts, you may be required to generate as-builts of your completed designs. Increasingly design contracts call for a routine transfer of CAD files to the owner, but in some situations as-builts should be considered an additional service, so you may charge for time spent both verifying field conditions and inputting them into CAD.

Area take-offs. Clients who lease out their space, as well as institutions and corporations that assign charges to departments or end users based on their real estate, need accurate takeoffs of their spaces. CAD is a perfect tool for this task, since it can easily assign area counts, with extreme precision, to spaces defined by complex boundaries (see Chapter 7).

Facilities management. Many corporations and institutions cannot afford (or don't want) their own in-house facilities management departments, are downsizing what they do have, or are too swamped with work to handle their project loads in-house. Such organizations often farm out work to design firms. Facilities management work can include programming, move coordination, space planning, and design (see Chapter 7).

Document management. As discussed in Chapter 10, organizations are becoming more explicitly aware that they have to more effectively organize and manage their data, whether on paper or in computers. Technical document management (TDM) is becoming a major concern in many organizations. Autodesk and other software companies now market TDM computer software. Clients may hire you to help establish and maintain a design- or property-oriented TDM system. Data management tasks might include overseeing the process of converting drawing archives to electronic form and setting up CAFM databases with the goal of reducing paperwork.

Methods of Setting Fees

Establishing fees that are both reasonable and competitive often requires estimating fees in various ways and arriving at a fee that is a reasonable average of the estimating methods. There are several typical approaches.

Person-hours

Fees generally are calculated by person-hours, by multiplying estimated hours by the billable rate per hour for the CAD drafter. As mentioned previously, clients nowadays are disinclined to pay a premium for routine CAD drafting, and if your CAD operations are well established, you should expect the projected person-hours to be the same or less than they would be for hand drafting, especially for floors plans (custom detailing is often done directly by the designer who normally bills out at a different rate anyway).

Charging by the hour for CAD work poses a problem for designers. If CAD really is more productive than handwork, then a specific level of work should be completed in less time, that is, fewer hours. Greater productivity reduces the number of CAD hours that can be billed to the project and requires that you add more projects during the same time frame, bill for other tasks, or increase your hourly rates. The person-hour billing approach also forces firms to compete solely on a fee basis. A growing number of design firms are thus backing away from charging for any CAD-based work on an hourly basis. They find that not only does charging hourly CAD fees create employee stress, it devalues design services. Indeed, the industry standard for CAD-based design fees in some sectors has dropped to under $1 per square foot, which, translated into hourly fees, barely allows for any *design* whatsoever.

Note that if CAD is truly integrated into the design process, you need not bill out separately for CAD drafters. However, you may be able to justify higher hourly fees or more hours of work.

Flat/Lump Sum

Particularly for architectural projects, many design firms charge a flat fee, with a contingency for situations that might require additional fees. A flat fee approach for CAD works particularly well if your CAD system enables you to approach the project efficiently; if the project poses complexities that challenge your CAD system and require more hours to complete, you risk losing money. You must have a clear understanding of the project itself and of your CAD capabilities to know if a flat or lump-sum fee will work for you.

Square Footage

Some firms, particularly those that do a lot of space planning, estimate fees per square foot. A 60,000-square-foot office space, for example, at 10 cents per square foot would cost $6,000. This method works best with simple, routine drafting projects and is advantageous on projects where the building floor plate is uniform throughout and consists of simple, repetitive elements. Charging by foot is often done for CAD as-builts and other forms of documentation rather than for design work.

Quantity of Layers

Another fee schedule is based on the number of CAD layers involved. A base CAD floor plate that shows only columns, exterior walls, core, interior partitions, windows, doors, and room names, that is, seven layers of information, might take one day to produce. If information on ceilings, electrical and data systems, furniture, landscaping, signage, and other specialties is also to be entered, additional time should be charged.

As Figure 5.2 shows, CAD changes the relative amount of work or person-hours required during each design phase. With CAD, the bulk of the work can fall into the initial phases rather than the final drawing production phases. Moreover, as Figures 6.1 and 6.2 show, CAD can transform design from a linear process to a circular one. Your fees should reflect this transformation, particularly if you set different fees for different design phases. Most likely you will want to shift more hours to the earlier design phases. You should also carefully review how CAD changes not only the relative number of hours during each phase but also project staff requirements. If CAD can reduce the routine drafting time needed in favor of more time being applied to design, the balance of hours needed for drafters and designers is altered, and, in turn, the mix of hourly rates that you will bill for changes.

Guidelines for Estimating Fees

The fees you set reflect different methodologies but also must reflect realistic trends and precedents. Otherwise, you cannot compete effectively with other design firms. You can obtain guidelines from several sources as discussed below. Many firms test estimates against several fee approaches and then average them out.

Recent History. For CAD-experienced offices, recently completed projects of a similar type and scope provide the most useful guidelines for setting fees. Team members, particularly those responsible for CAD work, can provide estimates for production times, person-hours, and so on. Although recent history offers valuable guidelines, you must anticipate factors that are unique to the current project (for a more in-depth discussion see Chapter 5) as well as changes in the profession that can affect fees.

Industry Trends. Major design magazines such as *Architecture* and *Interior Design* publish industrywide annual studies of fees charged by design firms, and for many years they have tracked trends in CAD-based fees as well as salaries and billable rates for CAD drafters. These studies are most useful in establishing trends, but you will need to make adjustments for regional differences and design specialties.

Peers and professional organizations. Many local, regional, and national chapters of design-related organizations offer panels, roundtables, and publications that discuss new trends in and approaches to fees. The American Institute of Architects (AIA), the International Interior Design Association (IIDA), the International Facilities Management Association (IFMA), the Society for Marketing of Professional Services (SMPS), and other design-related professional organizations all offer useful information on trends in fee ranges and scope. In some cities, financial officers, marketing staff, CAD managers, and other design professionals hold their own roundtables, and such peers are probably the source of the most pertinent fee guidelines.

AutoCAD in Design Contracts

Now that CAD has become ubiquitous in design work, especially for routine production, it need not be classified in contracts as a special service. However, design firms find that certain aspects of CAD work are becoming increasingly complex and can cause confusion, tension, and fee overruns if certain issues are not addressed up front with clients. Therefore, you would do best to spell out certain CAD-related issues explicitly in your contracts.

In your contract, you must clarify whether and when CAD is a routine service imbedded in various design phases, or if it is a supplemental service. *The AIA Standard Form of Agreement between Owner and Architect for Designated Services* (AIA Document B163), for example, lists thirteen categories of computer applications as supplemental services. These services include "Production of Drawings," but many design firms would not consider CAD a supplemental service if they use CAD as their standard form of production. On the other hand, an unusual analytical or design task that requires purchasing, installing, and training staff in the use of special software may warrant classification as a supplemental service.

Part of knowing when to charge standard or additional fees for CAD-related work involves having a clear understanding of the client's expectations. It may also require knowing what your competition does. CAD-aggressive firms that have already absorbed the costs of learning special computer applications may incorporate these services into routine work, thereby increasing clients' expectations of what you should provide as a matter of course.

CAD Deliverables

Most corporate, institutional, and governmental clients routinely expect designers to submit CAD files to them along with a set of prints at the end of a project. Many contracts spell out this expectation with a simple phrase to the effect that a CAD file will be provided to the owner upon the project's completion.

As both design firms and their clients become more computer sophisticated, however, both parties realize that more specific guidelines are often needed so that the CAD files that you submit to clients will be satisfactory. Many firms have gone over fee or have had to ask for additional fees to provide an acceptable file because the client's CAD expectations weren't clarified initially. Therefore, you may want to, or expect to have to, address the following issues in your contracts:

The preferred file format. Some clients simply want a *.dxf* file; others can read a *.dwg* or an *.iges* file. As Chapter 10 explains, however, your client will need to do more than merely specify a drawing file format; for the files to be of use to your clients, you will also need to know what version of AutoCAD your clients use. A *.dxf* file from Release 13, for example, cannot be read by a computer using AutoCAD Release 11 or 12 unless interim conversion steps are taken. Moreover, the process of converting AutoCAD files to other formats may involve choosing among a number of options; the options you choose can also affect the resulting files. Under some circumstances, you might have to purchase special third-party file conversion utilities to produce files that will match the client's specifications. Therefore, at the onset of the project you should encourage clients to be as specific as possible about their expectations for file formats. Bear in mind that for projects of great duration, this information may become obsolete; either you or your client could be using a new version of the software (or another software altogether) by the time the project is finished.

Media of exchange. Although most parties still exchange CAD files on diskettes, other forms of transfer such as modem, CD-ROM, and Bernoulli disks are also feasible. Even with diskettes, you may need to clarify whether the disks are to be formatted for PCs or Macintosh computers, or for standard or high-density drives. Clients may vary in their willingness or ability to accept compressed ("zipped" or "stuffed" files). They may or may not allow files to be sent to them by modem. Being unclear about this issue up front can prove costly down the line. As with software, new versions of exchange and storage media constantly appear, and for long-term projects you may have new transfer options by the time the project ends.

Guidelines on file content. Clients don't normally expect or want the information, such as architectural targets and dimensions, that you use in your construction drawings to get your designs built. But clients are becoming increasingly particular about layer names, blocks, and graphic standards for drawing borders and titles. If clients have special requirements in this regard for CAD files, these expectations should be spelled out in your contract. If written descriptions of the requirements are available, append them to your contract. If the client's requirements change over the course of the project, you may be able to charge additional fees for any effort you have to make to accommodate these changes.

Reimbursables. Design firms must distinguish what is chargeable as a reimbursable and what is a routine cost of doing business, that is, overhead that should be built into your design fees. Normally, plotted sheets are reimbursables as are blueprints; however, to be fair to the client, you must have a means of ensuring that the client is charged only for completed plots, and not for test or failed plots. Reimbursables, however, should not normally include general supplies such as plotter paper, pens, and ink, although your standard rates for plots should absorb these costs.

Copyright Ownership

As discussed in greater depth in Chapter 10, you need to confirm in your contract who actually owns your computer-generated work. Increasingly, your clients may ask you for copies of your work, including not only CAD drawing files but also databases generated during the course of the project for programming, specifications, and other tasks.

Designer's Liabilities

In addition to the need to confirm what services you'll provide, your contract with owners should include clauses to protect yourself from liability (see Chapter 10). Your firm's attorney can offer some guidance, especially vis-à-vis local precedents, but generally your contract might include the following:

1. A disclaimer that states that
   - CAD files are provided to the client only for informational purposes and that you are not responsible for postproject use of those files.
   - your client should agree to waive all claims to, and not to hold you responsible for, any losses, damages, or costs to him or her arising from the future use of any computer files you provide.
   - if your client wishes to use your CAD files for purposes other than reviewing the client's own space and facilities management needs, the client should request your permission.
2. A time frame—usually sixty or ninety days—during which you agree to correct or reissue CAD files owing to defects with the media on which the files were transferred or because the client somehow lost or destroyed the files.
3. A refusal to accept responsibility for CAD media that erodes or can't be read because of subsequent changes in technology.

Consultants

The issues you address in contracts with clients often reappear in contracts with consultants. As designers and their consultants increasingly exchange information by electronic means, clarifications about shared computer files becomes more important, and ownership of these files and guidelines for transferring files should be spelled out when you negotiate contracts with both consultants and clients.

In your contracts you should clarify whether you and the consultants turn CAD files over to the owner separately or whether you as the designer are responsible for merging all the project parties' work into composite CAD files, and if so, whether you are expected to review the consultants' files for accuracy and compliance with client standards.

Making AutoCAD More Affordable

If one reason were to be selected to explain why AutoCAD is not on every designer's computer, it would be cost. AutoCAD Release 13 ships with a list price of $3,995 (discounts are available for large orders, government agencies, and educational institutions). To run AutoCAD Release 13 properly requires a Pentium computer with 32MB of RAM, a 1GB hard drive, and a large monitor; the hardware alone costs at least $3,000 (without any add-ons). So at $7,000 minimum per workstation, AutoCAD is a significant investment, especially for small to medium-size design firms (which constitute the majority of design practices).

There are, however, ways to maximize your investment in AutoCAD while minimizing your financial outlay.

Use AutoCAD More Extensively

Many designers in essence make AutoCAD a more expensive outlay because they use only a fraction of its capabilities. The more you use AutoCAD in all phases of design (see Chapter 6), the greater the return you'll get on your investment. You won't reduce the initial outlay, but on a per task or per project basis, the costs will fall. In addition, if you aggressively explore AutoCAD's less obvious capabilities, you may find that you can save money by *not* having to purchase software that duplicates functions that AutoCAD also offers.

Market More AutoCAD-based Services

Many design firms find AutoCAD an expensive proposition because they acquire AutoCAD solely for producing ordinary construction documents, and, as discussed previously, such use no longer justifies high fees and taps only a small percentage of AutoCAD's capabilities. If you incorporate AutoCAD into special services, ranging from ADA studies to facilities management to modeling and 3D, which command premium fees, you can earn a better return on your AutoCAD investment.

Run AutoCAD Concurrently with Other Autodesk Software

Even design firms that use AutoCAD's more sophisticated features, such as 3D or SQL programming, often may not use these features all the time. Therefore, they may not need to have full-fledged AutoCAD installed on all their CAD stations. So one cost-saving option is to run full-fledged AutoCAD on a small number of workstations and to run another, more limited but affordable, AutoCAD-compatible product on all the others. The latter product can be used for routine drafting and analysis. When users need AutoCAD-specific features, they can import the drawing files into AutoCAD.

For design firms operating on the Microsoft Windows platform, the most applicable Autodesk product today, is AutoCAD LT for Windows. This software, introduced several years ago and now in its second version, runs only under Windows. It offers an interface similar to that of regular AutoCAD but with a more limited set of features; standard AutoCAD programming tools such as AutoLISP routines, for example, are not usable. Because the command menus, icons, and other features appear similar, however, any experienced AutoCAD user can quickly learn to use AutoCAD LT and produce baseline drawings. AutoCAD LT drawing files can be exchanged with AutoCAD, although features unique to AutoCAD will not translate to AutoCAD LT.* AutoCAD LT also requires less memory and is simpler to install and configure than regular AutoCAD. Another advantage is that AutoCAD LT lists at only $495 and need not be purchased through a registered dealer—it can be acquired through regular discount software retail stores and mail-order houses at an even lower "street" price. Autodesk produces symbols libraries for AutoCAD LT as do a growing number of third-party developers.

Using a compatible program like AutoCAD LT makes sense in offices where the bulk of CAD production consists of basic 2D drawings and where there is minimal demand for high-powered customization or 3D features. Project managers and designers can use AutoCAD LT to view, "doodle" on, develop, and redline drawings. When the capabilities of AutoCAD LT are exhausted, drawing files can be moved to a station with AutoCAD.

The one risk of using other Autodesk products such as AutoCAD LT is that you have no guarantee that Autodesk will produce future versions of such software to parallel future releases of AutoCAD, and thus you have no guarantee of continued file-exchange compatibility. In addition, AutoCAD LT's price and ease of purchase have angered registered AutoCAD

---

*As of this writing, the new version (LT 2) of AutoCAD LT for Windows does not read AutoCAD Release 13 drawings directly. You have to save AutoCAD drawings in Release 12 format or as .dxf files to be able to read them in Windows LT 2.

dealers; Autodesk apparently has sought to appease them by limiting the number of features that AutoCAD LT offers. However, if you weigh the low acquisition costs and the potential benefits of using such software against the possibility of someday having to switch to another product, you may find the risks worth taking.

Run AutoCAD Concurrently with Other Manufacturers' Software

Many design firms invest in AutoCAD because they want assurance that the files they create can be used by clients, engineers, and other consultants, and because they feel more comfortable using the world's predominant CAD software. For their own design and drafting purposes, however, some designers may prefer other software that they've already used or that has features that better meet their needs. Firms that have limited financial resources but want to put a CAD station on every desk often find that they can realize this goal only by using other, cheaper, software packages. Still others prefer software that offers less customization options but greater ease of use where 3D and database exchange are concerned.

Over the past decade, AutoCAD has continued to receive competition from CAD software that is both more affordable and targeted more to the architecture and interiors professions. And many of these products, while more limited in their abilities, are easier to learn and are more user-friendly. Moreover, many of AutoCAD's competitors make a point of assuring users that their files can be saved in a *.dxf* format that AutoCAD can read.

For firms that favor other CAD software yet wish to maintain compatibility with AutoCAD, a hybrid approach is perfectly feasible, provided that all CAD software meet certain criteria, including the following:

- All the CAD software you use must be able to share a common drawing file-exchange format. For CAD software, this is usually the *.dxf* file format (see Chapter 10).

- The software should be able to automate the drawing exchange process. This usually means creating automated procedures that write and read *.dxf* files—in AutoCAD they would be AutoLISP routines or Script files; in other programs they would be macros or other automated routines.

- Each software program should be able to adhere to officewide standards for drawing layouts, architectural targets, and layer names. The process of customization as described in Chapter 4 must not only be applied to all CAD software but also be implemented in tandem.

- All of the software programs should be able to use the same peripherals, such as printers and plotters.

- You need to have guidelines for determining which CAD software should be used for which types of projects or during which project phases. You might, for example, use AutoCAD for large long-term building or multifloor space-planning projects and use other CAD software for small short-term renovations. These guidelines must be based on a good understanding of the various features and relative strengths and weaknesses of each software program. You will need to modify your procedures if clients indicate a preference for a particular type of software for their project.

If your firm cannot develop or adhere to the criteria listed above, you will likely find that running multiple software standards is more costly in the long run, as you will in essence be operating two or more separate CAD systems. The smaller your firm is, the more costly such an uncoordinated multiple-software strategy will be.

Streamline Your AutoCAD Setup

As described in Chapter 4, AutoCAD permits you to enter commands by several methods: via typed keyboard commands and macros and pointing devices such as mice and digitizer pads. With Release 12, AutoCAD began to emphasize mouse-driven menus, dialog boxes, and toolbars. Increasingly, digitizers have become optional for ordinary command input; they are used more often for special command input or special functions, such as tracing.

Since they are optional, digitizers can be eliminated from your office standard CAD stations, saving you money. Many digitizer pads sell at $350 or more, so if the number of CAD workstations in your office is likely to increase, you may realize substantial savings by not purchasing digitizer pads to begin with.

You can realize additional cost savings by avoiding purchasing special input devices such as pucks with anywhere from twelve to forty programmable buttons. Most CAD users never use more than four buttons, and many find that having so many options is annoying; most users usually work quite productively with an ordinary Windows-compatible mouse such as the ones that come standard with each personal computer, or with one of the more ergonomically sound, affordable devices available at retail stores or through mail-order houses. CAD users are also more productive if they can use a single input device for *all* the Windows applications they work with; that way they do not have to learn different mouse input techniques for each separate program.

Selectively Upgrade

As Chapter 10 discusses, periodic upgrades to new releases of AutoCAD and other software are inevitable for committed computer users. Even so, you have options for how and when you upgrade. With careful planning, you can schedule and implement upgrades so as to minimize cash outflows and disruptions to office procedures. With a well-configured, powerful, and customized office system, you can also spread out the results of each upgrade over a longer time frame, thereby reducing the frequency with which you really need to upgrade.

Buy Uniform Workstations

Since AutoCAD runs on a wide variety of machines, many design firms can easily wind up using computer systems, monitors, and keyboards produced by many different manufacturers. This setup can potentially prove very costly, since not all computers may be able to share the same peripherals and components; you will need to spend time determining the optimal setup for each machine; and users may take time familiarizing themselves with each different model of computer.

Thus you can often save money by developing specifications for an office standard computer station; such specifications direct your purchases until the time comes to revise the standards. There are some advantages of creating standard computer station specifications.

- You may be able to realize bulk discounts by ordering multiples of the same system.

- You achieve economies of scale in shipping, installation, upgrades, and training.

- You increase the likelihood that all or most peripherals, cables, and especially memory boards can be used with more than one computer. Also, you don't have to discard all the related components when you discard a computer.

Buy Good-Quality Equipment

Your CAD workstations and related peripherals are among the most expensive capital outlays you will make. In reviewing products, you can easily be tempted to purchase equipment that either offers the most high-powered, state-of-the-art features or else comes at the lowest price. Careful consumerism is required to avoid ending up with equipment that very quickly proves unsatisfactory. For major purchases, look not only at the product features themselves but also at the manufacturer's track record, warranties, service capabilities, even financial stability.

Personnel Administration

The emergence of AutoCAD and other CAD programs has brought a new dimension to hiring and developing design firm personnel. Traditional job descriptions need to be modified, and new methodologies for evaluating prospective employees are required. Many design firms have felt "burned" after hiring CAD drafters because managers did not accurately describe their CAD needs, nor were they capable of evaluating prospective candidates.

CAD Job Descriptions and Responsibilities

To hire drafters/designers with the appropriate CAD skills, you must modify your traditional job descriptions to properly specify the technical skills that candidates need to possess so that they can use AutoCAD effectively in your firm. To determine the requisite skills, review your current projects and their projected needs. Consult, too, with the present staff and your CAD manager, if you have one. Review job listings in newspaper ads or in professional newsletters to get a good indication of what your peers are doing.

Some of the typical components of a job description include the following:

- a bachelor's or master's degree in architecture, interior design, or related discipline
- two or three years' minimum work experience in a design office
- experience producing full sets of contract documents, schematic design presentations, and other drawings. Indicate if experience is desired in a particular market, such as health care, education, or retail projects, because you can benefit from hiring a person with experience in using an AutoCAD system tailored to a particular design specialty.
- experience using a well-known third-party software add-on, such as Softdesk Auto-Architect or LandCADD.
- the assumption of some of the responsibilities of a CAD manager if your office does not already have one. Indicate that expectation in the job description as well. Do not assume that someone with basic CAD skills has the interest or abilities to qualify as a CAD manager.

> ### Job Titles: CAD "Operator" versus Drafter/Designer
>
> In the early days of AutoCAD, CAD users were typically called CAD *operators*. This nomenclature resulted in part because the earliest software versions and hardware were so difficult to use that just turning on the machine was truly an operation. Now, increasingly, as the typical AutoCAD user is a trained designer or engineer who uses the software for design as well as production, the term "operator" is inaccurate and often offensive.
>
> Therefore, in specifying a job title for an opening, avoid using the term "operator" unless the person you hire really will be nothing more than a technician.
>
> Instead, try to use titles such as "CAD Drafter" or "CAD Drafter/Designer." If your goal is truly to integrate CAD into your operations, drop "CAD" from the title altogether. After all, designers who work by hand probably have never had titles like "Pencil Operators" or "Pen Drafters."

AutoCAD Personnel: Interviewing and Hiring

Existing versus New Staff

One of the most critical strategic decisions that heads of design firms make is whether they should always go outside to find designers with CAD skills or whether they should train existing staff in AutoCAD. Both approaches have their advantages and disadvantages. From a long-term point of view, however, every firm benefits from educating its current staff in CAD, regardless of how many AutoCAD-proficient people the firm hires in the interim. Not only does training in-house staff increase your ability to "mainstream" CAD into your general operations; you end up with more versatile employees, who can produce both by hand and by computer.

Following is an outline of some of the pros and cons of hiring new people versus training existing staff in AutoCAD.

Table 9.2

Hiring New Staff	Training Current Staff
Pros	
General expertise in AutoCAD can be available immediately.	Your staff are already knowledgeable in the workings of your office, procedures, and projects; they have less need for training in office operations and procedures.
New hires can bring along useful tips for AutoCAD obtained through experience in other firms or in past CAD classes.	Your present staff can help incorporate AutoCAD into existing procedures.
	You can keep salary structures in line, because AutoCAD training is a firm benefit.

Table 9.2, continued

Hiring New Staff	Training Current Staff
Cons	
You must spend time and funds advertising, interviewing, and hiring people to fill the position.	You will need to spend time and money on training, whether internal or external.
New employees may not understand office routines, project procedures, and office culture.	Training staff may distract you from, or conflict with, your other duties.
Regardless of your new employees' skills, you still have to train them to use *your* customized AutoCAD system.	
Salary requirements may be out of sync with those of other positions in your firm.	

Interviewing and Evaluation

Hiring a new person specifically to do AutoCAD-based work requires care because the traditional methods of interviewing and evaluating won't work. Moreover, because project deadlines are so tight in the 1990s, few firms can afford to underestimate the skills of job candidates. As an employer you can face potential liability if you find that a new employee is unsatisfactory and therefore seek to dismiss him or her. So for hiring AutoCAD-proficient staff, a new hiring approach becomes essential.

Depending on your needs and requirements, an interview of a prospective CAD-using employee should include the following components:

1. Review of resume and portfolio. As with interviews for other positions, resumes and portfolios help to gauge a candidate's interests, abilities, and experience. Use these tools to obtain information on the types and scopes of projects the candidate has worked on; the extent of responsibilities he or she has assumed; and the degree of decision making and responsibility the candidate has had in developing CAD-based graphic standards and drawing content.

2. Review of a CAD disk. Traditional paper-based portfolios may show *what* the candidate has produced using AutoCAD but not *how*. A growing number of design firms now request that job candidates include in their portfolios a typical AutoCAD file they've worked on. This file should be reviewed by your CAD manager or by another CAD-proficient member of your staff who would keep an eye out for the following drawing elements and characteristics:

   - basic drawing setup, including drawing limits, extents, and layers.
   - degree of CAD customization, as evidenced by the presence of blocks, text styles, layers, linetypes, and other customizable features.

You can often gauge customization when you are opening one of the files; if you are asked for custom AutoCAD menus, External References **(XRefs)**, typefaces, AutoLISP routines, and other features that AutoCAD can't find on your system, then the file was created on a customized system. Another sign of customization is opening the **Rename** dialog box (Figure 4.10) and seeing many items that are available to be renamed.

- use of blocks. Analyze blocks for content, appearance, names, insertion points, and general suitability. How many blocks are created?

- memory and speed management. Do a test Purge to see how many defined but unused blocks, layers, and linetypes remain in the drawing. If the drawing file size seems exceptionally large relative to its contents, the candidate may need training in managing drawing memory.

3. AutoCAD tests. Some firms administer tests to evaluate the abilities of AutoCAD drafters. Products such as AutoCAD Evaluator, by Profisee Software, provide electronic multiple-choice tests. Once a candidate has entered his or her answers, the software automatically scores the test and analyzes the results by category of questions. Such tests are useful for determining candidates' general knowledge of basic AutoCAD, terminology, and operating context; the tests cannot evaluate drafters' knowledge of design production or of CAD as it pertains to architecture and interiors.

    In addition to administering tests yourself, you can find out whether the candidate has completed the AutoCAD Certification Exam, which is administered by Autodesk Training Centers. This is a two-level electronic exam consisting of multiple-choice questions and drafting exercises. As with the testing software described above, the exam is oriented to generic drafting applications and favors those who are proficient in taking this type of test. The exam is, however, a good indication of a candidate's seriousness about mastering AutoCAD, in that it can reflect whether someone has a solid *general* understanding of AutoCAD.

4. Training and credentials. The places where a candidate received training in AutoCAD may also serve as an indication of his or her CAD skills.

    If the candidate's primary training was received at design school, you need to familiarize yourself with the school and its resources in order to evaluate the significance of that training. If the design school happens to be a certified Autodesk Training Center (ATC), then the school must meet certain criteria for instruction and computer resources and should use the most current version of AutoCAD. Many design schools are limited in the resources they can allocate to CAD training, so some may choose to emphasize other CAD software, or, often as a matter of educational policy, avoid CAD and computers altogether.

Candidates' training and skills will also be affected by previous employers and work experiences. In small design communities, one can obtain, either through industry gossip or from AutoCAD users groups, some sense of which firms are aggressive computer users and have a CAD system from which your firm could benefit.

In 1997 the National Council of Architectural Registration Boards (NCARB) will present a new version of the Architectural Registration Exam (ARE). This revised ARE will rely heavily on computers for grading both multiple-choice questions and graphic exercises. Design problems will be solved using a unique CAD interface developed for the NCARB. While successful completion of the new version of the ARE will not in itself indicate whether someone is skilled in AutoCAD or in any other particular CAD software, the prospect of having to take this new exam will provide young architects with additional impetus to be computer-proficient, and as a result may help expand the supply of designers who are also CAD-competent.

Salaries and Benefits

In most cities, the supply of experienced design-oriented CAD drafters is still somewhat low relative to the demand for them. Therefore, unlike other design specialties, CAD drafters can expect to receive multiple offers and above-average salaries. Expect to pay for whatever skills and experience you are seeking in a new employee.

Because of supply-and-demand factors, many design firms have felt forced to develop separate salary levels for CAD drafters. On the one hand, this practice helps firms recruit CAD drafters, but on the other hand, it may be one more factor that isolates them from mainstream operations. Often design firms may find that training existing staff in CAD would be more cost-effective. A good training program in CAD is a welcome employee benefit.

Temporary Staff

Uncertain project schedules and workloads have made design firms leery of hiring too many people. Many firms prefer to use temporary staff when project crunches necessitate additional helping hands. Increasingly, design firms, just like many other types of employers, rely on consultants or other forms of short-term help as a matter of course.

While some temporary help is recruited through word of mouth or job postings, many design firms also rely on a growing number of placement or "temp" agencies to locate, interview, and screen CAD drafters. These firms keep a stable of experienced CAD-proficient employees on call; the agency handles all the paperwork involved in paying the drafters. By using placement agencies, design firms can hire temporary help without having to worry about complying with the myriad of state and federal regulations that affect short-term employees.

Even if a placement agency screens employees effectively, you should always try to interview candidates before inviting them on board, especially for long-term assignments. Where possible, try to interview and hire candidates with experience in architecture and interior design firms. CAD tends to be used somewhat differently in engineering, industrial design, and manufacturing firms.

Like most consultants, temporary CAD drafters generally are paid an hourly wage at a premium above ordinary hourly wages for regular staff. But you usually do not have to provide them with the typical employee benefits, such as vacation time, health insurance, and sick leave, nor do you have to carry them when billable work ceases. Temporary help, nevertheless, should always be treated with the same consideration as full-time help; good employee relations with "temps" help guarantee that they'll come on board eagerly the next time you face a work crunch.

Many firms find that using temporary help is a good way to locate permanent staff. As with dating, both parties have a chance to get to know one another before considering a long-term commitment—and hard feelings are less likely if the relationship does not become permanent. If, however, you ask someone hired through a placement agency to join your firm permanently, you may have to pay the agency a special fee for taking the consultant "off" the market.

Service Bureaus

Normally, temporary help work at your office or possibly at their own office, but another option is to farm out CAD work to service bureaus. Often these bureaus are reprographics houses that offer contract drafting as a way of augmenting other CAD services, such as plotting and scanning drawings.

The pros and cons of service bureaus resemble those of external plotting services. However, using service bureaus for drafting work generally requires more supervision and input than hiring plotting services. And external CAD production generally requires completely passive CAD input, with minimal interaction between CAD users and designers. Your design staff will need to compile extensive instructions before they issue drawings to a service; your staff then must check the completed CAD files when they are delivered to you.

Using service bureaus may be necessary to survive a short-term production crunch. Design firms that must continually rely on this option, however, either are perpetually understaffed or else have not effectively incorporated CAD into their basic design processes. Note also that many service bureaus do not use local help but instead rely on technicians in cities—not necessarily in the United States—where labor is cheaper. Apart from the impact that this practice may have on accessibility and turnaround time, relying on an anonymous labor force and particularly taking advantage of the cheaper labor rates available in developing countries, raises serious professional and ethical issues for designers.

Training in AutoCAD

Most design firms are too small, and operate under too many time and budget constraints, to provide formal training in CAD or, indeed, most other aspects of professional development. But you cannot implement a successful CAD operation unless you are prepared to offer some combination of funds and time to train your existing staff in AutoCAD. Even when you are fortunate enough to hire a skilled AutoCAD drafter "off the shelf," you will need to spend some time showing him or her how your firm operates; this type of training must be planned for as a matter of course if a new employee is to be productive in a short time frame. Proper training in AutoCAD involves not only AutoCAD itself, but also training in Windows, DOS, and other software that you use in conjunction with AutoCAD. You also need to ensure that someone shows the new employee how to use and maintain printers, plotters, and other peripherals.

There are several approaches to AutoCAD training. Since learning computer programs is a form of language learning and people learn languages differently—some by text, others by hands-on experience—it is wise to offer a choice, so that individual temperaments and schedules can be accommodated. For training sources and ideas, see the section on "Learning Resources" in *The AutoCAD Resource Guide*.

 Most design firms find that training in AutoCAD is successful only if it can be put to immediate use on real design projects. Using an actual design project is one of the best ways to teach AutoCAD to your staff and to get them excited about the software's capabilities.

Built-in AutoCAD Options

AutoCAD Release 12 and Release 13 ship with a self-paced tutorial package that consists of sample drawing files and documentation to guide users through the software. This is a good option for those who have the time and patience for this approach. It can also be a particularly good introduction for users who plan to take formal courses later on; such users then have a chance to become familiar with the software interface and the basic commands, which will help them keep up in a formal classroom environment and possibly qualify them for a more advanced course. Typically the tutorial requires one computer per user and at least ten hours (assuming one hour per lesson) of dedicated time.

Third-Party Software

Several third-party software developers have developed products to facilitate the learning of AutoCAD and other Autodesk software. These are listed in the "Learning Resources" section in *The AutoCAD Resource Guide*. Some typical products are

- self-paced tutorial guides, including texts designed specifically for interior designers and architects. These guides usually include a disk with tutorial files to use in conjunction with AutoCAD.
- video-based training materials.
- software add-ons that offer "on-line" tutorials that run inside AutoCAD. Some software packages, such as the AutoPack Tutor Builder by AutoPack, not only provide a given set of exercises but also let you modify the tutorial.

Formal Training Programs

Some people require the discipline and support of an omnipresent instructor and do better in a formal classroom environment. They also find that they can concentrate better when they are not constantly being distracted by colleagues, phone calls, and project deadlines. Formal training is also helpful for people who have learned AutoCAD informally and need to understand it in a more structured context while filling in gaps in their knowledge.

There are several options for formal training in AutoCAD:

AutoCAD Training Centers (ATCs). Autodesk sponsors ATCs in major cities in the United States and abroad. ATCs tend to be associated with technically oriented schools and universities, or they are independent CAD centers. They offer courses in topics such as AutoCAD Levels I and II, Production, Customization, AutoLISP routines, and 3D. Centers that offer training in advanced AutoCAD topics, such as ADS and AutoLISP routines, are designated Premier ATCs. One advantage of attending courses at such centers is that Autodesk requires that each center comply with the company's high standards for training and resources in order to become an ATC.

Schools and universities. AutoCAD is taught in classrooms throughout the country. Institutions that offer AutoCAD instruction range from design schools to community colleges and continuing education centers. If you have a choice, choose a course that focuses on architectural drafting rather than on generic AutoCAD.

Autodesk University. Over the past few years, Autodesk has sponsored a "university," which is held in various major U.S. cities. This event is a combination educational and professional development conference, offering classes in over two hundred topics related to AutoCAD and other Autodesk products. Many third-party developers are featured as well. The content is directed toward special applications and more experienced users, so the "university" is most appropriate for staff who are already AutoCAD-proficient and want to take their skills and knowledge to a more advanced level.

AutoCAD dealers. A number of Autodesk dealers offer courses and special product demonstrations.

CAD consultants. A number of independent CAD consultants include training in their bundle of services. The training they offer can range from one-on-one tutorials to larger group sessions.

In-house trainers. If your CAD operations are extensive or are growing, designating a member of your staff to train others in CAD is often most effective, especially when you need to train employees in using your in-house customization. Of course, you may need to provide training for your in-house trainers before they can train others.

In-House Training Programs

Although in-house training often seems difficult to organize, creating a program tailored to the needs of your firm can be the most useful route in the long run, especially if you have gone to great efforts to customize your system to the point where you have gone well beyond "generic" Auto-CAD. As mentioned earlier in this chapter, training should be provided not only for CAD but also for Windows applications and any other software that is heavily used in your office.

Because training typically involves nonbillable time and staff resources, it should be planned very carefully. Successful CAD training programs generally share the following characteristics:

- They address typical design issues that the firm contends with on a regular basis.
- They are scheduled at times convenient to both the trainer and the trainees.
- They protect the participants from work-related distractions.
- Ideally, they have a user-to-computer ratio of 1:1, 2:1 at most.
- Sessions cover specific and finite sets of topics, with stated objectives for each lesson.
- Training is conducted in English, not *computerese*.
- They are tied into personnel reviews and promotion policies.

Hands-on basic training should be supplemented by ongoing seminars in new developments and high-level applications. Training is also required when AutoCAD or Microsoft Windows is upgraded or when any software is substantially reconfigured in your office. More implicit but equally important forms of training include officewide presentations, ideally conducted during brown-bag lunches or happy hours, during which staff present successfully completed CAD-based projects and describe how they were designed using CAD. Combining CAD with food often seems to encourage the ingestion of CAD knowledge.

CAD Performance and CAD Parity

Although effective CAD use depends on an individual's interest, training, and initiative, it also depends on having a suitable CAD station. CAD is one of the most demanding computer applications that a design firm uses, so CAD users must have CAD stations that meet at least the minimum requirements for computers and RAM. Without such machines, CAD

What Makes a Good CAD Drafter/Designer?

With CAD, skill in drafting and design involves far more than what appears on paper. The way in which the CAD process is approached and the care with which data is entered into a computer is a form of craftsmanship, albeit a sometimes mysterious one.

Here are some signs of CAD craftsmanship:

1. CAD users draw as little as possible from scratch. Rather, they develop a few basic pieces and then use the commands **Copy**, **Rotate**, **Array**, **Mirror**, and **Block** to create additional entities. They use most CAD commands with a view to avoiding having to reinvent the wheel.

2. CAD users are precise. Unless they really should draw in a freehand mode (often done more easily in a "paint" or "sketch" program), CAD users create each line, arc, and so forth by using specific dimensions, angles, and coordinate points, or by using Object Snaps (**Osnaps**) to specify a relationship to another entity.

3. CAD users enter information not only according to how it will plot but also with attention to how the software will read it. On 1/16" scale plot, you may not notice or care that an apparently orthogonal line is off by 1°, but a skilled CAD drafter will care, because he or she knows that 1° degree can create distortions later on. By the same token, good CAD drafting entails having the lines that represent two perpendicular walls join at the ends, even if you cannot tell on a plot whether the lines meet or not.

4. CAD users plan for future changes. Skilled CAD drafters enter data, create blocks, write scripts, and perform other routines not just for immediate output but also with a view incorporating and facilitating whatever changes will occur in the future, making the work of others easier, and reducing the need for extensive quality control.

5. CAD users are also memory managers. On a daily, weekly, and monthly basis, they implement routines designed to free up computer memory so that they can work faster while reducing the possibility of computer crashes, downtime, and loss of data.

6. CAD users are considerate of others and are good team players. They never assume that they are the only ones working in a particular file or that they should be the only ones capable of doing so. They conform to office standards and procedures and document their work for others' benefit.

users cannot perform as you might expect them to. Before you can fairly evaluate their use of CAD, you must provide your CAD-proficient staff with the proper tools.

As important as providing individuals with the CAD tools they need is providing computer parity. In some organizations, employee rank may determine who gets the corner office. In CAD-intensive design firms (which tend to use the open-studio approach), ranking instead often focuses on who has the most powerful computer. In some cases, an exceptionally powerful computer or oversize monitor may be warranted for a specific employee, such as a designated 3D and animation specialist. In general, however, CAD users who perform substantially the same work should have comparable machines, just as project managers, secretaries, and others who perform like functions should have the same standard of equipment.

CAD Managers

Once a design firm acquires one or more AutoCAD workstations, it may also find that it needs to hire someone to assume the functions of a CAD *manager*. For very small practitioners, CAD management functions typically are assumed directly by one or more CAD users, who may often be one of the firm partners. For larger firms, CAD management functions generally are allocated to one or more junior or mid-level staff members.

A full-time CAD manager may seem an expensive proposition for many firms because many of the CAD manager's responsibilities, while essential to an office's efficiency, are not directly billable to projects; rather, they are absorbed in the firm's general operating overhead. And since experienced CAD managers can command mid-level salaries, their overhead rates are high. If no one assumes formal responsibility for CAD operations, however, especially in offices where CAD use is growing, a firm is likely to experience time-consuming operating inefficiencies and disorganization, which in the long run can exceed the costs of a CAD manager's annual salary.

In many design firms, the CAD manager may also assume responsibility for *all* computer operations in the office. In these circumstances, the title of director of management information systems (MIS) is often warranted.

Assigning Responsibilities

Design firms generally assign CAD management responsibilities in one of the following ways:

A full-time CAD manager. This option is usually selected only by firms with at least six to eight workstations, with a complex network or configuration of terminals, and/or with a highly customized system or the use of special applications that require programming expertise. In such situations, firms can realize certain economies of scale by consolidating non–project-based responsibilities into one job. Full-time CAD managers tend to offer a more sophisticated background and/or interest in computers in general and may be less likely to be formally trained in a design discipline (although a design background is definitely helpful). The full-time CAD manager may assume all, or almost all, of the CAD manager tasks listed later in this chapter. He or she may work on billable projects only when a particular CAD expertise is required.

A part-time CAD manager. This option is used by small to medium-sized firms with a small number of workstations or with a very reliable user-friendly computer setup that requires minimal care and feeding. A part-time CAD manager is expected to spend only 40 to 60 percent of his or her time on general CAD operations and to work on billable design/drafting functions the rest of the time.

A variation of this approach is to split the responsibilities among several people, each of whom assumes particular responsibilities, and who together constitute a "computer committee." For example, one person might manage the computer network, another might train staff in Auto-CAD, and yet another might develop and maintain officewide CAD drafting standards.

A CAD consultant. This option is used most often by individual design practitioners and small to medium-sized firms that either can't afford the overhead entailed in hiring a CAD manager and/or do not have sufficient work to keep an experienced technically proficient person on board full-time. CAD consultants are on call to handle software and hardware crises as they occur, to set up and expand office networks, to assist in training, and to provide general technical information. Although their rates are not cheap, they may be more affordable on an annualized basis than carrying a full-time staff person whose time is rarely billable.

CAD consultants come from various backgrounds. Some are computer-proficient architects and designers who have recognized that helping other firms improve their CAD operations is a lucrative business. Others are dealers in computer software and hardware; some are registered Auto-CAD dealers. Regardless of their background, ideally CAD consultants should have an intensive knowledge of *both* computers and design, and ideally, broad experience in installing and using AutoCAD. They also should be on call all hours during which your office is in operation.

Establishing the Need for a CAD Manager

The point at which a CAD manager becomes a necessity often creeps up suddenly on a design firm. Some firms try to determine proactively the break-even point at which a CAD manager becomes financially justifiable.

This break-even point occurs when the explicit costs of a CAD-manager—salary, benefits, research expenses, and so forth—are less than the explicit and implicit costs of not having one. Explicit costs include fees for CAD consultants and long-distance telephone bills for calls to vendors. Implicit costs include lost billable time and missed deadlines, both of which occur when CAD drafters can't use their computers or are spending hours installing or fixing software; suboptimal use of hardware or software; and the growing realization that somehow CAD isn't producing the results it should.

One way a firm can determine the break-even point is to ask staff to record on their time sheets the time they spend dealing with computer downtime or with tasks that a CAD manager would assume. If this information is recorded in the firm's accounting system through a separate nonbillable account, it becomes easy to determine at what point hiring a full- or part-time CAD manager is justifiable.

Hiring a CAD Manager

The CAD manager is a new position for many design firms and filling it is often challenging for several reasons.

- Most firms want a CAD manager who is expert in *both* computers and design, yet relatively few people are trained in both sets of skills. CAD managers working in design firms therefore tend to be either trained architects or interior designers who have picked up computer know-how on the job, or computer whizzes who don't necessarily understand the particular issues that the design profession works with.

- Compared to corporations or larger institutions, few design firms can offer competitive salaries, benefits, and job stability. Salaries for CAD managers in design firms, while often set at the middle of the range designers earn, are often no higher than entry-level information management salaries in industry.

- Design firms rarely offer a defined path of professional development for a CAD manager. Whereas many corporations recognize the importance of computer operations and offer career tracks leading to senior-level positions, many design firms offer little opportunity for promotion, except possibly to an operations management position, but rarely to associate or partner.

- Many design firms have only fairly basic technical needs and do not provide an ongoing set of new professional challenges; they rarely have the budget for research and development. Other industries tend to offer more interesting professional challenges for people with strong technical backgrounds.

- Partners and general managers, particularly those who don't use or understand CAD, often fail to give CAD managers the support—in terms of time, budget, staff, or respect—necessary to realize the firm's expectations for CAD. Thus many CAD managers feel that they are in the untenable position of having to produce results that are not possible with the scant resources they have been given.

For these reasons, the supply of skilled CAD managers tends to be outweighed by the demand. Therefore, firms that wish to hire someone who already has the requisite skills must be prepared to pay competitive salaries and provide opportunities for professional development. Otherwise finding a current employee with the technical skills and interest, encouraging him or her to assume the responsibilities of a CAD manager, and then providing financial and professional rewards are often a more effective approach.

To acquire and to keep a competent CAD manager, you must offer a well-defined job description, with the scope of responsibilities and required skills clearly spelled out. The job description should be part of the manager's personnel file and should be the criteria used for annual performance reviews. Many a person has been assigned the role of CAD manager without prior discussion or without a clear, written outline of job responsibilities; often the result is low employee morale, below-average performance, and high job turnover. CAD managers should also expect job-specific benefits, such as opportunities to attend conferences and training programs, ongoing technical education, and a budget for membership in professional organizations and subscriptions to professional periodicals.

Design firms that are fortunate to have a responsible and effective CAD manager rapidly find themselves dependent on that person. Should the CAD manager suddenly resign from the firm—or even go on vacation—staff may find themselves unable to carry out their CAD operations effectively. You need to protect your firm from the possibility of such a crisis by pursuing a twofold strategy: First, keep your CAD manager as satisfied as possible with his or her job. Although there is no way to guarantee that an employee will never quit, an employee with whom you have good relations is more likely to provide you with sufficient notice so that you can find a new manager before the current one leaves, or at least so that you can make temporary arrangements. Second, knowledge about your CAD system should be diffused throughout the office, not dwelling

Table 9.3 Checklist of a CAD Manager's Responsibilities and Qualifications

Responsibilities	*Qualifications*
Hardware Selection, Installation, and Maintenance	
research, recommend, and oversee purchase and installation of new computers and related peripherals	knowledge of IBM-compatible and/or Macintosh computers and ability to set up new machines and install operating systems
monitor hardware performance and attempt to diagnose problems; oversee hardware repairs and replacements	knowledge of and experience with major network systems (Novell, Lantastic, Ethernet, AppleTalk)
plan, oversee installation of, and maintain a computer network (if such is used in the office)	
Software—General	
research, recommend, and oversee the purchase and installation of new software and software upgrades	knowledge of typical software installation and configuration requirements, memory requirements, and general troubleshooting abilities
develop and maintain a system for naming computer files and directories	
develop and maintain routine procedures for file backup and archiving	
Software—AutoCAD	
oversee purchase and installation of AutoCAD on computers	experience in using and installing AutoCAD
customize each installation in accordance with office standards	ability to configure the AutoCAD *Config.sys* and other files
	ability to install and configure other Autodesk or third-party AutoCAD add-ons
Design Projects and Related Tasks	
set up CAD-based files for new projects	a basic understanding of architectural (or engineering) drawing conventions, techniques, and issues, both CAD and non-CAD
develop and maintain electronic libraries of CAD symbols and templates	
develop and maintain CAD-based officewide graphic standards	
research and develop technical solutions for CAD applications as they arise	
perform CAD-based design and drafting as required	
Training and Personnel	
participate in the hiring and training of CAD-skilled personnel	ability to work well with people
help project managers to schedule and staff CAD-based projects	good judge of personalities and skills
	good communications skills
provide training on an ongoing basis as required by project needs and other developments	patience and common sense
Budgeting, Planning, and Administration	
develop and monitor an annual budget for computer acquisitions, upgrades, training, and maintenance	ease of use with numbers; experience in the use of spreadsheets and database programs
research and recommend affordable options for hardware and software purchases	research abilities; contacts with computer vendors; familiarity with major types of hardware and software
maintain an inventory of computer software and hardware	ability to schedule and juggle multiple tasks and projects

entirely and exclusively in the CAD manager's head. Information about CAD operations should be documented in a CAD manual and dispersed through training sessions, memos, and other means of communications.

Mentoring and Apprenticing with AutoCAD

Traditionally, designers have sought to pass on their knowledge to younger members of their profession. Indeed, apprenticeship has a long tradition in architecture and is somewhat formalized in the requirements that enable young architects to become eligible to sit for their registration examination. (Interior designers have similar prerequisites for the National Council for Interior Design Qualification (NCIDQ) examination, but because of the inconsistent licensing situation for interior design in the United States, these standards are not so uniformly enforced.)

CAD presents new challenges for design professionals who wish to serve as mentors and trainers to younger, inexperienced designers. Most senior designers practicing today graduated from design school before CAD entered the curriculum, and many have been able to carry on professionally without having to learn CAD themselves. Yet in offices where CAD is the standard mode of production, these designers still must train and communicate with drafters who use CAD. So a good many firms find themselves in need of new methods of transferring knowledge from one professional generation to another.

Mentoring CAD users will be easier in the future, once more CAD-proficient designers take over the reins in design firms. In the interim, mentors must stretch their teaching abilities by trying to transfer information through a medium with which they are not entirely comfortable. Some suggestions for facilitating mentorship in the CAD era are listed here; note that these are really just electronic versions of good mentorship techniques that you may have applied in the context of manual drafting.

- Learn to redline in the CAD context; see Chapter 6.
- Require junior drafters to use CAD for *all* kinds of drawings—plans, elevations, details, and even specifications. Encourage them to put projects together on CAD.
- Don't just hand redlines to CAD drafters and let them enter the work into the computer passively. Rather, give them problems to solve on the computer. For example, have them design a staircase that meets parameters that you specify; then sit down together and analyze the results.

- Make sure that drafters aren't 100 percent dedicated to CAD. Get them out into the field, to client meetings, and into researching and writing product specifications and checking shop drawings. Make sure they get a fully rounded design experience (as both professional registration programs and many cooperative design schools require).

Finding the time to serve as a mentor is challenging these days, when both project fees and schedules limit the amount of time that you can expend on professional growth. But if designers don't make the effort to educate the newer members of the A/D professions, there will be no one to whom we can pass the professional torch. Use CAD to fulfill your professional responsibilities for educating future designers.

The CAD Standards Manual

One of the most important training and personnel tools is the office CAD standards manual. Such a document describes comprehensively your office's CAD configuration, standards, policies, and resources. The office CAD manual can take several forms:

- a completely separate book
- a section in your existing office graphic standards manual
- a part of your office personnel handbook (Even if you don't use this form, personnel handbooks should nonetheless describe your firm's policies for computer training, expectations for computer proficiency, and policies for using and maintaining computers and related equipment.)

The order and contents of CAD manuals vary considerably among design firms, but at minimum your manual should include these items:

- a table of contents
- guidelines for organizing and naming computer files and directories (see Chapter 10)
- A CAD layer names drawing matrix (see Chapter 7)
- printouts of prototypical CAD drawings
- printouts and explanations of CAD block (see Figure 4.18)
- printouts, or listings, of CAD-drawn detail (Many firms keep detail libraries separate from other libraries; even so, a listing in the CAD manual can be useful.)
- general CAD drafting guidelines

- plotting procedures and guidelines and default plotter pen menus (see Chapter 8)
- guidelines for preparing CAD files for issuance to consultants and clients
- guidelines for reading and writing files in *.dxf* and other file formats
- tips on using AutoCAD with other Windows software
- procedures for routine maintenance, including tape and disk backups, antivirus checks, and hard-drive optimization
- tips on managing memory and speeding up CAD drafting
- guidelines for archiving computer files (see Chapter 10)
- information on training options for CAD and on in-house reference manuals and other learning devices

Ideally, the CAD manual should be paginated and presented in such a format, for instance, in a three-ring binder, that you can easily make periodic additions and revisions (not unlike the binders that house building codes).

While the primary purpose of a CAD standards manual is, of course, to document and circulate information on the office CAD system, don't overlook the other benefits of creating such a manual: it can be a design opportunity and a desktop publishing exercise. A well-designed manual not only will be more effective as an internal personnel tool, it may also impress prospective employees and consultants and current or prospective clients. The mere presence and overall appearance of such a manual says a great deal about your firm's approach to CAD operations and quality control.

The CAD Technical Committee

Although paper is essential in order to communicate and document information about CAD standards and procedures, a verbal forum is also crucial in circulating guidelines on CAD use as well as collecting feedback. Design firms that are large enough to warrant establishment of committees of any sort often form a CAD technical committee.

CAD technical committees, which often meet over brown-bag lunches, typically are charged with developing and documenting the materials that constitute the office CAD standards manual. They also review feedback on existing CAD standards and procedures, determine ways to improve productivity, and develop proposals for acquisition of new hardware and software. CAD technical committees also may review training programs, organize office presentations of CAD projects, and circulate an office "technical" newsletter (which may be supplemented by e-mail announcements and memos). Such committees serve the critical role of dispersing CAD-related information throughout the office and act as a clearinghouse for staff views on how CAD is—and should be—used in the office.

David Packard, the cofounder of Hewlett-Packard (makers of a brand of AutoCAD workstations and pioneers in inkjet printing technology), writes in his memoir, **The HP Way: How Bill Hewlett and I Built Our Company** *(HarperBusiness, 1995), that much of his company's success can be attributed to an emphasis on* **verbal** *communications and instructions. Although Hewlett-Packard provided written guidelines for its operations, Packard found that employees tended to understand and follow instructions better if they heard, not merely read, about them and if verbal instructions were followed up by ongoing personal contact.*

Even design practices that are too small to organize a formal committee will find that scheduling regular times to meet and review CAD procedures is a helpful way to ensure that all staff are operating on the same wavelength where CAD is concerned.

CAD Ghettos and CAD Burnout

Despite the power and increasing affordability and ease of use of AutoCAD and other electronic drafting programs, many design firms have had limited success in integrating CAD into their overall operations. One major reason that AutoCAD has not lived up to its full potential is that many designers and drafters worry that they may be pegged as "CAD techies." And many design firms inadvertently justify this anxiety by segregating—both physically and functionally—CAD users from their design colleagues. This enforced isolation creates what is often called the "CAD Ghetto," and it can accelerate professional burnout.

Listed below are some symptoms to look for in diagnosing the emergence of a CAD ghetto and burnout, as well as some ways to combat these ills.

CAD Ghettos

CAD ghettos tend to emerge in firms where CAD drafters are physically or psychologically separated from the rest of the design team. Typical indicators of ghettos include

- a particular area of the office which is specifically designated for CAD drafting and is separate from other design and drafting functions.
- staff and managers who use terms such as "CAD operator" or "technician" in lieu of "drafter" or "designer" in general conversations, personnel reviews, and job descriptions.
- senior staff who fail to assign to CAD users any type of work other than CAD work or to promote staff with CAD skills to positions that do not explicitly require CAD knowledge.
- CAD users who are not invited to participate in project meetings, do not receive project memoranda, or are not included in other routine procedures.
- CAD users who are treated as *passive* operators, responsible only for inputting the designs and redlines of others, rather than as *proactive* team members who are encouraged to use the powers of CAD to develop and refine designs.
- designers, project managers, partners, and other non-CAD users who have little knowledge of computers.

If your firm has a CAD ghetto, here are some steps it can take to bring CAD users in from the cold.

- physically place CAD users with other designers and staff, within project pods, thereby disbursing them throughout the office (see Figure 5.5b).
- include CAD users in all phases of a project, from project start-up to job meetings and communications, client presentations, and site visits.
- schedule projects so that CAD users can also assume *non*-CAD responsibilities.
- require every designer to do his or her work on CAD.
- unless it is essential, avoid using the term "CAD" in job titles or descriptions for positions such as drafter or designer.
- put a computer on every desk. (For financial reasons, this is not always immediately feasible, but it should be a goal.)
- if your office operates a computer network, tie *every* computer into the network. This configuration will help disseminate computer skills and responsibility throughout the office and minimize the degree to which CAD users are expected to bear responsibility for computer operations.
- if your office does have a network, emphasis e-mail (both internal and external) as a way of managing officewide communications and

planning projects. Encourage staff to use modems to send faxes and distribute computer files and to use an electronic time sheet to record their hours. Get employees into the mode of thinking of computers as a natural and easy way to complete tasks.

- Provide CAD training for all the staff—at times that are convenient for them—and integrate CAD proficiency into the firm's policy on professional development and advancement.

- Maximize your computer system's user-friendliness and accessibility. Ideally, this involves installing AutoCAD on a Macintosh or Windows platform; if DOS is in use, purchase add-ons that make negotiating the computer easier. Greater ease of use will encourage non-CAD users to explore computers and expand their skills in this area.

The CAD ghetto concept applies not only to CAD users but also to CAD software itself. Firms that think of AutoCAD and other CAD software solely as a production tool and not as part of a family of interconnected computer tools are severely underutilizing their resources. What is so exciting about the Windows version of AutoCAD is that it promises to increase the ease of sharing data between AutoCAD and other applications.

CAD Burnout

Given the fast schedules and limited fees for design projects in the 1990s, professional burnout has become a strong possibility for all members of the design professions. CAD users are at particular risk for burnout, however, because CAD is one of the primary tools that enable design firms to keep apace with tighter schedules and more demanding clients. Whereas other staff may have responsibilities that enable them to take periodic respites from the computer, many CAD users are assigned to CAD tasks for forty or more hours per week. And the risk of burnout increases when CAD users work in CAD ghettos, as described above.

Some indicators of current or potential CAD burnout are

- loss of interest in, and lack of enthusiasm for, specific projects or design work in general.
- increased lateness to work or calling in sick more often.
- inability to meet project schedules and deadlines.
- declining work quality, as indicated by mistakes, omissions, and inconsistencies.

- unwillingness to assume additional responsibilities or to expand either CAD or non-CAD skills.
- withdrawal from the office social scene.
- outright refusal to use CAD.
- complaints about physical problems, such as pain in the back or arms, eyestrain, or headaches. (Such complaints may indicate an actual physical problem brought on by a preexisting medical condition, repetitive motion stress, poor HVAC or lighting, or some other reason, but staff who are happy in their work tend to complain less about, or even ignore, minor physical problems.)
- actual or threatened job resignation. (Burnout is often a factor when a person quits one job without first lining up another one.)

The following are some steps design firms can take to lessen CAD burnout in their firms:

- Eliminate any signs of a CAD ghetto, as described above.
- Set project fees and project schedules so that CAD users are under less duress and have time to diversify their responsibilities and skills (this may mean that you accept less competitive fees up front, but you will reduce the risk of project burnout at an inopportune time).
- Limit the number of hours (especially consecutive hours) that CAD users sit at CAD stations on a daily or weekly basis.
- Provide ergonomically sound, physically comfortable workstations for CAD users.
- Meet with CAD users on a regular basis to review their professional goals and interests, and take steps to help them meet their goals. Provide real opportunities for professional growth.

Routine Administration

Memory and Speed Management

For AutoCAD to be a responsive and effective design and drafting tool, it must run both quickly and reliably. Therefore, your CAD configuration must reflect an optimal blend of RAM, hard disk space, and coprocessor power. As Chapter 3 discusses, the *minimum* AutoCAD configuration already demands a powerful computer; the *optimal* configuration requires even more. Acquiring the necessary hardware, however, is only half the story. The other half is maintenance, consisting of routine computer housekeeping that must be done on an ongoing basis, whether daily, weekly, or monthly. If you don't pay constant attention to memory and speed management, your high-powered CAD station resembles a Triple Crown racehorse that is never exercised, fed, or groomed properly and thus cannot perform to its highest potential.

Memory and speed management involve several tiers of maintenance, all of which center on three goals:

1. to free up hard disk space,
2. to increase operating speed, and
3. to reduce the possibility of system conflicts, crashes, and corruption.

Some maintenance falls under the province of CAD or network managers, who are responsible for periodically reviewing the office's overall system requirements and performance; clearing the system of excess software, project files, and obsolete utilities; and "tweaking" the system's memory configuration as developments dictate. Various tools are available to help computer managers perform these tasks, particularly for networks. Officewide tape backup systems and diagnostic software, for example, can help optimize the network (see Chapters 3 and 10).

Other maintenance functions fall into the laps of individual computer users. Although some users may fall back on CAD managers or support staff when they encounter problems, they will have less need to do so if they are properly trained in basic memory management techniques. They should be expected to implement these techniques as part of their routine responsibilities. It is especially important that individual CAD users be responsible for site-specific hardware and software needs that cannot be diagnosed and addressed by an office network file server.

The following section contains a list of some memory maintenance routines for both Windows in general and AutoCAD Release 13 in particular. The value of these routines will be clear to any experienced AutoCAD user, so the rationale is not explained here. Project managers do not need to have a detailed understanding of why these techniques are important, but they should make sure that information about the techniques is circulated to all computer users; indeed, such information should be summarized and regularly updated in an office CAD manual, as described earlier in this chapter. Such documentation is especially useful as neither Microsoft Windows nor AutoCAD provide comprehensive information on memory and speed management in their documentation; much is often learned by word of mouth, in computer publications, or by experience. Therefore, CAD users in your office should be encouraged to share what tips they have with their colleagues.

System Memory

Management of your system's memory means optimizing your overall hard drive and the underlying operating system, for example, DOS and Windows. The condition of your hard drive affects both AutoCAD and any other software you run on a given machine. Note that in some cases, managing system memory may involve reconciling conflicting demands that various software applications and utilities are making on your hard drive. Some of the basic tasks include the following:

- Check the configuration of your DOS and Windows software (see Chapter 3). If you installed the operating software according to the software *defaults*, you may need to adjust the memory to reflect the actual needs of each application that will be run on the computer. If the computer is part of a network, then it needs to be configured in tandem with the network's requirements. Recommended frequency: every three months or more, depending on your system's level of change and activity.

- Partition your hard drives. This involves configuring your hard drive into different partitions, each of which is named differently. Different families of software applications and files are stored on different drive partitions (see Chapter 3).

- Run utility software to "optimize" your hard drive. Utilities, such as Norton Utilities by Symantec, search through your computer for bad files, suboptimally organized directories, fragmented drives; in essence they clean up after you. Recommended frequency: weekly or monthly.

- Run utility software to check for "bad" files, such as error files, viruses, and other problematic items that can at best slow down your system and at worst cause your computer to crash or lead to permanent corruption of your files. Recommended frequency: daily or weekly. Antivirus software should automatically check every new file that is copied to your computer hard drive.

- Minimize the number of applications and files that you keep open simultaneously. Although "multitasking" is a useful Windows feature, it drains the computer's memory and increases the potential for software conflicts.

- On peer-to-peer networks or networks where users can "log *on*" to different file servers at will, "log *off*" from file servers that are not needed.

- Keep all computer directories free of extraneous files. Such files include sample files, tutorials, unused font files, and production demonstrations. Remove unnecessary files from the hard drive. You can also remove application files and their directories for software programs that are no longer used or have been superseded by a newer version. Recommend frequency: daily or weekly for project files; monthly or quarterly for global overhaul or whenever a project is completed. Note that such purges not only free up hard drive space but also save space on backup tapes or disks. You may also want to invest in utilities that allow you to safely "deinstall" applications, or portions of them, without inadvertently removing essential operating files.

- Keep AutoCAD subdirectories free of extraneous files. These include both extraneous AutoCAD files, such as sample or tutorial files, as well as project files, which instead should be stored in project subdirectories (see Chapter 10). You can also delete plot files (*.plt*) for completed drawing plots; backup files (identified by the file extension *.bak*), which provide essential backup during a drawing session but are unnecessary once the drawing session is finished; and error files that are created during a "crash." Leaner AutoCAD subdirectories reduce the time AutoCAD needs to check its configuration each time it starts up.

- Archive completed projects promptly. Just as designers archive selected drawings then clear out drawing flat files once a project is completed, so they should archive CAD files for a completed project (see Chapter 10) and then remove project drawing files from computer drives once they're no longer needed.

AutoCAD Memory Management

In addition to keeping their hard drives free of excess files of any sort, AutoCAD users need to get in the habit of keeping their AutoCAD drawing files "lean and mean." The commands suggested below help speed up the drawing process by making better use of AutoCAD's resources, particularly by speeding up screen regenerations, and, in some cases, by reducing drawing file size, which in turn frees up hard drive space, which can then be used as swap file space.

- Routinely **Purge** drawing files of unnecessary definitions of blocks, layers, linetypes, and type styles.
- **Turn off** and **Freeze** unused layers.

- Keep drawing **Limits** as small as possible, ideally no larger than your drawing area.
- Use External References **(Xrefs)**.
- Create **Blocks** for any entity that occurs more than once in a drawing file.
- Turn on Quicktext **(Qtext)** when you don't need to read text blocks.
- Turn off the **Blipmode** option.
- Turn off the **Fill** option.
- Lower the **Viewres** (which dictates the resolution or fineness of objects on-screen).
- Use basic AutoCAD typeface fonts, such as standard or Roman, when using text during heavy-duty drafting sessions; save fancy outline fonts for printing or use External References to link the text that contains them.
- If you use TrueType fonts, set **Textfill** to off and select a low value for Text Quality **(Textqlty)** until you need to review the text prior to plotting.
- Minimize the use of **Hatches**, place them on a separate layer that can be frozen when unneeded, or else use **Xrefs**.
- Minimize the use of complex **Linetypes** and **Mlines**, except where needed.
- Work in 16 rather than 256 AutoCAD colors for routine "black-and-white" drafting and adjust your monitor accordingly.
- Use the **Appload** command to unload ADS and AutoLISP routines that you don't use.
- Delink connections to external databases when you don't need them.
- If you do create links to external databases, minimize the number of defined attributes and the amount of accompanying data in your drawing file; keep as much data as possible in the external database.
- Use preset **Views** to move around the drawing screen.
- Disable the **Use Menu in Header** option in the **Preferences Misc** dialog box. This will stop AutoCAD from loading a new *Menu* file when it opens a new drawing.
- Create custom **Menus** (see Chapter 4) for different types of drawings, including one for heavy-duty high-speed work; this latter menu should involve loading as few menu items as possible.

TIP Some of the techniques listed above can be imbedded in the defaults of your customized AutoCAD configuration as described in Chapter 4.

Controlled Obsolescence

The only certainty after God and taxes is that the computer you just purchased is (or shortly will be) technologically obsolete. The technology curve remains sharp, and the newest invention that you just read about in a publication is already being superseded. If you are determined about using computers aggressively, you are committing yourself to an ongoing investment in new machines and software. With the technology moving along as quickly as it has been in recent years, you will be lucky if you can get good use from a machine for three to five years.

Although computers rapidly become *technologically* obsolete, that does not mean that they are automatically *functionally* obsolete. With careful planning and purchasing, you can stave off the need to replace machines. Moreover, you can make the process of upgrading less costly and less painful.

Postponing Obsolescence

You can keep your machines productive for a good long time after you purchase them by having a deliberate and rational policy for making purchases, upgrading, and discontinuing the use of certain equipment or software.

Buy Good Quality Machines for Your Major Workstations

Computer firms come and go; when a manufacturer discontinues production of hardware, finding replacement parts and repair services can be difficult. Good quality machines purchased from manufacturers that will support their discontinued products may be less costly to maintain and may hold their value longer than top-of-the-line exciting new products whose manufacturer may go belly-up six months later. When considering a major hardware purchase, look at the manufacturer's financial statements as well as its product literature.

Buy Expandability

You may not need 64MB of RAM or a 1000MB disk drive *now*, but with the next release of AutoCAD or Microsoft Office, you soon might. It can be foolish to purchase excess capacity for a machine that you expect to discard within the next twelve months, but for longer-term investments, a reasonable amount of excess capacity can help extend a machine's utility for you now and its resale value later.

Note that expandability includes not only RAM and hard drive capacity but also the ability to add peripherals, memory boards, and any number of other devices.

Play "Musical Computers"

CAD, desktop publishing, and other graphics-related software, as well as heavy number-crunching programs, are memory hogs, and your fastest machine will soon grow sluggish when you use these programs. But routine spreadsheet and word-processing functions require much less power. So when it comes time to upgrade to new CAD stations, give the older machines to project managers, specifiers, and secretaries; in turn, their old machines can become designated for even less memory-intensive functions. Many firms, for example, use long-obsolete PCs solely to run plot files or a simple database program for their architectural library.

Track Depreciation Schedules and Salvage Values

Your accounting department should be able to track the depreciated value of your equipment and advise you on what the tax and cash flow consequences are of holding on to aging equipment. At every stage, accounting considerations should be discussed in the context of other decisions about retaining machines.

Choose Carefully between Purchasing, Renting, and Leasing Machines

Some equipment will become obsolete so quickly or will be so little used that it is not worth owning. Other equipment, such as heavily used PCs, you can justify purchasing. Making careful decisions about leasing, renting, and purchasing will save you money, especially when you consider financing costs.

Unloading Old Computers

Design firms that have lived through a cycle or two of upgrading computers can rapidly accumulate a computer "graveyard." This is a sad and unprofitable use of office space and resources. There are several options for getting rid of an old computer product.

Sell It

Even if it is five years old or more, a good machine may have some market value, and selling it may be preferable if you need the cash. Markets for secondhand equipment can be found in a number of places.

- Computer equipment vendors often take trade-ins and sell secondhand or reconditioned computer equipment.
- High-school and university students often are good markets for secondhand computers. They usually require a machine that has just enough power to produce term papers, and often they are unconcerned about long-term usefulness; they will probably sell the machine when they graduate.
- Local newspapers and other publications, such as various want-ad publications, offer space for computer advertisements.
- The Internet offers many electronic bulletin boards and sales exchanges for all kinds of products. One advantage of using the Internet is that users can address needs very precisely in minimal time; however, the various Internet exchanges are national and even international, so you may end up dealing with a prospective buyer on the opposite side of the country. Both the AIA and other design-oriented professional groups offer Internet forums (often local or regional) that may include listings of computer equipment; these forums can be especially useful since your professional peers are more likely than the general public to need CAD-related equipment.

One challenge in selling computer equipment, especially if you do so informally and some bargaining is involved, is determining a fair market value. You can collect information from pursuing the preceding options. You'll also find prices listed in a number of publications (as well as through the Internet), usually supplied by the United Computer Exchange (telephone: (800) 755-3033), which is a national clearinghouse for used microcomputer equipment; it functions much like a computer-oriented version of the New York Stock Exchange.

Be careful that the costs of trying to sell equipment—such as placing want ads or shipping the product cross-country—do not offset the value of your equipment to the point where it might be cheaper to give it away.

Sell It to Your Staff

Although computers in general are now quite affordable, a CAD station is still a significant investment for many designers; often even contemplating such a purchase is intimidating. You may find, therefore, that your own colleagues may be interested in buying a machine that they are familiar with, having worked with a similar setup. You will save time in marketing and dispensing of such equipment. Note, however, that you cannot legally include an AutoCAD license with the equipment unless you hand over that license yourself, in which case you forfeit your rights to the software.

Donate It

When a machine no longer has significant market value, or, from an accounting viewpoint, no salvage value, you may wish to donate it to a worthy institution. Many educational institutions and other nonprofits need secondhand equipment, provided that it runs fairly current, mainstream software and is in good condition. Contacting local institutions—perhaps some that you've done design work for or that you support in other ways—is a good way to find prospective donees. A number of organizations, such as the East-West Foundation (telephone: (617) 542-1234), act as clearinghouses for nonprofits that need computer donations; some also ship computer equipment to developing countries.

Give It to a Technical School

When a machine is so obsolete that it barely operates, it does not even constitute an acceptable charitable donation. Don't just throw it into a landfill, however. Give it to a local technical school or college that offers courses in electronics. The students can learn something by taking the electronic dinosaur apart.

Security

As computers proliferate in your office, they become more and more fundamental to your daily operations. As you become more dependent on them, you also become more vulnerable to any situation that prevents your computer from working. Some potential catastrophes include computer crashes, power outages, theft, flood, fire, or even vermin. When the worst happens, you must not only arrange to repair or replace damaged or missing equipment, you also have to find some way to ensure that you do not lose productivity because your staff could not access their normal production tools.

Computer-dependent offices need a systemwide strategy for protecting their computers from all possible causes of downtime. Such a strategy should address potential problems with your computers, but it should also apply to all your equipment and operating assets. Security involves different levels of protection from different types of damage or loss.

Physical Destruction

Physical destruction can occur when an office is subject to fire, floods, or deliberate physical mutilation, such as might occur during a robbery. Equipment can also be damaged if an office has inadequate HVAC and can't maintain proper temperature and humidity levels. Officewide measures are required to protect computer equipment from physical destruction or at least to limit its effects. Some precautions include

- having a security alarm system installed in the office.
- having a fire alert and fire extinguishing system, such as sprinklers, installed in the office.
- having upgraded, code-compliant electrical power as well as electrical surge protectors installed at each machine.
- having a properly maintained HVAC system.
- keeping equipment away from pipes, sprinkler heads, and direct sources of heat.

Theft

To a potential thief, PCs and small computer peripherals, such as printers or scanners, are probably the most desirable pieces of property that your firm owns; the hot-goods market for drafting boards and diazo machines, for example, is far less extensive.

The most important form of theft prevention starts at the exterior of your office. Prospective thieves are less likely to try to enter a building with good security and limited access. Having sturdy locks and even an alarm system in your office also helps. Another deterrent is making sure that if your office entrance contains any form of storefront, no temptations are visible from this vantage point.

Should thieves succeed in breaking into your office, you can still deter them from actually removing the machines. One strategy is to purchase security kits, which generally consist of cords that wrap around both the computer and a piece of nearby furniture and contain locks for holding the machine in place. Some computers have ports specifically for computer locks. Note, however, that locks may prevent a thief from running off with equipment, but they don't stop the thief from damaging the machine in the processing of trying to move it.

System Damage

Considerable damage to your computer system can occur not only from external physical activities but also from computer users themselves. Some of this may be completely unintentional, but some damage can be quite deliberate. Forms of computer system damage include

- faulty system configuration,
- improper file management procedures,
- poorly maintained computer drives,
- computer viruses, and
- deliberate erasing or corruption of computer files or directories.

With a well-configured operating system running on robust machines, you can provide protection against inadvertent or minor damage by installing utilities such as ones that can retrieve and repair "trashed" files. But you must also take conscious steps to protect against extensive deliberate damage. This includes limiting access to networks and to important directories and files by requiring log-in passwords. If you are on the Internet, you want to make sure that no external party can access your network without first being "cleared."

Under ordinary circumstances, a responsible, honest employee is unlikely to present a hazard to your system's integrity. Design firms by the very nature of their practice are prone to abrupt layoffs, however, and some employees may leave under less than cordial circumstances. Many employers have found that disgruntled employees can, at the very minimum, take with them software or project files that they don't own; at worst, they can deliberately set out to destroy a computer system as a parting gesture. If you must dismiss an employee, make sure that your computer system is protected from the consequences of negative feelings. Changing the network log-in password is a quick form of protection.

System Backups

As Chapter 10 discusses, computer system backups are essential in today's computer-based office. If done on a daily basis, backup routines enable you to reconstitute data that has been lost because of a computer crash, damage, or theft. Depending on your backup system, you may not be able to retrieve all your work, but you should be able to minimize your losses to no more than a day's or hour's worth of work.

Computer-savvy clients rely on sophisticated computer backup systems themselves, and nowadays they might have less sympathy with designers who failed to meet a project deadline because they did not implement routine backup procedures. Indeed, backing up is so important that contracts in many fields now require that it be done routinely.

Insurance

Since they are major assets, computers should be covered under your insurance policy. Be sure to provide your insurance agent with a current inventory of your computer assets. This inventory should be updated whenever you make a major acquisition, not when your policy is renewed. If you rent computer equipment, you will probably need coverage for that as well.

Discuss with your insurance agent what steps you can take to insure your computer system properly without paying more in premiums than necessary. You may be able to reduce your premiums by implementing special safety measures. Some carriers offer special coverage for computers that guarantees replacement based on current market value, not on historical book values. Although it is rarely offered, some insurers may sell special insurance to cover the time spent reconstituting lost data or renting computers while your existing ones are being repaired.

Space Planning for the Ergonomically Sound Office

When it comes to their own office spaces, many design firms resemble the proverbial cobbler's children, whose parents were too busy making shoes for others to provide proper footwear for their own children. Design firms are often so busy planning space for others that they forget their own workspaces. Yet properly designed CAD workstations are essential for the productivity and health of your employees.

CAD workstations require more than merely removing a parallel rule and making space for a computer terminal and monitor, keyboard tray, and mouse pad. Any workspace that houses a computer must offer appropriate seating and lighting as well as HVAC and noise control. Figure 9.4 shows some basic features of a proper CAD station. If your firm frequently

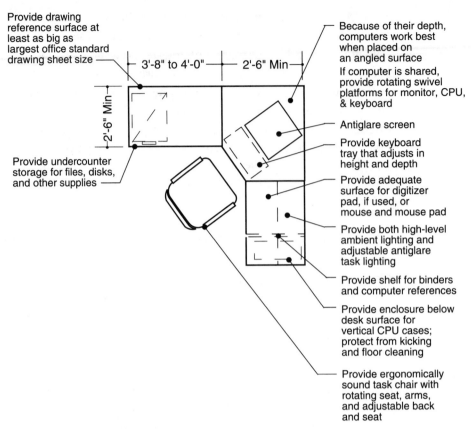

Figure 9.4 A plan layout of the basic components of a typical CAD workstation

specifies office systems furniture, you have immediate access to products and guidelines for such stations.

Although design offices are generally considered to be low risk in terms of worker safety issues, anyone who spends significant amounts of time in front of a computer terminal is vulnerable to physical injury. Potential injuries include carpal tunnel syndrome, eyestrain, and back problems. The prevalence of problems among VDT workers has sparked national concern, and in some communities, such as San Francisco, there have been attempts to mandate safety guidelines for VDT workers. Although there is no national legislation (yet) requiring that employers provide specific ergonomically sound workstations, a growing body of standards and guidelines from various organizations, including the American Industrial Hygiene Association, the American National Standards Institute (ANSI), and the National Safety Council, are now circulated, and these offer ammunition for any employee who experiences work-related physical problems. These standards are augmented by the guidelines in the Americans with Disabilities Act, which more and more states now incorporate into their regulations. This act requires employers of a certain size to provide accessible workspaces that conform to specific standards. While reconfiguring workstations is not an inexpensive project for design firms, it may be less expensive in the long run than paying out excessive amounts of workers' compensation.

Since the workstation has the most direct impact on an employee's physical well-being, it is rightfully the primary focus of office space planning. Careful attention should also be given to space planning for other pieces of equipment that constitute your CAD system or computer network. For plotters, for example, you need to provide sufficient floor space for the machine itself, but you also want to ensure that people using it don't impede general circulation paths. A storage area for plotter pens, ink, and paper should be located nearby. Adequate lighting, HVAC, and electrical and data power need to be specified. Similarly, accessible and sensible spaces for printers, scanners, and other devices are essential. For strategic equipment, such as a network file server or backup system, a secured area, such as a closet with a lockable door, is worth considering.

10 Managing AutoCAD Data

Architects and interior designers have always been data managers. Although they may envision themselves as designers, conceptualizers, artists, or craftsmen, in the routine scheme of things they have always had to manage data: data on clients and their needs; data on buildings, materials and construction, codes, and budgets; and data concerning other components of design documents. Moreover, any designer with responsibility for any aspect of office administration is involved in the circulation of paper and information through the office and to and from clients, consultants, vendors, and other external parties.

Computer-based data requires the same attention as paper-based data. The myriad forms of computerized information, however, force one to become more conscious of data management. Despite the vision of a paperless society, computers have simply allowed us to generate more and more paper, faster and faster, giving us more hard copy to manage alongside more data. Since no one can feel entirely comfortable being dependent on data stored on disks, tapes, and other electronic storage media, we are further compelled to supplement our computer-based databases with paper copies.

When this plethora of paper is not rationally managed, the costs of managing information can far exceed the benefits of generating it. Moreover, the data management issues for designers become more complex, as Figure 10.1 suggests. First, a single computer file may have the capacity to generate output in infinite forms; a typical AutoCAD drawing file, for example, can produce fifteen separate floor plans. Second, a single paper document may actually be a composite of files from several different computer applications; the ability to copy and paste between computer applications and to link multiple files to produce output is increasing. Since the relationship between electronic files and paper ones is not 1:1, the ability to track data, that is, to produce an electronic trail alongside a paper one, is essential.

The data management issues that computers raise have made data and document processing a major concern for computer-dependent organizations. A growing number of software packages are now available precisely for the purpose of managing computer files and documents. Indeed, Autodesk has recently been repositioning itself from a producer solely of CAD-related software to a maker of technical (or total) document

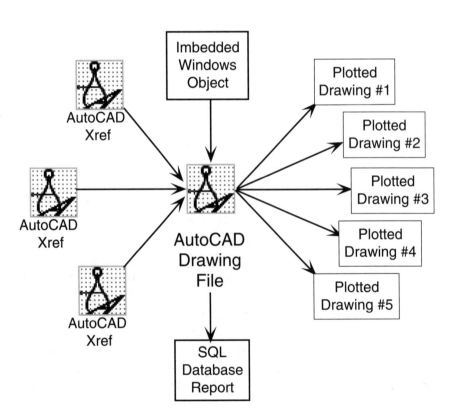

Figure 10.1
Advances in computer software increase options for computer output, but they also obscure the relationship between computer files and hard copy.

management (TDM) products, as evidenced by the introduction of its WorkCenter and View software.

The emergence of TDM software also reflects a growing body of government and corporate regulations and standards that seeks to bring some control and uniformity to the flow of both paper and electronic data. The federal Paperwork Reduction Act of 1995 attempts both to standardize and to reduce paper generation, and to rely on electronic communications where possible. International production standards for quality control such as ISO 9000 include specific guidelines and standards for document and data formats and circulation. Although many of these standards are more applicable to large corporations, industry, and government agencies, they have begun to affect design firms as well.

Trying to manage data can seem overwhelming at times, especially since technology is constantly evolving and most design firms are too small to be able to afford to hire staff who can focus solely on these issues. Data management is less onerous, however, if you keep in mind its overall purpose and benefits and establish a systematic approach. While a well-planned system may need to be updated every few years, it does not need to be "state-of-the-art" to control your data effectively.

As Figure 10.2 shows, you should think of data management as a systematic flow of information that reflects a particular principle of computing: that data should never be created from scratch more than once. You can copy, convert, link, and back up data, but if you have to continually retype or redraw the same work, you are not using, organizing, or protecting your data efficiently.

Data Organization: Directories and Files

The most important step in organizing data—for computer software in general and AutoCAD in particular—is to establish a computer-oriented filing system for your office. Your filing system should specify how computer files are named and where they are stored on individual computers and networks. A well-designed computer filing system saves you enormous time and headaches in storing and retrieving data, and reduces the likelihood that your work will be lost or damaged. Your computer filing system will also shape your procedures for sharing work, as well as doing backups, archives, and other routine data management functions. And effectively planned computer filing can reduce the costs of computer and paper filing and archiving.

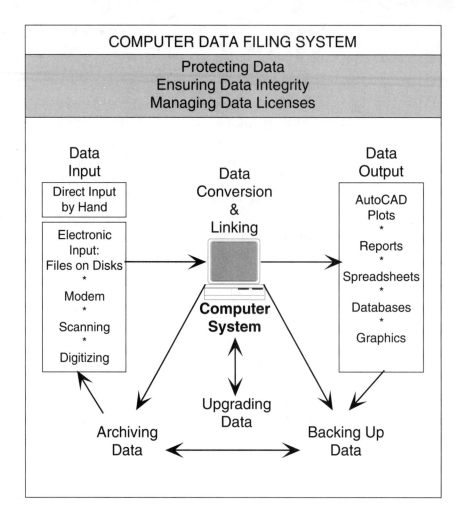

Figure 10.2
Effective computer data management produces a logical flow of information, which can be recycled as necessary.

In the early years of AutoCAD, when it was DOS-based, CAD data generally was filed according to its own separate criteria, without regard to how the project files produced by other software programs were stored. The emergence of sophisticated network software as well as user-friendly and flexible file-management systems such as Microsoft Windows File Manager and the Macintosh hierarchical file system has encouraged users to organize files by *projects* rather than by computer *applications*. Now, in many offices, you will find AutoCAD project files stored in project directories rather than in separate CAD directories. As both AutoCAD and other Windows software increasingly break down the barriers between applications and project files, and as data exchange becomes easier, organizing computer files by project rather than by application becomes inevitable.

As more and more of your work goes on computer, your computer filing system may end up mimicking your paper filing system. Indeed, many design firms use file directories and individual file names that directly parallel their paper-based filing system. When a computer file prints, it includes the computer filing address; if the file relates to the standard file-naming system, then filing the paper version becomes easier.

Directory Organization

The first step in establishing an AutoCAD filing system is to have a general policy regarding the location of different types of AutoCAD-based files. In general, AutoCAD *application* files and directories—of which there are many—should be separated from AutoCAD files *you* have created. In addition, you want to separate prototypical files, such as office-standard drawing sheets and symbol libraries, from project-specific files.

As Chapter 3 discusses, with a small CAD system on a single workstation or a small peer-to-peer network, you can use a single hard-drive designation to store all sets of files. For larger systems, especially with centralized network servers, you should consider partitioning your hard drive into several drives or server identities. DOS and Windows application files would be on one drive, AutoCAD application files on another, project directories on yet another.

Although the degree to which you assign files to directories and subdirectories will vary, you may want to adopt a general policy of trying to group files by like characteristics. And while you don't want too many subdirectories, you also don't want too many files under one directory umbrella, as this will make locating a specific file more time-consuming. The optimal file directory structure evolves as an office evolves, so it is not possible to recommend set guidelines, but you can derive clues by looking at how your hard copy files are organized in hanging folders and file cabinets.

File Access

To establish directory and file organization, you need to decide whether and to what extent you want to restrict access to particular files. Normally, you want employees to have access to their project files and to office templates. You should, however, restrict access to the AutoCAD application

Figure 10.3 *A typical approach to organizing directories for AutoCAD application files and project files. This example shows directory names restricted to eight digits, as Windows 3.1 DOS requires. With Windows 95, directory names can be much longer.*

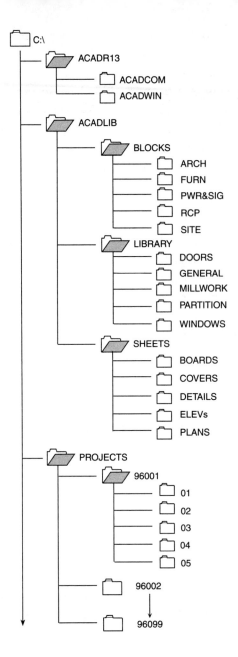

software and to any other files that will affect your default AutoCAD configuration (this holds true for any major application software on your system). In addition, you may want to specify prototypical drawings and any other office-standard templates as *read-only files*, so that users can view or copy these files but cannot change them.

382   MANAGING AUTOCAD DATA

AutoCAD enables you to protect files in several ways (see Chapter 4). As part of your configuration of AutoCAD you can require that users enter a log-in name to use specific files. You can also control whether users in a network can access files that are opened, that is, in use, by someone else. AutoCAD is not the only software that can determine who will have access, however. In Microsoft Windows File Manager you can specify file attributes that allow certain files to be read only or hidden. In the Macintosh 7.x operating system, you can lock a file through the **Get Info** box. Your options increase if you use network software, which can usually offer a wider range of security and access levels.

There are both practical and political aspects to who has access to computers. Computer network managers, for example, can exert enormous control in an office through their ability to control access to files and applications, even whether someone can log on to a computer. By the same token, if there is unrestricted access and as a result work is lost or damaged, tension may erupt among colleagues. Therefore, policies on accessing and restricting files should be established not by computer users alone but by upper management, and those policies should be established with an understanding of the *sociology* of data circulation in your office as well as the basic technical requirements of managing access.

AutoCAD File Names: Typical Approaches

A logical approach to file naming is essential to building an effective CAD system; a successful approach enables you to address one of the most challenging aspects of computer data management, which is simply to *locate* a particular computer file. In addition, an official file-naming policy saves users time thinking up file names while simultaneously limiting the possibility that each computer user will apply personal, often inconsistent, criteria to naming files.

Design firms have different approaches to file naming depending on their needs and on how they use AutoCAD. You should, however, keep the following considerations in mind:

- Since a CAD file can produce hard copy in infinite formats, CAD file names in and of themselves are not particularly meaningful. The file pathname *C:\Project\96001\Client \LT010596.doc* might be a fair indication that the file is a letter written on January 5, 1996, to the client contact for project 96001. But the AutoCAD file *C:\Project\96001\Plan-Flr1.dwg* can generate hundreds of different drawings. Having the file name print out on a CAD plot is only part of the story; you also have to know what layers, plot parameters, and pen settings were used to produce it. Thus, as stressed elsewhere in this manual, you can see the importance of documentation.

- Historically, names for DOS and Windows files have been limited in length to eight digits, followed by a "." and a three-digit file name extension. Spaces cannot be used. Windows 95 will permit much longer file names; Macintosh file names already can contain up to thirty-one characters, including spaces and all characters except a colon (AutoCAD for Macintosh file names are restricted to nineteen digits). Even if you operate solely on a Macintosh or Windows 95 platform, or perhaps on both platforms, most of the rest of the world still does not, and won't for a while. Therefore, for the foreseeable future, you may want to employ an eight-digit file-naming system. Even when you work with longer file names, a disciplined approach to naming files is still necessary.

- You can have two files that share the same file name, provided that they are stored in separate directories. But if you place them in the same directory, one will overwrite the other. Therefore, you cannot be cavalier about renaming, copying, and moving files, especially where networks and multiple users are concerned.

When your firm develops guidelines specifically for CAD file naming, your approach should be shaped by the overall officewide computer filing policy. Following are some sample approaches for naming CAD files.

File prefixes that begin with the project name. One advantage of this approach is that it is intuitive and usually easily recognizable. The disadvantage is that many clients and projects have names at least eight characters or more in length. In DOS and Windows 3.1 you will be forced to abbreviate the client names if you want to include other relevant information. And when acronyms are used, they often are used inconsistently so that duplicate directories and files appear. You might find, for example, that files for your client Gotham Bank appear under names ranging from "Gotham" to "GotBnk," and so forth. Therefore at the beginning of a project, set file-naming guidelines that clarify a standard approach.

Bear in mind that using client names can conflict with a client's right to privacy. You may wish—and in some cases you may be required—not to publicize that you are working on a top-secret project for a government agency or a new residence for a celebrity, for example. With the client-name filing approach, your computer system can reveal the nature of your clientele. For this reason, many plotting services do not like to receive client names with plot files or orders; they much prefer client project numbers.

File names that use the project number. Most firms use project numbers to identify project file directories and often project files themselves. Project numbers tend to be used more consistently than names, and often they require no more than four or five characters. As mentioned above, project numbers are more discreet than project names; consider this approach to file naming if confidentiality is a concern.

Many projects are split into subprojects. Although these subprojects may be given completely separate project numbers, many firms instead add on suffixes, such as 01, 02, and so forth, to the original project number, as shown in Figure 10.3. Your computer filing system can work with either approach.

File names that relate to drawing titles. Another approach is to relate drawing file names to the drawings they produce. For example, *Flrplan1.dwg* would indicate an AutoCAD drawing file that produces a plan of Floor 1. The drawback of this approach is that most AutoCAD files, particularly those for construction drawings, produce *multiple* drawings. It is generally more useful to use the drawing title approach for CAD files, such as details and elevations, which generally only produce one formal plotted drawing.

Basic numeral or alphabetized names. Some users simply apply sequential numbers or letters to AutoCAD files and then cross-reference these names with written drawing file logs. This approach only works if CAD users are meticulous in documenting their work. It is wise to add a prefix, such as a project number, for each project, so that file *1.dwg* for project A won't override *1.dwg* for project B if the two accidentally end up in the same file directory.

The ConDoc approach. The ConDoc system (referred to in Chapter 6) attempts to standardize all aspects of architectural drawing production, including the order and naming of drawing names and numbers (something contractors would appreciate). Adherents to the ConDoc approach can apply it to some extent to naming CAD files. The ConDoc approach is based on hard copy, however, and you have to remember that no direct 1:1 relationship between drawing names and CAD files necessarily exists. A typical CAD file-naming system based loosely on the ConDoc system is shown in Figure 10.4.

Figure 10.4
A hybrid approach to CAD file naming inspired by the ConDoc system

ABC Design Inc.
File-Naming Matrix - Project #9605

AutoCAD File Name	Drawing Number	Drawing Title	Script
9605A001	A-001	General Information	N/A
9605A002	A-002	Partition Types & Schedules	
9605A003	A-003	Doors, Frames, Hardware Schedules	N/A
9605PL1	D-101	Demolition Plan\Floor 1	Demo.Scr
9605PL2	D-102	Demolition Plan\Floor 2	Demo.Scr
9605PL1	A-101	Construction Plan\Floor 1	Const.Scr
9605PL2	A-102	Construction Plan\Floor 2	Const.Scr
9605A201	A-201	Building Envelope	N/A
9605PL1	A-301	Reflected Ceiling Plan\Floor 1	RCP.Scr
9605PL2	A-302	Reflected Ceiling Plan\Floor 2	RCP.Scr
9605PL1	A-401	Power & Signal Plan\Floor 1	P&S.Scr
9605PL2	A-402	Power & Signal Plan\Floor 2	P&S.Scr
9605PL1	A-501	Finish Plan\Floor 1	Fin.Scr
9605PL2	A-502	Finish Plan\Floor 2	Fin.Scr
9605PL1	A-601	Furniture Plan\Floor 1	Furn.Scr
9605PL2	A-602	Furniture Plan\Floor 2	Furn.Scr
9605A701	A-701	Detail Plans	
9605A702	A-702	Detail Plans	
9605A801	A-801	Elevations	
9605A901	A-901	Details	

A hybrid approach. Inevitably, the particular needs of individual firms result in customized file-naming methods that combine various approaches. An example can be found in the AIA CAD Layer Guidelines (see Chapter 7), which shows a case study where the drawing files naming system is based on a blend of A/E/C specialties, client number, and drawing names. For example, an architectural floor plan numbered A2 for client 9601 would be named *A9601-2.dwg*, an electrical lighting plan numbered E3 would be named *E9601-3.dwg*, and so forth. As with the ConDoc approach, a hybrid file-naming system, such as the above example uses, will not fully account for all drawings that a single CAD file can produce, but it can be a useful guideline, especially for multidisciplinary design/engineering firms.

External References

A growing number of design firms use AutoCAD's External Reference (**Xrefs**) feature to link different CAD files together to create drawings. Thus, what you might see on paper as a single floor plan may be composed of many different files for building exterior walls, columns, core, furniture, and so forth. Because AutoCAD will look for External References with specific names in a designated path, not only the names of linked CAD files but also their locations in particular subdirectories become critical.

AutoCAD Prototypical Drawings and Utilities

Although project drawing file naming is perhaps the most important aspect of file naming, the standard template files that comprise your customized AutoCAD system should also be named with care. Apart from retrieval and identification issues, many of these files may contain links to other AutoCAD files and are programmed to look for those files in a particular path. You might see an AutoLISP routine that inserts a particular library symbol (or block drawing). To work properly, the routine has to find that block in a particular file directory. If you move that block drawing to another directory or rename the directory, the AutoLISP routine will be unable to function unless it is programmed to give you an opportunity to direct it to the moved or renamed drawing.

AutoCAD Plot Files

Although you can plot or print AutoCAD drawings directly to an output device, many users find it faster and more efficient to write plot files first; those files can be stored in a plot file directory until they are ready to be plotted in-house or sent out to a plotting service.

Plot names should be named with care, because, as previously mentioned, a single CAD file can generate multiple plots. If you are plotting an entire set of construction plans from one CAD file, you might want to name each plot file in a way that reflects both the source of the CAD file and the expected output. Since most drawing plots have a drawing number, such as D-1 or A-1, you might use that number. A plot file for a demolition plan (Drawing D1) from CAD file *9601Fl1.dwg*, for example, could be named *9601-1D1.plt*. Using such nomenclature is helpful when you have to monitor the status of multiple files as they are being plotted.

A strategy for plot file names also proves critical if plot files are stored in a shared folder on a network file server. If everyone writes plot files with, say, the name *Plot1.plt* to this shared folder, users will end up writing over (in effect, erasing) one another's plot files. Every plot file, therefore, should have a distinct name.

Special AutoCAD File Managers

The complexities of CAD files combined with file names restricted to eight characters can make understanding the exact contents and status of a drawing file difficult. The file directory information provided in most computer operating systems generally tells you no more about a file than its name, byte size, and last date of modification. Yet it is often necessary to know when and why a file was created, who created it, the purposes for which it was last modified, and so on. It is also useful to know what the file contains graphically, yet opening up large CAD files simply to check their contents can be time-consuming.

With Release 13 for Windows, AutoCAD has addressed this last concern to some degree by providing a preview function in the **Select File** dialog box (Figure 10.5). In addition, Release 13 provides a **Browse/Search** capability, which enables you to search through a collection of bitmapped "snapshots" of all the CAD files in a file directory (Figure 10.6). Although these features can be helpful, you can only see one view of a CAD file. For large complex files, a single view is a weak indicator of a file's full contents.

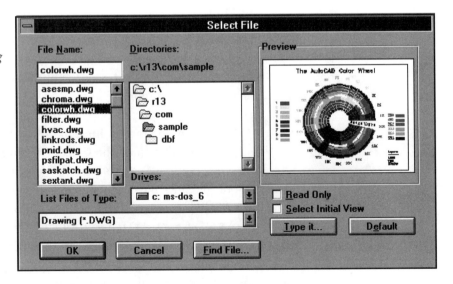

Figure 10.5
The AutoCAD **Select File** *dialog box showing a "preview" of a particular drawing file*

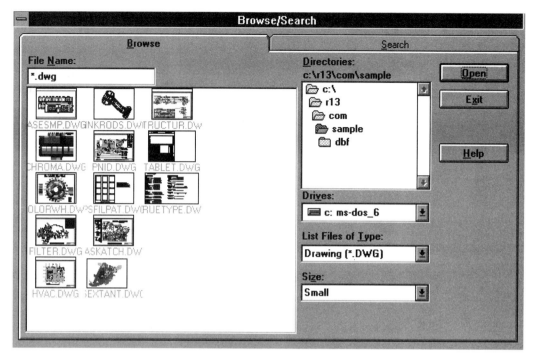

Figure 10.6 The AutoCAD *Browse/Search* dialog box showing "snapshots" of a collection of drawing files

In many offices, CAD users are required to document on paper what each CAD file contains and its current status. But ongoing maintenance of such logs is time-consuming and disruptive. Many firms often find it useful, therefore, to invest in third-party file management software—whether or not it is specifically designed for AutoCAD—to track the evolution of drawing files. Figure 10.7 shows a sample screen of Autodesk WorkCenter, a new electronic TDM product that manages all kinds of files, with particular attention to drawing-driven projects. File management software typically allows you to keep track of when a drawing was originally created, by whom, and for what purpose, as well as whether it was approved or intended for a particular deadline. Some of the data is automatically provided by the software; other information must be entered by a user. One advantage of software packages designed specifically for AutoCAD, whether by Autodesk or by third-party developers, is that they address functions particular to AutoCAD. For example, they can permit redlining and file reconciliation, and track External References **(Xrefs)** files. As projects grow in size and complexity in your office and become the products

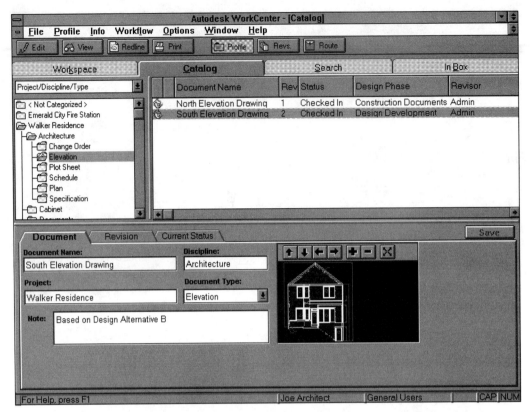

Figure 10.7 TDM *software, such as Autodesk WorkCenter, shown here, helps illuminate the contents and characteristics of CAD files without your having to open them or maintain extensive logs by hand.* Computer screen image provided courtesy of Autodesk, Inc.

of larger "workgroups," these electronic file managers are often the only way to leave a full "paper trail" of your CAD operations.

Data Input

Once data has been entered into a computer system, it is easy to use and reuse that data in an almost unlimited array of formats. Once it is input, the data should never have to be reconstituted manually from scratch; instead, it can merely be modified.

An increasing portion of the built and buildable environment has already been put into an electronic form that CAD users can access from many different software packages and operating platforms. Thus design firms can obtain base project information, such as geological site characteristics and building floor plates, in formats that AutoCAD can read, without your having to enter data manually.

Where CAD-ready information is not available, design firms are confronted with the choice of either having a drafter redraw the information in CAD from scratch, which can sometimes take hours, or finding less labor-intensive ways of getting information into the computer. Current technology offers several alternatives, none of which is free, but some of which may be cost-effective, depending on the situation.

Files on Disks

If you have the opportunity to acquire existing CAD files that can be read in AutoCAD, you can receive the files on several forms of media. Media technology is constantly evolving, and when exchanging data, each party must confirm what media can be used by the other. Storage technology reflects an ongoing trend of storing greater amounts of memory in ever smaller forms.

Diskettes

The traditional mode of computer file transfer within and among offices is the computer diskette. Earlier versions, often called floppies, were thin, flexible 5 1/4-inch squares. These were gradually replaced during the late 1980s by 3 1/2-inch disks that were stiffer and more resilient. Initial versions, called double-density disks, stored up to 800K; these then were gradually replaced by high-density disks, the current version of which contains 1.4MB of information. Specialty 3 1/2-inch optical disks that can store far more megabytes are available, but these are used primarily for file backup and archives rather than for routine data transfer.

To accommodate advances in disk technology during the past few years, most IBM-compatible personal computers ship with disk drive bays for both 3 1/2-inch and 5 1/4-inch diskettes, although drives for 5 1/4-inch diskettes are rapidly being phased out. Macintosh computers normally ship only with 3 1/2-inch disk drives. On older machines, not all 3 1/2-inch disk drives can read high-density disks.

Although the top-quality 3 1/2-inch diskettes are sturdy products, they should be kept away from magnets, and every office should maintain an inventory of cardboard diskette mailers, which protect disks in transit via mail or messenger.

CD-ROM

CD-ROMs (Compact Disc, Read-Only Memory) are flat disks that can store up to or even more than 650MB of information—more than four hundred times the data capacity of a high-density 3 1/2-inch disk. Given this capacity, CD-ROMs are the logical choice for memory-intensive storage needs, such as libraries of photographs and font files. They can also store encyclopedias, dictionaries, and thesauruses. CDs are increasingly used for electronic versions of manufacturers' catalogs; you can use CD versions of the *Sweets' Library Manufacturers* and the *AIA Master Systems*.

Just five years ago, CD-ROM drives were expensive, external, slow devices. Now you wouldn't think of purchasing a PC without at least one if not two CD drives. You also will find that many software manufacturers offer their products on CD; this medium saves enormous storage space and installation time. Some manufacturers, such as AutoCAD, will give you discounts if you accept CDs rather than diskettes; software producers find that CDs are cheaper to produce and distribute. And users find software installation faster and easier when only one CD, as opposed to ten or more diskettes, is needed.

Traditionally, CDs have been considered passive repositories of files. You could read and copy files *from* CDs onto your hard drive, but you could not write or copy files *to* them. Increasingly, writable CD drives (or CD-Rs) are now available, allowing you to copy files to CDs. Although still expensive, they enable you to store a great deal more information than you can on standard diskettes and tape drives. The CDs can be written on only once, but the information on them is claimed to be readable for a hundred years. Design firms are starting to use CDs to transfer CAD drawings to clients; to store multimedia presentations; and to archive complicated projects.

When selecting CD drives, a critical choice is the drive speed, which is usually expressed as a transfer rate measured in kilobytes per second. The top of the line used to be 300K/sec ; now 600 or 900K/sec is becoming the standard transfer rate (the rate may also be expressed in multiples, such as 3X or 6X speeds).

If you do invest in a CD-R, you will find that you have a number of options to review. In order to ensure that the CD-ROMs you write can be read by other parties, acquire a CD-R that records files in formats that adhere to internationally recognized standards such as ISO 9660. The CD-R should also be able to produce CDs that can be read by various operating systems (DOS, Windows 95, and Macintosh) and store the kinds of files (graphics, multimedia, audio clips) you envision writing to the CD. You can invest in one of a growing number of CD "authoring" or "premastering" software, such as Corel CD Creator by Corel, which help you organize your files, create CD directories and indexes, and maximize access to the data you've stored in the disk. Yet another feature to consider is whether your CD-R device and software will permit multisession recording or whether you only have one "shot" at making your disk.

 In computers, CD drives function not only as repositories for information but also as a stereo system component: they play music. If your computer has a sound board, you can present an on-screen AutoCAD slide show while using your favorite music CD as background music.

Tapes

Computer storage tapes are mainly used for performing regular backups and archiving computer files (as discussed further on in this chapter). Tapes are not normally used for transferring working files on a routine basis. One reason is that tape drives are not integral to personal computers; rather they tend to be peripheral devices, separately purchased, that can only be accessed using special software. They also tend to reflect proprietary tape sizes and formats. They can store large amounts of memory, however, from 90MB to over 1 GB. Unlike CDs, they can be written over more than once.

 Equipping every computer in your office with the most current file transfer technology is an expensive and time-consuming proposition. One cost-saving strategy is to designate one or two machines for file transfer; these would be equipped with all types of drives and would be used to transfer files, via the hard drive, from one type of disk media to another.

Modems

Modems are rapidly becoming essential in office computers, because they enable computers to quickly transmit information via telephone lines. Successful transmission via modem is faster than overnight express services or messengers. Modems enable different offices to exchange not only computer files but also electronic mail via the Internet and internal networks.

Design firms often use modems to send CAD plot files to external plotting services, thus eliminating a step (the messenger) and cutting down turnaround time. Many plotting services provide software to facilitate transmission and to generate an equivalent of an order form for documentation and billing.

Other uses of modems include distributing CAD files to engineers and other consultants and, still to a lesser extent, to clients. Design firms with offices in several cities find modems an effective way to share project work. Modems also enable firms to connect up with the "information superhighway" via on-line services such as America Online, CompuServe, and Prodigy. You can also obtain upgrades and updates from Autodesk by logging onto one of its several CompuServe forums. Some electronically aggressive design firms have already begun to market their services over the Internet.

Selecting a modem can seem intimidating because its features sound so technical. But for most design firms, it is essential that you understand only a few criteria in order to make a selection.

- The speed at which a modem transmits information. Often referred to as the baud rate, transfer speed is measured in bytes per second (BPS). Most plotting services prefer that files be transferred at the maximum baud rate available, but other parties may only be able to send or receive files at a lower baud rate. So aim for modems with the highest BPS available; the modem software should automatically adjust the modem to receive at lower speeds when necessary.

- The software that comes with the modem should enable you to transmit files as well as faxes and e-mail. Not all modem software permits file transfer, and making the financial stretch needed to purchase software that has file transfer capabilities is worthwhile.

- The modem should work with user-friendly modem telecommunications software that facilitates file transmission. Such software ideally will store frequently used modem telephone numbers, provide electronic transmittals, automatically redial numbers if the initial try didn't work, and notify you when another party has sent you e-mail.

- If your firm runs a computer network, you can purchase modem software that allows everyone to access the modem. Users can share the modem not only for transferring files but also as a direct, "paperless" fax machine. The more sophisticated software queues multiple faxes and redials numbers that were previously busy; this cuts down the line of people waiting to use the central office fax machine.

- Whether the modem software itself or your computer network provides it, you must have a means of controlling how external parties can access your office network through a modem. You certainly want off-site employees to be able to log on to your system, but you usually will want the opportunity to screen "strangers."

In offices where modems are *infrequently* used and/or used only at specific times, you can borrow an existing telephone line to transmit data; simply attach the modem to the phone line with a two-headed phone jack and consider investing in a line switching device. If you use the modem more frequently (at least daily), then you can justify installing a dedicated phone line for the modem.

Modems can be great time-savers when you need to distribute information, but, along with the fax, they increase the acceleration of work in the design office of the 1990s. The presence of a modem can increase expectations on the part of clients and consultants regarding turnaround time. Think carefully about who you inform about your telecommunications capabilities.

Computer Networks

Among a computer network's most valuable features (see Chapter 3) is its capacity to transfer files from one computer to another without users needing to resort to disks, CDs, or any other media of limited storage capacity. Given that typical AutoCAD drawing files can easily exceed the typical diskette capacity of 1.4 MB, and can grow to as large as 10 MB or 20 MB, networks are often the only quick and easy way to transfer drawing files. The office network is "wired" into the general public via modem as described above, then the need for diskettes and other transfer media drops significantly.

Communicating by modem now costs significantly less than using other forms of communications. While postage rates have increased over the past decade, the cost of electronic data transfer has dropped by 80 percent over the past few years.

Other Forms of Electronic Input

Data exchange via disks or networks is the most common way of transferring data. But often the information you want in a CAD file is not yet available in electronic form. You can pay to have someone input the data by hand, but for some applications, other methods are worth exploring.

Scanning

Scanners enable computer users to copy graphic images from paper into electronic form. These devices work by reading the images and re-creating them as computer files consisting of pixels. The higher the scanner's resolution, the more dots per inch the computer image will contain and the finer the resulting image will be (the resulting file will, of course, also be larger).

Participating in the Information Superhighway

These days, much of the media is crammed with news about the "information superhighway" or the Internet. Using a PC, a modem, and a telephone line, you can log on to any number of on-line services, such as CompuServe and America Online. Through these services, you can "download" computer files, send and receive e-mail, participate in forums, and tap into library catalogs and news services from around the world.

Design firms increasingly face strong professional incentives to participate in the Internet. Among them are the following benefits:

- You can transfer text and drawing files to engineers, clients, and reprographics services via the Internet, thus reducing turnaround time and saving time and money that would be spent using a messenger or overnight express service or the mail.

- With most e-mail services and fax modems, you can use a single command to "batch" a memo or file to multiple parties; this capability saves you enormous time handling communications for large project teams.

- You can communicate and share work with colleagues without having to be together in the same office. e-mail allows staff to work at home while tending an infant or being housebound because of bad weather or an infirmity; on a business trip, or at a construction site.

- You can reduce the inflow and outflow of paper, particularly for informal messages that don't warrant a paper trail.

- You can access general information services, such as libraries, newspaper clipping services, and research databases, without having to physically travel to the source or wait for a periodical to arrive in the mail.

- You can perform focused searches to find information without having to collect and sort through large amounts of extraneous information.

- You can communicate with anyone anywhere in the world without having to worry about different time zones or long-distance phone bills. In New York, you can e-mail a message to a client or colleague in Tokyo without worrying about establishing the proper time to do so. And your regular e-mail rates will absorb the costs of these international communications—contacting a colleague in Brazil costs the same as calling a consultant on the other side of your town.

- The American Institute of Architects (AIA) runs an on-line service called *AIA Online* ((800)864-7753) that focuses on the needs of designers and provides electronic access to a number of highly useful sources, such as the *Commerce Business Daily*, *Construction Market Data Early Planning Reports*, and manufacturers' product information. You can also participate in member roundtables, log on to professional directories, market your own skills and services, and obtain demonstration software for new computer programs and free computer "shareware." Several local AIA chapters sponsor on-line services as well.

Participating in the Information Superhighway (continued)

- A growing number of software manufacturers, including Autodesk and Microsoft, now provide technical support, product documentation, technical forums, and other services on-line. Because many software manufacturers find it cheaper to service customers through the Internet, they are encouraging its use. Consequently, you can download software upgrades, AutoCAD "patches," debugging utilities, and installation and instructional information, and even register your new computer products and upgrades via modem. It is often easier to reach manufacturers via e-mail than by endlessly dialing an 800 number, only to be placed on hold for half an hour or longer.

- Several design firms are starting to use the Internet as a marketing outlet, alongside standard approaches such as printed materials. Many publish their own "Web" page or post electronic "portfolios" in Autodesk's CompuServe Showcase Forum. It is too soon to tell if the Internet will become a lucrative source of clients and design commissions, but it is already an additional way for design firms to publicize what they do.

- Many designers and other prospective employees use the Internet, both to advertise their skills and to find job listings. *AIA Online*, as well as such as regional AIA affiliate forums as the Boston Society of Architect's ***BSA Architects Online***, serve as employment referrals and may allow participating forum members to attach resumes to their on-line "member" profiles.

- The federal government prefers it. The Clinton administration has promoted the information superhighway from the outset; through its internal directives and the passage of the Paperwork Reduction Act of 1995, the White House has encouraged the government to process information electronically, whether that means sending messages to the White House or filing taxes.

Although many aspects of the Internet offer enormous appeal, some design firms view the information superhighway with caution, and rightly so. Like all computer-generated products, e-mail inevitably has a downside if it is not implemented and managed properly.

- Internet costs can add up. Most services charge a flat monthly fee for a certain number of hours of basic services, and then a per-minute charge for additional time and for use of "premium" services. Costs can add up if use is not carefully tracked. However, Internet charges may be offset by savings in mailing, messengering, and other charges.

- E-mail, much like the fax, is yet another phenomenon that feeds into our national compulsion to accelerate all aspects of life, personal and professional. Why wait for mail or a personal delivery when it can be instantaneously "e-mailed"? Once you get into the Internet mode, you'll find it hard to slow down, and your clients and consultants will also speed up their expectations once they start to communicate with you this way.

> ### Participating in the Information Superhighway (continued)
>
> - Although Internet offers access to an overwhelming amount of data, not all of this data is useful, and much of it is still easily accessible in other formats, such as daily newspapers. If you fail to establish a criteria for what your office should use the Internet for, you could easily suffer from "information overload" and find yourself paying twice for the same data.
>
> - E-mail can be a distraction. Although it offers many professionally useful capabilities, e-mail also encourages fun, non-essential electronic bantering (or "electronic logorrhea") back and forth. You may think that your colleagues are using e-mail for serious project communications when, in fact, they may be participating in an e-mail forum that is not remotely work-related.
>
> - Inevitably, e-mail potentially brings on a range of problems, such as copyright, security, and privacy violations. Particularly when e-mail is used in the somewhat public environment of an office computer network, you should pay close attention to security and protocol issues.
>
> Like every other electronic invention, the Internet does have a potential downside, but if used carefully, it can enhance your office's operations. It can be particularly helpful for smaller design firms, which tend to have limited staff to handle communications and information services. e-mail can help take the load off receptionists, secretaries, librarians, and other support staff. Whether it can produce additional revenue is debatable, but used effectively, the Internet could *save* you money in the overall scheme of things.

Most software that works with small-scale scanners, such as Adobe Photoshop, saves scanned images in bitmapped file formats such as *.tiff* or *.pict*, which can be read by any software that supports these formats. Scanned images of this nature can be used in desktop publishing materials, marketing brochures, and programming reports. You can, for example, import construction and furniture product cut sheets for inclusion in a spec. For large sheets, such as D- or E-size drawings, you need access to a direct image scanner, which looks like a plotter but instead reads hard copy and translates it into computer files.

Scanned images normally become bitmapped files composed of color or black-and-white pixels. Although AutoCAD can read bitmapped scanned images, it only recognizes them as single graphic entities, not as a collection of editable layers, entities, colors, and other components. These images, therefore, cannot be manipulated as entities in a standard AutoCAD drawing file can. If you want to manipulate the images, you must first "vectorize" your scanned bitmapped images into entities that AutoCAD can work with. This process requires special software, which is available from many third-party developers.

Vectorizing is useful for providing manipulable CAD files, but it has its limitations, which every designer should be aware of before undertaking this process (or having it done by reprographics or digitizing services, as is the norm). Some of the limits of digitizing follow.

- Even though special software exists to convert raster images to vector files, the vector files still have to be converted into a format that your AutoCAD system can make sense of. Conversion software can't produce attributed blocks or place entities onto separate layers that correspond to your office-standard layering system. The software may define an entity as a line when you would prefer that it be a polyline; it can't make judgments. Vectorization is just one step in the conversion process; you still must budget time to convert the drawing file to a file your office can use.

- The quality of the vectored image reflects the quality of the original. If the original drawing is faded or incomplete; has creases, folds, or tears; or is itself a reproducible and therefore slightly distorted, the scanned image will mirror these flaws. Some "raster-to-vector" software does offer features that allow you to make adjustments for distortions and omissions, but your human judgment will still be necessary.

- Scanned images are inherently inaccurate. Even if you work from an accurately scaled paper drawing that is in perfect condition, you will encounter inaccuracies in your CAD files. One reason is that a hand-drawn pencil line representing a wall on a 1/8-inch scale plan only *approximates* the wall's location; written dimensions indicate where the wall actually goes. In CAD, you expect that wall's location to be exact, to a fraction of an inch. Digitizing cannot provide such accurate placement, and in this example there are several inches of latitude in terms of where the wall will be located. Therefore, a scanned and digitized image of a floor plan would be a reasonable general graphic facsimile of the floor's appearance but not a truly accurate representation, and as such the image would require considerable editing to make it a useful CAD file.

- Often you'll need to accept a trade-off between a vectored file's accuracy and its size. The better the resolution at which the file is scanned and vectored, in terms of greater dots per inch (dpi), the greater the quality of the resulting file. The standard is 200 to 300 dpi; 800 to 1000 dpi is the high-quality rate. While higher rates of dpi increase the quality of drawing files, they also produce much larger drawing files, which can be more unwieldy to manage.

- The process is expensive. Vectorizing architectural drawings requires either purchasing a large format scanner (still expensive, although prices are dropping) and software, or using a digitizing service, which charges expensive rates per hour or per scanned drawing. Although you may be able to negotiate bulk rates for large quantities of drawings, you can often expect to pay upwards of $100 for a 42" × 30" sheet to be screened, as well as charges for file conversion and for handling extra-large drawing files. If you then have to spend several hours of your own staff time reshaping the resulting file to conform to office CAD standards, the process can cost more than if you had drawn the information from scratch in AutoCAD.

Although scanning and vectorizing is still expensive and problematic, it does serve some important purposes. It is often the best way to replicate in CAD conditions that would otherwise be impossible to document accurately. Intricate stone or wood carvings on a building facade, a complex angled polygon floor print, and a corporate logo are difficult to measure and input from field work, but a scanned image will be a good approximation. As Chapter 7 discusses, scanning has useful applications in historic preservation. Scanning and vectorizing can also have archival and storage benefits. A number of facilities departments, for example, can justify spending thousands of dollars scanning their paper drawings into files because the cost of digitizing is less than the cost of storing them on-site. Even if the resulting images are not inherently accurate, you can retrieve them more easily via electronic TDM software (described earlier in this chapter) than if you had to pursue hundreds of tubes of full-size drawings. Digitizing also gives organizations a way to "back up" drawings that are prone to fading and other forms of physical damage and deterioration. Finally, even if they are not accurate, digitized images can be useful as CAD underlays to guide you while you trace over them or reconcile written plan dimensions with field verifications.

There are variants on digitizing that are more accurate than straightforward scanning and file conversion. As the section on historic preservation in Chapter 7 mentions, you can use desktop photogrammetry to create accurate CAD files from photographs. A range of site-surveying tools can help to replicate existing conditions. While these tools are accurate, to use them effectively you need some training, technical skills, professional judgment, and time. They are definitely worthwhile for long-term, complex projects where extreme accuracy is paramount and project fees and the time frame can incorporate the costs. For simple, fast-track jobs, however, direct input by hand may be faster and cheaper.

A number of third-party vendors have developed software specifically designed to convert scanned images to AutoCAD files. These are listed in *The AutoCAD Resource Guide*. Many offer a range of digitizing options, such as redlining features or the ability to produce hybrid raster/vector drawing files. Even if you leave digitizing to external services, this guide can help you determine whether a service uses digitizing software that will produce a file that meets your needs.

If you use a local reprographics firm to create digitized files, you may think your files are being digitized locally. But these days, there is a good chance that your files have been modemed to technical shops in another state or country as far off as India or China, where technical skills come at far cheaper rates. It is wise to find out in advance where the actual digitizing operation is, as this factor can affect turnaround times as well as output quality. Although technicians abroad may be very skilled, they are not necessarily able to make judgments about data conversion that reflect standards in the American design community.

Scanning drawings is just one of many ways to convert noncomputerized information into a computer file. You can use scanners not only for paper but also for photographic slides. A technology called Optical Character Recognition (OCR) converts text from paper and faxes to text files that can be edited in word processors. New and exciting technologies emerge constantly, and over time software and devices become more affordable and easier to use. Keeping up with all the developments is possible only for professional computer gurus. But bear in mind that if you want to enter something into your computer, you can probably find a technology to help you do so without re-creating the data from scratch.

Public Domain Data

A number of companies now compile and sell CAD-ready files containing plans of public domain sites and properties as well as graphic images. In addition, both state and federal agencies have compiled Geographic Information Systems (GIS) databases of government-owned public spaces and buildings. Some examples of public domain data are floor plates and elevations of major urban office buildings and suburban office parks, geological site surveys, and CD-ROM libraries of copyright-free details and images. A growing number of software companies provide electronic maps of the entire United States; refer to *The AutoCAD Resource Guide*. Note that in 1995 Autodesk, already heavily used in geographic information systems, announced the acquisition of Automated

Methods Ltd. This change should enable Autodesk to market its GIS products directly instead of merely supporting third-party GIS developers.

When contemplating the use of such sources, be sure to find out where the information comes from. What techniques were used to develop GIS maps? With CAD floor plans, for example, did the plate come from original architectural drawings, site surveys, and/or the property owner? Did the company field verify existing conditions and, if so, when? How accurate are the dimensions likely to be?

Using public domain data can save enormous time you would otherwise spend replicating information. If your client knows that you are using public domain data, you may avoid being liable for whatever defects or inaccuracies may result from that data. But, as always, you should at least do a cursory field walk-through to ascertain the accuracy of the files. You also want to check to make sure that the data truly is "royalty" free. Recently a number of legal disputes have arisen over the ownership of databases that document public space; ownership of the underlying data normally is not contested, but the ownership of the database itself may be.

Data Conversion

A growing concern for users of AutoCAD and other software packages is the ability not only to input data and receive output but also to share information *electronically* with other parties, regardless of which software each party uses. Electronic file sharing can save design firms tremendous time in obtaining and distributing information, and it also helps them dispense with unnecessary duplication of input.

AutoCAD files, however, cannot automatically be read by most other software, so designers must spend time (and often money) producing usable file conversions. Moreover, the ability to translate information in and out of CAD files is increasingly critical in landing and keeping design projects and in determining how much of the design fee must be spent on CAD-based activities. Many design firms have seen their project profits reduced because they underestimated the time required to convert computer files into a workable format. Other firms have failed to qualify for projects altogether because they could not guarantee electronic compatability. Therefore, understanding the issues inherent in CAD data conversion is essential for setting fees, marketing services, and scheduling projects.

As a design manager you do not need to master all the intricacies of file conversion, which in any case are constantly evolving. But when planning projects that involve data file conversion, you do need to address the following issues:

- What will you want to do with the information that will be shared? Do you need a single graphic image, an ASCII text file, or a CAD file with individual entities that can be modified?
- What file formats can AutoCAD and other software programs you own use to import and export data?
- How long will the entire conversion process take, including preparing the file before conversion and editing it afterward?
- Can you convert the files using your existing software, or do you need to purchase conversion utility software?

Converting AutoCAD Data: Major File Formats

To share AutoCAD-generated files between your computer system and another system requires having identical software or the ability to convert the files to a "universal" file format.

Every software application saves its files in a particular, often proprietary "normal" file format. In addition, almost all mainstream applications can "import" files from, and "export" files to, universally recognized file formats. Because of the importance of file exchange, a number of national and international organizations have sought to develop international standards to which software applications can conform. Autodesk participates in some of these efforts and has been particularly involved with the promotion of *.iges* standards (discussed later in this chapter).

Universal file formats vary in terms of their flexibility, robustness, and reliability. Some formats can work with almost any application; others are of very limited usefulness. File formats also evolve in response to software developments, so you sometimes have to specify a particular version of a file format. To import a graphics file, for example, you may have to specify whether the file is saved in the *.tiff* file format and also whether it is *.tiff* version 4, 5, or 6. Moreover, many file formats offer a number of conversion options, requiring that you choose between black and white or color, for example . You need to understand the implications of using one option over another.

The AutoCAD Drawing Format—*.dwg*

AutoCAD creates files in its own drawing file format, defined by a *.dwg* extension in the file name (as in *Plan1.dwg*). All AutoCAD files must contain this extension in their file names in order to be read by AutoCAD application software. The *.dwg* extension also helps you distinguish drawing files from other sorts of files, such as word-processing documents (with *.txt* or *.doc* extensions) or spreadsheets that might otherwise have identical file names.

In shaping AutoCAD for various computer platforms, Autodesk has traditionally emphasized the ability of an AutoCAD drawing file to move between different computer operating systems without any need to alter the file. Thus you can open a Macintosh-created AutoCAD drawing file on a Windows-based machine without having to modify it. However, even working within AutoCAD, you should be aware of some limiting factors when exchanging files across different platforms or among different versions on the same platform.

- As with most software, AutoCAD drawing files are 100 percent compatible *upward* but not *downward*. AutoCAD Release 13 will keep everything in a file created by Release 12, but Release 12 will not recognize features unique to Release 13. For example, Release 12 will ignore Release 13 features, such as Groups or Multiline styles (AutoCAD Release 13 will display a text window informing you of features that will be lost when a file is converted to the Release 12 format).

- With Release 13, AutoCAD for Windows has embraced more Windows-specific features, such as editable text boxes and OLE objects. These items will be ignored by the DOS version of AutoCAD Release 13.

- AutoCAD Release 13 can read AutoCAD files created in earlier AutoCAD releases. Through the **Saveasr12** command, Release 13 files can be saved in Release 12 format (with the limitations described above), and this format can be read in both Release 12 and Release 11. You can also use Release 13 to edit a drawing file created in Release 11 or Release 12 and then save it again using the **Saveasr12** command. If you need to open up a Release 13 drawing file in Release 10 or earlier, however, you have to save your drawing file as a Release 12–compatible *.dwg* file and then use an external AutoCAD utility called Dxfix to convert the drawing to formats that earlier AutoCAD versions can read. There are a number of third-party utilities available that can perform this process more easily and more reliably.

- When drawing files are moved from one computer to another computer that has a different AutoCAD configuration installed on it, unique features of the original drawing environment may be lost. On the original computer modifications may have been made to AutoCAD software files, such as the *.cfg*, *.pcp*, or *.mnu* files, as well as to special font files, External References **(Xrefs),** and plotter configurations, but these settings will not be picked up in other machines unless they are copied over. When the drawing file is opened on a new machine, AutoCAD will highlight most lost features with special "alert" dialog boxes.

Because of the predominance of AutoCAD in all CAD markets, a growing number of other CAD drafting packages now read files with a .*dwg* extension. This feature eliminates the need to convert AutoCAD files into .*dxf* or other interim file formats although, as with different AutoCAD platforms, some data may be lost.

The Drawing Interchange Format—.*dxf*

The .*dxf*, or Drawing Interchange Format, is an ASCII text file format that many other software packages (not just CAD programs) can read. Files with a .*dxf* extension are then converted into the particular file format of the other software package you want to use. Developed by Autodesk, .*dxf*, thanks to AutoCAD's popularity, is now a dominant drawing file-exchange format for CAD drafting packages (as well as other graphics software) used in the design professions; indeed, the ability of other CAD products to read and write .*dxf* files affects their potential usability and competitiveness. Using the .*dxf* format can also affect *your* marketability, as it enables you to share information with clients and other parties regardless of what computer software you use. Indeed, in their request for proposals, a growing number of potential clients require that at the end of a project the design firm provide a .*dxf* version of the CAD files used to generate a drawing.

To convert a .*dxf* file into an AutoCAD .*dwg* file, or to do the reverse, requires using the AutoCAD commands **Dxfin** and **Dxfout**. For most ordinary drawings files, the .*dxf* conversion process itself takes at most only several minutes. The process may take longer if the program that created the .*dxf* file also imported data attributes.

What can be time-consuming is not the .*dxf* conversion process itself but rather the work required to edit files at either end of the process. Many design firms have been surprised to see their fees eaten away and deadlines missed because they underestimated the time needed to convert files created from the .*dxf* file compression process to workable AutoCAD files. Indeed, here is a good rule of thumb: Budget one hour, minimum, for converting a .*dxf* file to AutoCAD.

Designers must also recognize that the .*dxf* file conversion process is just one step in the total conversion process; even though your software can read the file, it doesn't make the file usable in the context of your AutoCAD system. A typical *.dxf* file from an external source, for example, will load with layer characters, block definitions, graphic conventions, and typefaces that may be completely different from your office standards. Information on drawing units and page layouts may be lost. Moreover, although the *.dxf* conversion process is fairly reliable, it cannot completely reconcile differences between AutoCAD and other CAD software. Therefore, experienced CAD users normally expect to have to spend time addressing issues such as the following:

- There is a possibility that part of the file may be corrupted in the process of importing drawing files. You can configure AutoCAD to perform an automatic **Audit** of errors in new drawing files that have been created by the **Dxfin** command. An audit detects and can correct errors, saving you hours of time and frustration.

Apart from errors identified by a .dxf audit report, other signs of potential difficulties include the presence of layers that can't be renamed or unlocked and the inability to purge definitions for layers, blocks, and other customizable features that can't be found in the drawing.

- The *.dxf* files can lose definitions for layers, blocks and block attributes, custom linetypes, text styles, and dimension styles.

- The *.dxf* files do not necessarily retain information on drawing units and scales. Moreover, AutoCAD draws in "true" scale; many other CAD packages draw in "rulers" and generate *.dxf* files that have model-space entities drawn at a fraction of the "true" scale. A floor plan drawn at 1/8" scale in another CAD package, for example, may come into an AutoCAD file at 1/96 of its "true" size and will have to be scaled up to work in AutoCAD model scale.

- AutoCAD is an extremely accurate software program that takes measurements to one millionth of a unit. Some less powerful CAD programs are not as accurate, so after conversion subtle distortions can crop up in the dimensions of entities, and small "rounding" errors can accumulate into considerable distortions. These potential problems can be identified by using the **List** or **Dist** commands to measure items, such as the distance between column grids or a typical door swing, whose dimensions you are certain of. Address these distortions immediately (usually by carefully rescaling the drawing) or they can cause problems further along in the project.

- As described in Chapter 7, AutoCAD specifies layer colors and line weights differently from the way many other CAD packages do. With AutoCAD, layer color and linetypes can bear an important relationship to on-screen entity characteristics and plotted pen weights; in many other CAD packages, color is merely a way to distinguish objects on-screen and has little impact on plotted output. Thus, many *.dxf* files import into AutoCAD with color assignments for layers and/or entities that in no way reflect the color or line weights at which they should plot. Therefore, considerable time may be spent relating another system's colors and pen weights to yours for plotting purposes. To reduce the time spent on this task, request information on pen weights, layer standards, and colors from the person providing the original *.dxf* file. If you repeatedly work with *.dxf* files from a particular source, you can automate the conversion process by writing AutoLISP routines or Script files.

- As many other CAD packages, AutoCAD supports its own unique typefaces. These typefaces will be recognized in *.dxf* files as text, but special text attributes, such as type style, height, width, and alignment may be lost, and the text will have to be reformatted.

- During the conversion process, *.dxf* files will not read data attributes unless they are attached to defined blocks. Other CAD packages often store data differently, and they may not export or read attributed blocks.

- AutoCAD defines shapes differently than do other programs. What it defines as an arc, circle, or ellipse may import into other programs as a set of polylines, for example.

- The *.dxf* file format does not support many elements found in paint or desktop publishing programs, including paint and picture objects, pen and fill patterns, grayed layers, and pen sizes that vary in height and width.

- The *.dxf* file format is evolving over time to reflect new AutoCAD features. If another CAD software package's *.dxf* format relies on an earlier version of AutoCAD, the conversion process may produce files that won't work well in Release 13. Alternatively, a *.dxf* file created from AutoCAD Release 13 may have trouble translating smoothly into another program which reads only earlier versions of *.dxf* files.

In addition to understanding the quirks involved in the *.dxf* process, designers need to understand the different options for creating *.dxf* files which both AutoCAD and other software provide. Most AutoCAD users resort to AutoCAD's default options, which are to write a *.dxf* file to six decimal points of accuracy and to select all entities. However, you may elect not to **Dxfout** the entire file or to change the level of precision; alternatively, your clients may require *.dxf* files be written using particular options. It saves time if you determine this in advance.

The Initial Graphics Exchange Specification Format—*.iges*

The *.iges* format is an ANSI-standard file exchange format developed by a committee of interested parties in government and industry who sought a universal exchange format for CAD programs. Like the *.dxf* format, the *.iges* format translates all aspects of a drawing file into a text file, but since it was developed by an independent group rather than by Autodesk, it handles non-AutoCAD entities as well. The format tends to convert information "asymmetrically," meaning that if you import an *.iges* file that you just exported, the information would come back slightly differently. This factor can cause problems when precise dimensioning is required. Also the *.iges* format does not always recognize entities that are AutoCAD-specific.

Prior to Release 13, AutoCAD included in its base software the commands **Igesin** and **Igesout** for importing and exporting *.iges* files. With Release 13, you must purchase a separate translator package from Auto-CAD called the IGES Translator V 5.1.

The *.iges* format has never been adopted by the A/D community to the degree that the *.dxf* format has, partly because most architecturally oriented CAD software packages include *.dxf* file translators but rarely *.iges* ones. In addition, *.iges* files tend to be larger and therefore more time-consuming to use than *.dxf* files, and early versions of *.iges* have flaws. If you exchange files with government agencies, manufacturing and industrial designers, or foreign organizations, however, you may be required to use the *.iges* file format. Note that even within the *.iges* format there are different versions and formatting options, as well as foreign-language versions, access to which can help you to comply with different international requirements. The U.S. Department of Defense, for example, requires that its contractors comply with its Computer-Aided Acquisition Support (CALS) standard, to which *.iges* can conform.

Other Common Exchange Formats

The *.dxf* and *.iges* file formats are used when a file needs to be shared between two different CAD programs or operating platforms and when information on layers, blocks, linetypes, and other base file information must be retained so that the user can apply standard CAD drafting tools. However, sometimes only the *graphic image* needs to be exchanged.

Other formats generally are used for desktop publishing and graphic presentations. Typical applications include using a client logo on presentation boards and program reports, reproducing an office floor plan in slide presentations or a magazine article, and adding graphics to headers for spreadsheet and database printouts. Using these file formats enables you to share a single image among many different applications, helps build consistency in your graphic output, and reduces your dependency on other sources of reproduction, such as photography, scanning, and photocopying, by allowing you to import images.

Along with graphic file formats, another common file-exchange format is the text (*.txt* or ASCII) file format. Almost any application can read text files, so you can use the text file format, for example, to open up word-processed specifications in an AutoCAD file or to edit an AutoLISP routine or a database file in a word-processing program. Depending on your word processor's capabilities, *.txt* files may import as unformatted lines of text, with or without page breaks. Import options such as the rich-text format (*.rtf*), however, allow you to retain the original file's tabs, pagination, and other formatting information.

If you are working primarily in Windows or Macintosh applications, you'll generally find that the Clipboard utility, which both operating systems support, is useful for copying graphics and small amounts of text between different applications as shown in Figure 10.9 (although the AutoCAD for

Figure 10.8
The AutoCAD **Export Data** *command (shown here) and* **Import Data** *command allow you to share AutoCAD-generated data with other applications.*

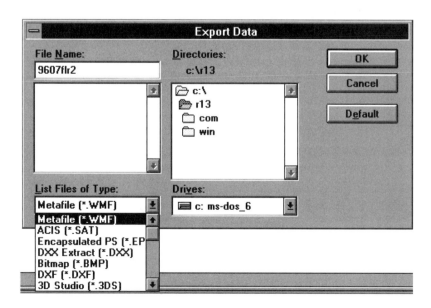

DATA CONVERSION 409

Figure 10.9
The Windows Clipboard Viewer allows you to see what entities you are "clipping" from one Windows application to another. The AutoCAD-drawn panel workstations, shown here in plan view, could be "pasted" into a word-processing document on client furniture standards.

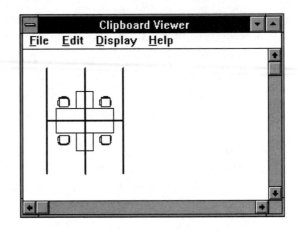

Macintosh clipboard works only in AutoCAD). The Clipboard utility creates *.wmf* files in Windows and *.pict* files in Macintosh; most applications on their respective platforms support these file formats. Saving files in the file format of the other application is preferable to using the Clipboard to copy text when the entire document, including formatting and underlying file features, needs to be transferred to the other computer application's standard file format (see Figure 10.8).

File-Exchange Documentation

During the past five years, vast improvements have been made in the options and the reliability of file exchange formats. Yet both computer applications and the file-exchange formats they support are so dynamic that you must also budget time to test different import and export options between any two applications. The knowledge gleaned from such tests should be shared with the rest of your office so that the learning curve for future file exchange is reduced. One way to share file-exchange format information is to present it as a matrix. Figure 10.10 summarizes the file formats that your major applications support; there is space for adding annotations about your experience applying these file formats in various applications. If you regularly export and import files from a client or consultant who uses a particular file exchange format, include information on that process in your matrix as well. You can also write macros or Script files—in AutoCAD as well as in other applications—to automate routine file exchange. Print-outs of these text files can be included in your documentation as well.

ABC Design Inc.
File Exchange Format Matrix

CODE:
√ = Import/Export
X = Export Only
M = Import Only

File Format	AutoCAD Release 13	Microsoft Word 6.0	Microsoft Excel 5.0	Microsoft Powerpoint 4.0	Microsoft Access 2.0	Adobe Pagemaker 5.0	Adobe Photoshop 3.0	Autodesk 3D Studio		Comments
3D Studio (.3DS)	√	√						√		
AutoCAD .DWG	√									
AutoCAD .DXF	√									
BMP	X									
EPS	X	√								
GIF	M									
IGES	√									Requires special IGES translator
PCT	M	√								
RTF (Rich Text Format)										
SYLK			√							
Text Only w/Line Breaks	M	√								
Text w/Layout										
Tiff	M	M								
Windows Metafile .WMF	√	√								

Figure 10.10 A file format matrix helps you document reliable file exchange procedures between computer applications.

Data Linking and Embedding

To exchange data, the traditional approach has been to convert computer files to other file formats. With Windows and Macintosh, however, the wave of the future may be linking files rather than converting and exchanging data. With features such as Windows Object Linking and Embedding (OLE), file conversion may eventually become a thing of the past. Users will be able to share information without regard to specific computer applications or file formats.

OLE enables users to share objects, whether text, graphics, or even sounds, between documents, regardless of the objects' origin (see Figure 10.11). You can, for example, link or embed a portion of a Microsoft Excel spreadsheet into a Microsoft Word document or an AutoCAD-generated drawing into a desktop publishing document.

Under OLE, the file that produces the object to be shared is called the *server*. The files that receive the object are called *clients*. Conceptually, OLE seems similar to the Copy, Cut, Paste approach of the Windows and Macintosh Clipboard and are intended to be used with equal ease. But OLE has much more power than the Clipboard utility. With ordinary Clipboard actions, no lasting relationship is built between applications; with OLE, you can create lasting links between software applications; this capability has important implications for document coordination and updating.

OLE actually consists of two processes: linking and embedding. Accessed in AutoCAD through the **Olelinks** command, *object linking* establishes an ongoing connection between a server and a client. When a linked object is changed in the server, any client document that contains that linked object reflects the change when it is opened. Linking is particularly useful when two files are used to create a document and both are being actively edited; linking offers built-in coordination.

With *object embedding*, the client does not automatically reflect changes that are made to the server. But if you embed an object from another application in the client, you can—in some Windows applications—access the server application by selecting the object in the client file and then editing the object in the original server application as necessary. This procedure spares you from having to open up the server application separately, or even from having to determine what the application is.

Although OLE is of growing interest to Windows application developers, it still needs refinement, and it is not yet available through all Windows

Figure 10.11
Depending on what software you use, you can use object linking to paste in a range of object types, ranging from graphics to sound files as shown here, into AutoCAD.

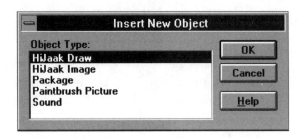

Figure 10.12
OLE objects need not be confined to graphics or text: you can paste in sounds as well. In this illustration, tapping the small microphone icon in the upper left side of the AutoCAD screen opens a Sound Recorder file that you can edit and play.

applications. OLE also is not always confined to object sharing: in some applications, selecting an embedded object will cause a video to run or a sound file to play.

Under Release 13, AutoCAD files can be both OLE clients and servers with other Windows applications. When two AutoCAD files need to be linked, however, using the AutoCAD **Xref** command, as described in Chapters 5 and 7, is preferable. In addition, OLE objects from other applications paste into AutoCAD as grouped objects, which can be scaled and rotated but not fully edited as conventional AutoCAD entities can be. Text will import, but fills and patterns will be lost.

OLE is analogous to the *Publish and Subscribe* feature in Macintosh System 7.x. And the concept of Windows servers and clients is reflected in various terms used in other applications and computer documentation as summarized below.

Data Object Producers	Data Object Users
Server	Client
Source	Destination
Publisher	Subscriber

OLE offers great potential for document exchange, coordination, and consistency. It does require careful organization of files, however, since linked objects build their relationships through specific directory paths. In addition, OLE has the potential to obscure the original source application of embedded objects if objects are merely "embedded." Careful file management is essential in order to keep track of the relationship between client and server.

Upgrading Data

Design firms that are committed to using AutoCAD—or any other software application—for the long haul are also committing themselves to upgrading their computer system on a regular basis. Most major manufacturers of software, including Autodesk, issue upgrades of their most popular software every few years. Companies do this not only to be responsive to user needs and to stay competitive but also to keep up with developments in hardware and operating systems that require adjustments to their software code. Design firms with a serious commitment to Windows and AutoCAD must plan on upgrading at least some portion of their computer system every two or three years.

Upgrades may be inevitable, but usually you have only one reason to rush out and purchase an upgrade immediately upon issuance. That reason is financial. Many software manufacturers offer discounts if you purchase their upgrade within a specific time frame. With the issuing of AutoCAD Release 13, for example, Autodesk offered users a $100 discount if they purchased the CD-ROM version of the Release 13 upgrade before January 31, 1995. If you definitely plan to upgrade within the near future, that savings may be worthwhile, especially if you have multiple AutoCAD licenses.

Purchasing an upgrade, however, does not mean that you have to install it immediately. Indeed, there are several reasons why you should postpone installing an upgrade for several months, if not a year or more (much of this advice applies not only to AutoCAD but also to Windows 95).

- Upgrades inevitably have bugs. Manufacturers usually are under pressure to release upgrades as soon as possible, so they often forgo the need for a complete beta test. And even the most thorough beta tests can't anticipate all the ways a user may use a software application. The initial version of AutoCAD Release 13 contained a number of bugs that had to be addressed by "patches."* By waiting to

---

*AutoCAD "patches" are designed to make changes to an existing installation of AutoCAD rather than having to install a newer version of the software entirely from scratch (although you *can* do the latter). These options are similar to repairing a hole in an automobile tire—you can either patch the existing tire or use a new one.

upgrade, you let other users discover the "bugs" and inform the software manufacturer about them.

- Upgrades can disable or harm your customization. For example, although AutoCAD Release 13 reads Release 12 drawings, it offers over fifty new commands while eliminating others. If some of the discontinued commands are embedded in your Script files or AutoLISP routines, you will have to rewrite these files. Changes in features and commands will also have to be reflected in your office standards, documentation, and other aspects of your customization. Moreover, new features that offer you more power and greater ease may compel you to entirely rework procedures so that you can take better advantage of the upgrade's power.

- The upgrade may not work with third-party software. If you use a third-party add-on for AutoCAD, you may have to wait until an upgrade for the add-on is available in order to use it with the newest version of AutoCAD. While many add-on developers aggressively parallel AutoCAD's upgrades with their own, a lag time is inevitable, and then you also have to deal with similar upgrade issues, such as bugs, in the add-ons as well.

- The rest of the world won't necessarily upgrade immediately. You will find that many of your clients, consultants, and other contacts who use AutoCAD will also delay in upgrading, whether or not for the same reasons. Only external plotting and digitizing services are likely to upgrade quickly, as they have a professional incentive to be compatible with the latest version of AutoCAD on the market. Until you perceive pressure from clients or consultants to use the newest version of particular software, however, you have no external incentive to rush to upgrade.

- Upgrades can be disruptive. Even if the new release offers many desirable features, the process of installing it and learning it takes time. Formal training in the upgrade's new features will be necessary. And even the most talented drafters must take time to *un*learn commands they can no longer use.

- Upgrades require downtime. Particularly if AutoCAD is installed on a network, the office will have to plan downtime to install the upgrade and test it. Testing includes not only using the application but making sure it works with plotters, printers, and other peripherals and with the overall network configuration. Sometimes installation can be done nights or on weekends, but often it may require computer downtime during normal working hours.

- When you upgrade, the list price of the upgrade software is only the start of the entire costs you assume. With AutoCAD, you almost always have to purchase more RAM to be able to continue operating with the same relative power and speed. You may have to purchase upgrades to third-party software and spend money upgrading an in-house customization. You have to spend unexpected staff time on installation and training. If the upgrade process does not go smoothly, you pay implicitly for the resulting problems, such as delays in producing work.

Planning Upgrades

Because of the costs and time involved in implementing serious upgrades, offices should schedule them carefully. Judicious planning not only will make the process go more smoothly but also will save you money over the long haul.

Upgrades can be done in various ways, depending on the software you are upgrading and the number of workstations involved. After conducting a study of firms and how they implemented upgrades, Autodesk* recently determined that four major approaches are used.

1. The big switch. Generally this involves a firm changing its general operating system, such as switching from DOS to Windows or Macintosh; making a major system upgrade, such as moving from Windows 3.1 to Windows 95 or Windows NT; and/or systematically shifting from 16-bit to 32-bit software applications. The big switch requires that you pay close attention to all aspects of your computer system. It often involves purchasing new machines or installing more memory as well as training staff and living with computer downtime. Such a major change requires careful scheduling so it can be accomplished properly and systematically without impinging unreasonably on ordinary office routines and major project deadlines. This move may involve the highest up front costs and risks, but over the long haul it can produce a more efficient system that offers the greatest benefits to the entire office.

2. AutoCAD Release 13 "2-step" upgrade. This approach involves switching to Release 13 now and upgrading the operating system

---

*Information on this study is provided by Amar Hanspal at Autodesk Product Marketing.

(Windows or DOS) later, or vice versa. The two-step approach requires less initial overhauling, disruption, and training, and allows users to become familiar early on with the most important aspects of the upgrade. It often occurs when a firm seeks to use specific new features that the software upgrade package offers. Experience in implementing the first step in the upgrade can help when your office plans installation of subsequent steps. However, you will have to wait until both upgrades are implemented to get the benefit of all the new software operating in tandem.

3. The rolling upgrade. With this approach, you upgrade your system as necessary or as you can afford it. This often means upgrading only a few workstations at a time, perhaps in "waves." Such a strategy may seem to save money up front, and often works best when a specific portion of the office really hungers for the upgrade while others would be disrupted. But it reduces the benefits that would be seen with an officewide upgrade and causes the office to operate with a "double standard" in the interim. In addition, it can produce friction if members of your staff feel that upgrades are arbitrary or produce inequities in the allocation of computer resources in your office. In considering this approach, you also need to look at the current allocation of work among project teams; upgrades should be scheduled with an eye toward who is working on which project team and how urgently a particular team might need the upgrade, or, alternatively, who is too tightly scheduled to be able to deal with the disruptions that an upgrade can entail. With rolling upgrades, special attention must be given to organizing CAD files and partitioning of hard drives, so that one set of software files doesn't interfere with the others.

4. The waiting game. With this approach, you wait and upgrade only when you feel strong pressure to do so, whether from your staff or from external parties or because a new project justifies it. You might upgrade one station just to have the upgrade features available or to be able to read and write a particular version of an application file. The waiting game approach does not make you state-of-the art, but it can save you from spending money before you absolutely need to. In addition, you can let others go down the learning curve involved in dealing with an upgrade and its likely bugs. Playing the waiting game, however, should always be done deliberately and systematically, using explicit criteria for determining when to stop waiting and to step into the fray.

Scheduling and Budgeting Upgrades

If you are seriously committed to using AutoCAD as your basic CAD software, you should build the costs and time for upgrading into your ongoing operation budget. As mentioned above, you should budget not only for the primary costs of upgrading AutoCAD software but also for the secondary costs:

- new computers or additional RAM
- upgrading operating systems if required
- upgrading any third-party software for AutoCAD (or switching to new products)
- upgrading your in-house customization (see Chapter 4)
- downtime for installing and configuring upgrades
- downtime and staff time to beta test the upgrade
- time for training users in the upgrade's new features

Historically, AutoCAD has released upgrades every two or three years since it was first released. An upgrade approach that many firms implement successfully is to upgrade to *every other* new release—not only with AutoCAD but also with other major software. This can be a highly sensible approach for firms that have gone to a major effort to customize and implement the software they already have. By deliberately delaying upgrades, they can spread the costs of acquiring and customizing their software over a longer time frame and minimize the disruptions that upgrades entail. Since most new releases of software packages provide some way of reading earlier versions, remaining compatible with users who have upgraded is not a major issue.

What are the costs of this approach? To upgrade from Release 12 to Release 13 costs $495 per license; upgrading to Release 13 from Release 11 or earlier costs $695. While upgrading to Release 13 from Release 11 costs $200 more per license than an upgrade from Release 12, this additional cost can be negligible compared to the costs of an interim upgrade from Release 11 to Release 12 and eventually to Release 13, not to mention the costs involved in disrupting your system twice to implement upgrades.

Managing Licenses for Data

When you purchase a copy of AutoCAD, you become a licensed and registered owner of AutoCAD software. A license entitles you to certain privileges and obligations; understanding the terms of your license can save you a great deal of money and headaches over time. Indeed, as firms use computers more and more, software constitutes an increasingly large share of firm assets, and thus software licenses are worth protecting carefully.

Many software purchasers do not bother to read the license agreement that accompanies the envelope containing AutoCAD disks. Nor do they seek to understand the finer points of the agreement. License agreements should be studied and obeyed, not approached cavalierly. Some key points should be noted as listed below; many of these points are applicable to *all* your software.

- Your AutoCAD license allows for *one* copy of AutoCAD to be used at *one* workstation. You can get around this rule by installing it on two workstations, but you have to be able to prove that only one installation is in use at any one time.
- AutoCAD Release 13 gives you licenses for DOS and Windows versions of the software. You can install both on the same machine, but you cannot install one of each on separate machines and run both versions simultaneously.
- You may make a backup copy of your AutoCAD software, but this is to be used for emergency purposes only, not for additional installations.
- If you upgrade to a newer version of AutoCAD, you can keep the older version on your computer for sixty days, then you must remove it. After sixty days, you should destroy the diskettes containing the previous version. Your license does not entitle you to run two releases of AutoCAD concurrently for the long-term.
- If you use a network version of AutoCAD, you may operate the software on a multiuser license provided that you purchase a License Pack for a specified number of licenses. You must purchase an additional license for each additional concurrent user. The multiuser license does not restrict the number of workstations that can access networked AutoCAD, but it does restrict the number of users that can *simultaneously* use AutoCAD.
- You should not make copies of your AutoCAD documentation. Any documentation that arrives on disk, CD, or on-line may be printed out, but it also should not be copied.
- Your licensed AutoCAD software may ship to you on several forms of media. Even if you cannot yourself access some of the software because it is on media you can't use, you may not give it to someone else; you may only use the media that meets your needs and your hardware capabilities.

- You cannot sell or give away your AutoCAD software whether in its original form or in a modified form. If you do want to transfer your license to another party, you should contact your AutoCAD dealer or Autodesk; by transferring the ownership, however, you give up your rights to that license.

- When using certain Autodesk multimedia software, you should review Autodesk's distinctions between "software files" and "user files." Certain files may not be used in commercial presentations without Autodesk's permission. If you use multimedia applications heavily, combining images and sounds created in a number of different software, you will need to become familiar with all the applications' licensing policies.

- Being a registered owner of AutoCAD entitles you to receive future releases of the software at significant discounts (often you save more than $2,500) over the standard list price. In addition, you are entitled to "patches," which address bugs in the earlier versions of your software. You also get access to technical support and Autodesk resources. So you should register your new software immediately, maintain copies of your purchase records and software agreement, as well as hold on to the original AutoCAD disks.

*Both your original AutoCAD software diskettes (or CD-ROM) and accompanying license and registration information should be filed away in a safe place, but not in the **same** place. By keeping the disks separate from other information, it will be easier for you to replace your software or to qualify for upgrades in case your original diskettes are lost, stolen, or damaged; you can use your license and registration information to "prove" your ownership should you no longer have your disks.*

- Some software manufacturers are benign about software copyright violations of their products, but historically Autodesk has not been. It has an internal department that functions specifically to deal with copyright infringements and software piracy (and will provide you with complimentary information). If you break the terms of the agreement, Autodesk can fine you and remove your license to use the product (although often the company will require that you simply pay the full price for a new license). And even if Autodesk is soft under certain circumstances, copyright violations can harm your relationship with your local AutoCAD dealer.

- Under U.S. copyright law, if you make or use an unauthorized copy of AutoCAD or any other Autodesk software, you can be fined up to $250,000 and face a maximum of five years in prison. Additional unauthorized copies can result in fines up to $100,000 per copy.

- Autodesk is not the only entity authorized to penalize copyright violators. You may be subject to an audit by private organizations, such as the Business Software Alliance, that can obtain legal jurisdiction to enter your office—often without prior notice—and inventory the software installed on your computers. Such organizations also have the jurisdiction to issue fines at their discretion; these fines have often been in the six figures and often constitute far more than the value of the illegally copied software.

In scope, intent, and specifics, Autodesk's license policies are not terribly different from those of any other software manufacturer and are intended to deter users from engaging in the prevalent crime of computer software piracy. Whenever a computer user installs a product, he or she must always consider the copyright issues inherent in using that product.

Licenses and Employees

Even when a design firm is meticulous in its policies regarding software licenses and copyrights, it must recognize the possibility that such policies may be either abused or unintentionally ignored by employees and consultants.

An employee may not know, for example, that he or she is violating the software's copyright by installing a copy of your firm's licensed AutoCAD software on a home computer in order to practice and improve his or her CAD skills. Someone else may not understand that sending third-party add-on software files along with CAD drawing files to engineers or consultants is also a violation. Even sending special typeface files to a printing service may violate copyrights (especially if the printing service forgets to erase the files from its computer system when your order is completed). Often unintentional violations can occur right under your nose, as when an employee copies a software installation from one machine to another when he or she switches desks, but then forgets to erase the first installation.

Both because of the practical and ethical need to adhere to copyright laws and because of the very real possibility that they may undergo an external "software audit," many firms are making more concerted and explicit efforts to inform employees about software and licensing issues. In addition, some firms undertake their own internal software audits, wherein they systematically inventory all the software that is installed on all their machines and compare it to their registration records. Any unauthorized copies of software are erased from the office system. For additional information on conducting internal inventories and related issues, you can telephone the Business Software Alliance Anti-Piracy Hotline at (800) 688-2721.

Licenses and Computer Inventories

Because software licenses, especially with AutoCAD, can have enormous financial and legal implications, every design office should track its computer software and hardware via an inventory. As Figure 10.13 shows, this inventory should contain current information on all the software products you own, including date purchased, cost, and origin of purchase; the version or model of each product; serial numbers for hardware and registration numbers for software; and phone numbers for vendors. The machines on which each program is installed should also be noted; indeed, you can

ABC Design Inc.
Computer Software Inventory

Software Program	Software Manufacturer	Version	Date Registered	Registration #	Machine Installed On	Support #
AutoCAD R13	Autodesk	R13.b	4/15/95	00105-013808-9400	PC#3	H. Smith, AutoCAD Dealer, 546-9999
AutoCAD R13	Autodesk	R13.b	4/16/95	00105-013808-9401	PC#4	H. Smith, AutoCAD Dealer, 546-9999
AutoCAD R12	Autodesk	R12.c	9/1/93	00105-013808-9402	PC#5	H. Smith, AutoCAD Dealer, 546-9999
Microsoft Office for Windows	Microsoft	4.2	10/10/94	00-045-0200-1111	PC#1	1-206-635-7200
Microsoft Office for Windows	Microsoft	4.2	10/10/94	00-045-0200-1331	PC#2	1-206-635-7200
Microsoft Office for Windows	Microsoft	4.2	10/10/94	00-046-0550-2211	PC#3	1-206-635-7200
Microsoft Office for Windows	Microsoft	4.2	10/10/94	00-046-0550-2212	PC#4	1-206-635-7200
Microsoft Office for Windows	Microsoft	4.2	10/10/94	00-045-0670-5656	PC#5	1-206-635-7200

Figure 10.13 A sample computer software inventory

"link" this data to an AutoCAD-based floor plan that shows the location of your office's equipment.

Computer inventories have uses well beyond documenting licenses and being prepared for possible license audits. They can be used by financial officers to track the firm's assets, calculate depreciation, and purchase insurance coverage; by CAD managers to plan and budget for upgrading and acquiring additional software and machines; by marketing staff, who may need to include an updated report of your firm's computer assets in a proposal; and for internal software audits, as discussed previously in this chapter.

Many computer operating systems contain utilities that print out detailed reports about the computer on which the system is installed; such reports include statistics well beyond basic features, such as RAM and CPU power, that are included in a purchase invoice. Such reports should be included in the computer inventory files and updated each time a computer has memory or other components added to it.

Backing Up Data

Having a system for making regular backups (making copies) of all important computer files is absolutely essential in any computerized design office. Backups provide recourse in case of unexpected disasters, computer shutdowns, and theft. Increasingly, provisions about backing up files are required in design agreements and other contracts where computerization is essential to the delivery of a design product.

Individual practitioners and very small firms with few computers tend to perform backups by copying select project files onto diskettes at the end of the day or at some other designated time. Such a system relies on the conscientiousness of each computer user to remember to do this; this system can also take enormous time and storage space since one AutoCAD drawing file alone can easily fill up a single diskette. Individual disks must be properly labeled if you are to have any ease in finding files on them.

A better, more efficient solution is to establish an automatic tape backup system. The advantage of such a system is that, first of all, tapes can store far more megabytes of information than diskettes can; second, you can arrange for tape backups to be performed on a regular basis, independently of whatever individual computer users are doing; third, if you run a network, one system can backup all computers on the entire network. You can often program backup software to automatically backup the system on a daily basis, say, after midnight.

Backup Media Options

Backup systems come with various forms of media. Like the formats for data transfer described earlier, backup systems are constantly evolving, and older media can become obsolete within a few years. Moreover, the long-term reliability of some types of media is always questionable. Therefore, every office should be prepared to shift all its archived files to a new backup system every several years.

The most state-of-the-art backup systems are always the most expensive. However, they are always capable of storing more information than previous systems, and are usually faster and more efficient in retrieving data. As Figure 10.14 shows, while the *total* costs of backup systems may remain steady, the cost *per byte* has dropped sharply. You can now purchase tape backup systems that store 1.5 GB (1,500 MB) for the same price that you would have paid five years ago for a system that could store only 100MB. With many backup systems, the industrywide cost per byte is down to 3 cents. Note, however, that this increased capacity is quickly eaten up by larger AutoCAD Release 13 drawing files and other application files.

Two common forms of backup media are tape cartridges and Bernoulli disks. Both require proprietary backup devices, cartridges, and software. There are also a growing number of 3 1/2-inch diskettes that look like

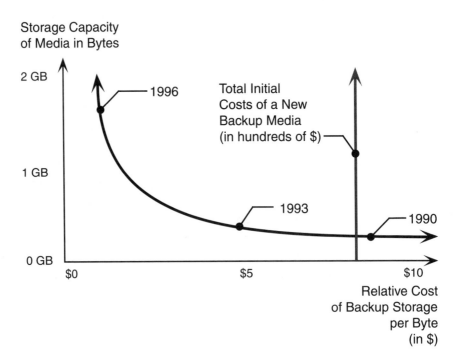

Figure 10.14
The upfront total costs of new backup systems remain constant, but you now get far more storage capacity for your money.

standard 1.4MB diskettes but actually can store far more memory, as much as 128MB or 230MB. Writable CD-ROM is another emerging option.

Backup Procedures

Design offices use a number of different backup methods, depending on the capabilities of their backup equipment, the quantity and size of their files, and the number of computers needing to be backed up. Whatever approach an office subscribes to, each person must understand how the backup system operates, how to adjust his or her computer to participate in the process, and what the risks are that data may be lost anyway. The last point is especially important, since under some setups, the network backup system may need to be augmented by individual backup procedures.

Single Daily Tape Backups

With this option, you store files on one tape over and over again, every night. One night's worth of backup files will be overwritten the next night. The tape is used until it starts to corrode. This method involves minimal administration and uses only one tape at a time, but by using only one tape you can only retrieve the previous day's work. You also have no backup, so to speak, of your backups.

Separate Tape Backups for Each Day

With this option, you use a particular tape for a particular day of the week. Thus you use the "Monday" tape every Monday night, the "Tuesday" tape every Tuesday night, and so on. This method involves more administration but provides at least five days' worth of backups, so that on Friday, you can retrieve Monday's version of a file that you just irretrievably changed or deleted.

Your tape size and your backup software also determine your backup procedures. To select the best routine for your office, consider the following options:

- Should each night's backup *overwrite* (that is, erase) the previous night's backup, or instead should each backup *add* the current version of a file to what is already on the tape? If you take the latter approach, you will have many versions of your file to refer to, but you will use up empty space on your backup tapes much more quickly (tapes, however, are ultimately much cheaper than scrambling to recover lost data).

- Can your backup software determine whether a particular version of a particular file is already on the tape as it performs a backup? You may be able to specify that the backup software only back up new files or newly revised old files. This feature saves space on the backup media and saves you time searching through duplicate files.

- Can the backup software do its thing while someone is working on a particular file on the computer? Some backup programs will ignore any files that are being worked on; others may abort the entire backup process if the program detects a user working on a file.

- What are your options for finding and retrieving older files? Do you have to scroll the entire tape, or can you issue a **Search** command? Does the software have sophisticated search options that enable you to find, for example, all files with a *.dwg* file extension that were saved on or before March 15?

- How quickly can the tape drive access information? Some models are extremely slow. The newer technologies can access all portions of a tape or other media extremely quickly.

Although overnight backups are essential and generally address most needs, computer users have discovered that backing up at the day's end does not help retrieve information that is lost when the computer system suddenly crashes at noon. Therefore, many firms are backing up their backups (!) with systems that store files during operations, using various methods to split data and to "mirror" it on multiple disk drives so that when one system or system component goes, another can be accessed. The most well-known kind of these systems is Redundant Array of Inexpensive/Independent Disks (RAID), which breaks up stored files onto multiple disk drives, so that users are not dependent on the integrity of one disk. This approach is analogous to a building that has a separate source of electrical power so that it can continue to use electricity during a power outage. For firms that feel that they cannot afford a minute of downtime or lost data, systems such as RAID are essential.

Protecting Backups

The purpose of computer backups is to protect your data. For backups to do this, they themselves must be protected from any sort of damage or corruption.

A critical element of protection is physical storage for backups. You should have one copy kept on-site, preferably in a fire- and moisture-resistant file cabinet. On a regular basis, you should also move a copy off-site. In some offices, the CAD manager takes a copy home. In others, an employee will make regular trips to a bank safe-deposit box.

Regardless of what backup method you use, it is often wise to make a periodic backup copy of your entire system and file it away for general safekeeping. Ideally, this should be done monthly or weekly for critical files, and it is a good idea to rotate tapes so that you can always retrieve earlier versions of a file in the event of a catastrophe. You can also use these backups to find older files that have been obliterated from your current system.

Apart from physical duplicates, backup protection involves keeping your system free of viruses and other potential sources of file corruption. Running file diagnostic programs, which look for problem files and suboptimal configurations and other potential sources of problems, should be done regularly. Many diagnostic programs, such as Norton Utilities by Symantec, also will optimize your disks, reorganizing files on hard drives so that data takes up less space and can be accessed more efficiently. This, in turn, helps your tape backup software store files more efficiently. Many of these utilities, as well as DOS and Windows, enable you to *un*delete damaged or deleted files.

If you upgrade your backup system to a new media format, you are well advised to make copies of all your backups and archives using the new media. This way, you don't have to worry about media obsolescence, and you won't have to reinstall the older backup system to retrieve archived files.

Archiving Data

Almost every design firm has a process for archiving paper copies of the projects they have completed. This process should be replicated for computer files as well. Bear in mind, however, that computer technology poses interesting challenges and modifications to the traditional archival processes for paper.

How to Archive AutoCAD Drawing Files

AutoCAD files generally are archived by saving a drawing file under a file name with a file extension other than *.dwg*; many firms use *.arc*, which is a logical abbreviation for "archive." With this file extension modification, drawing files cannot be opened up without first being renamed or copied under another file name that contains the *.dwg* file extension.

Before making an archival copy, purge the file of any unnecessary and unused blocks, layers, and other features. There is no point in preserving information that becomes irrelevant once a project has been completed.

Although recently completed projects can linger on computer hard drives for a while, immediately copy archive files to a diskette or backup tape and store them in a safe place, such as a fire-proof file or safe-deposit box. If you have off-site storage for archival tubes and boxes, consider keeping copies of computer files with them as well. All storage places should offer protection from moisture, fire, and other potential sources of damage or destruction.

When to Archive AutoCAD Drawing Files

The policy for archiving files may vary according to the scope and nature of a project, and particularly whether or not the client involved is an ongoing or repeat client.

Some firms archive a copy of the AutoCAD files that produced *all* sets of drawings that were formally issued and stamped. Others simply store the files that are current at the project's end. Still others archive each version of a drawing file during a project's evolution, and then selectively purge files when the project is completed. Whatever approach your firm takes, it is advisable to have a clearly stated, publicized policy for CAD drafters to follow automatically; this policy should be contained in the office CAD standards manual. If a particular project agreement calls for special modifications to this policy, the project manager should indicate that to the design team. Policies on archiving should state clearly whether archives should be created under any of the following conditions:

- when a formal set of drawings is issued, particularly when drawings are stamped,
- for each major phase or version of the drawings, such as at design development, bid documents, construction documents, and as-builts,
- for each CAD file that is turned over to a client as an as-built, and
- for each CAD file that is shared with a consultant.

Challenges with Archiving CAD Files

Although computer-based archival files can be created instantaneously, ensuring their usability over the long haul is difficult. In particular, the following characteristics of computer archives should always be kept in mind:

Computer storage formats constantly evolve. As described earlier, the media for using and storing computer files constantly changes, with the only determinable trend being that the media becomes smaller while storing more information. Because of ongoing evolution, you may easily find that AutoCAD drawing files that you archived only five years ago cannot be read by the disk drives of the computers you use today. You may have to do interim moving and copying of files to get the old file onto your current computer's hard drive. Some older diskettes or tapes may be completely inaccessible.

Computer software is always advancing. Competitive software developers like Autodesk, Microsoft, and others are always improving their products and issuing upgrades. Most make an effort to ensure that previous versions of their product can be read by the newest release. But some software changes so radically in subsequent versions that it is not really feasible to support the earlier versions, or the earlier versions are so skeletal by comparison that not much meaningful information is retrieved.

Computer media is not always stable. Although certain types of computer storage, such as CD-ROMs and many brands of backup tapes, may offer long-term integrity, the long-term stability of some media is questionable. Many older diskettes have been known to have data flake off the surface; others have been warped or scratched so that they are unreadable. Computer media may not have the same problems of paper archives, such as fading and tearing, but they are not inherently more stable.

Resolving Archiving Challenges

The issues described above are typical to all software and hardware setups. While they can't be avoided, they can be anticipated. Some of the precautions that you should take as a matter of course include the following:

Make hard copy archives. Just as you would for your hand-generated projects, archive paper plots of all your important AutoCAD files. Plot them at the proper scale on vellum, film, or other archival-quality reproducibles, and store them in acid-free tubes in a safe place as you would your traditional hand-produced drawings.

One archival school of thought holds that all files, regardless of the software that produced them, should be saved in ASCII file format (which is what .dxf files are). Since ASCII is recognized by most versions of most software, such file conversion might increase the odds that a very old computer file, made from a long obsolete program, could somehow be read by a current software application.

Store computer files in multiple formats. To stave off the possibility of finding that you can't read older computer media, store archival files on two or more different types of media, for example, use backup tapes as well as on 3 1/2-inch diskettes. You also may want to upgrade individual files to the newest versions of your software. For example, older drawing files created in AutoCAD Release 9 or Release 10 might be "*re*archived" as Release 12 or Release 13 files.

Periodically make upgrades to new file formats and tape backup media. Each time you make a major shift in your backup media, copy all your important CAD archives to the new media. This way, you won't have to rely on older equipment to access your archives.

Often CAD files of older projects are of great value for marketing purposes. Especially for marketing staff who are non-CAD literate, trying to access older CAD files can be a time-consuming experience. Marketing staff should use the CAD archiving process as an opportunity to collect materials for future marketing needs. They need not store AutoCAD files in their original AutoCAD drawing file format; instead, they can import what they need into other software, such as desktop publishing or presentation programs or store the images in a CD-ROM-based library of images.

Keep proper records of archived CAD files. Most design firms keep logs of project archives. These logs should be expanded to include names of CAD files, the date they were archived, the media, and what version of AutoCAD they were created under as well as the plots that they produced. Figure 10.15 shows an archival log form that reflects CAD media. You can also use TDM software such as Autodesk WorkCenter to help track, store, and prepare CAD files for archiving.

A very quick way to document the contents of an archival disk or tape is to print a file directory. You can do this easily, for example, using the DOS command **C:\Print dir c:** or the Macintosh **Print Desktop** command. These printouts can be attached to the pertinent archival form.

Contents	Drawings	Computer Files Disk Name*	Software	Media
ABC Design Inc. **Archiving Log Form** **Client:** ACME Corporation, Inc. **Address:** 999 Exchange Place Gotham City, MA **Project Name:** 999 Exchange Place Renovations Floors 9 - 16 **Project Number:** 9405.00 **Date Archived:** 1/10/96				
Construction Set dtd. 1/2/95	Tube #9405-1	Disk #9405-1A; Disk #9405-1B	AutoCAD R12	3 1/2" HD Diskettes
Construction Set dtd. 1/2/95	Tube #9405-2	Disk #9405-2	AutoCAD R12	3 1/2" HD Diskette
As-Built Set dtd. 1/1/96	Tube #9405-3	Disk #9405-3A; Disk #9405-3B	AutoCAD R12	3 1/2" HD Diskettes
Misc. Studies 1994-1995	Tube #9405-4	Disk #9405-4A; Disk #9405-4B	AutoCAD R12; Autovision	3 1/2" HD Diskettes

*See attached Computer directory printouts for computer file names

Figure 10.15 A sample archive form, showing one way you can document CAD archives and relate them to your hard copy

Protecting Data: Copyrights and Liability

The increased ease and speed with which CAD files can be exchanged inevitably raises concerns about the appropriate use of these files. Architects and interior designers must monitor the distribution of their computer-based work just as they do the distribution of hard-copy drawings and specifications. And the two faces of "CAD risk management" are copyright protection and liability.

Copyright Protection

Architects and interior designers traditionally protect ownership of their work with copyrights. A valid copyright is indicated by the word "Copyright," the abbreviation "Copr.," or the symbol ©, followed by the name of the copyright author and the year of first issuance. A designer's right to ownership of these drawings is protected by The Copyright Act of 1976, The Architectural Works Copyright Protection Act of 1990, and various federal and state acts, as well as by contracts between designers and owners, contractors, consultants, and others. Article 6 of AIA Document B141: *Standard Form of Agreement Between Owner and Architect* (1987 edition) includes an industry-standard wording for copyright protection. Copyright protection occurs automatically when an original work is created. A design must meet a test for "originality," however, to be eligible for copyright protection.* If you want to be able to bring suit for copyright infringement more easily or in any other way protect your copyrights aggressively, you must file your work with the U.S. Copyright Office—something most design firms do *not* do on a routine basis.

Copyrights traditionally have protected an idea's *expression* but not necessarily the idea itself. In the design world, this has meant that drawings, specifications, and other "tangibles" had copyright protection, but the resulting built design may not. The Architectural Works Copyright Protection Act expanded protection for design ideas but came with enough ambiguities to leave the extent of copyrights for design ideas still murky. Copyright issues, therefore, will continue to be worked out in contracts, arbitration, and the legal courts.†

As tangible instruments of expression of design ideas, CAD files are copyrightable materials. Yet many designers blithely give CAD files to clients, consultants, and other parties without being aware that they are distributing intellectual property. They may also be unaware that CAD files can be manipulated far more quickly (and invisibly) than can hand

---

*The Architectural Works Protection Act of 1990 protects the "overall form" of an architectural work as well as the "arrangement and composition of spaces and elements," but it does not protect "individual standard features." *CSM Investors, Inc. v. Everest Development, Ltd.*, 804 F. Supplement 1304 (D. Minn, 1994).

†The legal approach to protecting an idea's expression but not the idea extends to copyrights for computer applications as well. The courts have tended to protect the written code that produced a computer interface or product but not the interface itself. Thus in a well-known copyright infringement lawsuit brought by Apple Computer against Microsoft over the Microsoft Windows interface, the courts determined that Windows could approximate the graphic interface of the Apple Macintosh; if perhaps Apple could have demonstrated that Microsoft had cloned the underlying code that produced the interface, Apple might have had better luck.

drawings, thus obscuring copyright infringements. With a film or vellum drawing, considerable work is required to make changes while maintaining consistency in lettering, line weights, and other graphic elements. With an AutoCAD file, typefaces, line weight menus, hatch patterns, and architectural targets can be instantaneously modified, and various components of a drawing can be easily reordered. A fresh plot or screen image of a modified drawing file can look dramatically different even if the underlying design remains identical.*

Judges and other legal professionals may not uniformly accept that, or even understand why, CAD files or any other computer-generated work should be considered worthy of copyright protection. As with other legal issues, precedents for computer-related legal disputes vary among states and court systems. Always consult with legal professionals who are well-versed in the legal precedents of any state in which you practice.

Given the ease with which CAD-generated work—and most computer-based documents—can be copied, designers must vigilantly protect their copyrights. Methods of protection include the following:

- In all your contracts, make explicit your copyright ownership for CAD files (see Chapter 9), as well as any other software files that you produce for use with a project.

- Limit the circulation of CAD files solely to parties who have business using them and who themselves have incentives to protect their own work. Such parties generally include engineers and other consultants.

- Distribute only those *portions* of CAD files that other parties really need. Engineers, for example, do not need finish plan information. Facilities managers usually don't need construction notes, dimensions, and architectural targets. Use the **Wblock** command to produce edited versions of your drawings for distribution to others.

- Avoid distributing your own AutoCAD Scripts file, AutoLISP routines, block libraries, and other tools that you've created to produce your own drawings. Others may be able to reproduce with impunity the *look* of a simple CAD drawing, but tools are generally more

---

*For traditional drawing technology—on film or vellum—copyright disputes are often resolved by overlaying the plaintiff's and defendant's drawings and reviewing the extent of similarities. With CAD-produced drawings, the files themselves might need to be compared (they can be overlaid in one file). An expert witness on CAD might also review the files for similarities in layer names, blocks, and other elements that designers typically customize.

particular to individual CAD users or firms and therefore can be used as evidence that you produced a particular CAD file.*

- Document carefully and completely the parties to whom you distribute your CAD files. Whenever CAD files are circulated, note on an accompanying office transmittal the names of the files sent, the date and time they were created, and the byte size of each file. Back this up with archival copies—in both disk and hard-copy formats—of the CAD files you've distributed.

Modify your office-standard transmittal and other forms to encourage and facilitate documentation of transfers of computer-generated works.

- Create an AutoLISP routine that produces a copyright alert dialog box whenever someone opens your AutoCAD files.

CAD files are not the only computer files you may have an interest in protecting. Computer files that produce specifications, programming reports, budgets, graphic presentations, and other output are also worth protecting.

Firm policy about sharing CAD files should also be followed by your employees. While technically, a design firm's partners may own the CAD files that their employees create, the employees may naturally feel entitled to a copy of the CAD file as they would a print of a hand-produced drawing. Since CAD files may nowadays constitute an important part of a designer's portfolio, it is reasonable to expect to have to permit employees to make copies of CAD drawings they've produced. A stated policy regarding copies of CAD files can be included in your firm's employee or CAD standards manual.

Liability and Legal Disputes

Although designers want credit for their designs, they must also take responsibility for them. In today's litigious world, this means that design firms must see themselves as being potentially liable for any problems that computer usage produces.

---

*CAD files are a form of database. In copyright disputes over databases, the courts often ignore the amount of effort involved in creating the database and rather focus on what is unique in content. Under this approach, a database of potential clients which took five years to create would not be afforded copyright protection if it could be replicated by using telephone books and other sources of information generally available to the public. Extrapolating this to CAD, a CAD drawing of a public building which someone with good surveying skills could draw from scratch would not be copyright protected. However, the process of creating that CAD file might be.

Computer law is still a young legal specialty, evolving on a case-by-case basis, and few clear precedents have emerged. But, as computer-based design proliferates in the A/D professions, the potential for liability in this area expands as well. Based on legal disputes that have already occurred in the A/E/C field and in other professions, potential sources of liability and other legal disputes can be predicted. In disputes, however, CAD itself may not be the explicit issue under discussion. Indeed, it may never even be referred to; rather, the *results* of improper use of CAD will be the focus of disputes.

Possible Sources of Disputes Involving CAD

Disputes over ownership of work. As mentioned earlier, design contracts typically give designers ownership of copyrights and physical possession of their work. But clients who increasingly need CAD-based drawings for their own operations in leasing and facilities management may feel entitled to all CAD files relating to their properties. They also may expect to be free to distribute these files to other design firms that perform subsequent work on spaces *you* designed. If contracts do not explicitly spell out what clients get upon a project's completion, the distribution and ownership of CAD files may become a source of dispute or cause designers to "give away" more work than necessary.

A subset of such disputes occurs when design firms modify or develop software for a particular project. You might, for example, create a unique database to plan and monitor the design, construction, and move-ins for a major, multiyear building project. You believe that you own this database (because it is a copyrightable product and because your firm is the licenser of the physical software itself) and that the client is entitled only to the resulting building. The client, however, believes that it owns the database, because the software was developed for its project and probably at its expense and because the client can use the database for additional work on the building.

The flip side of ownership disputes occurs when the client wants a building that is unique and doesn't want the designer to reuse distinctive portions of the design on other projects. Yet many designers do develop their own "trademark" or "stock" designs, often using AutoCAD to develop signature elements.

Disputes over CAD-based work to be delivered. At the completion of a project, many owners expect to be given a CAD-generated "as-built" drawing for their records and future use. Contracts reflect this expectation but rarely specify what the CAD files should contain and what their format should be. As institutional and corporate clients become more sophisticated in their use of CAD, however, they become more specific in their needs. Some even request that the designer provide CAD files that reflect a particular layer-naming system, color or pen settings, and other standards. These clients may also assume that the designer has tested the files to ensure compatibility between the design firm's computer platform and theirs. If these expectations are not addressed at the outset of the project (preferably in a contract), the designer may not even be *capable* of providing the desired CAD file. And if creating this file involves considerable work or the purchase of a file translation package, the designer might then feel entitled to request an additional fee.

Disputes over unauthorized use of CAD files. Several cases involving a building contractor or other party accused of using an architect's CAD file without the architect's knowledge or authorization have already come to court. In some of these cases, even though the builder had violated the designer's copyright, the designer nonetheless was held potentially liable for problems with the design and had to prove ignorance of the copyright violation. In short, unless they can prove a CAD file was stolen, designers can be liable for any CAD file that credits them with a particular drawing design and production.

Disputes over accuracy of CAD file contents. AutoCAD's internally consistent coordinate geometry, powerful dimensioning capabilities, precise line-weight production, and controlled interface create the illusion that the AutoCAD file represents a flawless database. However, an AutoCAD file only reflects the quality and validity of the data that it receives, and given the realities of the built world, it can never be 100 percent accurate (see Chapter 7). Experienced AutoCAD users understand that the AutoCAD universe is only a platonic ideal. But a neophyte CAD user might find the computer interface compelling enough to use CAD files in total blind faith and make incorrect and potentially costly decisions on the basis of unverified electronic information.

The longer CAD is used throughout a project, the more CAD files become a mosaic of information from different sources: site survey and terrain information, field survey verifications, and other data that is supposedly accurate. As the project evolves, however, pinpointing flaws in accuracy in any source of information becomes more difficult.

Disputes over data in the public domain. Whether on a public forum on the Internet, a shareware program, or a software program you've purchased, tons and tons of data now is instantly available to computer users.

Sometimes this data *appears* to be in the public domain, but for one reason or another it may actually or potentially be copyrightable. For example, you can purchase a CD-ROM containing an image of Da Vinci's *Mona Lisa*. Since you can see this painting in a publicly accessible museum and since Da Vinci lived long before current copyright laws went into effect, you might think that you can copy this CD-ROM image at will into your own files. Depending on the computer software's license agreement, however, the Mona Lisa image may not be "royalty free." This has been a particular issue with products such as GIS databases, where the sites surveyed are public lands, but the surveyors feel justifiably entitled to copyright their computerized replications of it. As multimedia applications and new forms of Internet-produced data pervade the design world, tracking copyrights will become even more complex.

Liability for employees' mistakes. Traditionally, members of the professions have assumed responsibility for their employees' mistakes and omissions, provided that the employee's "misconduct was committed within the *scope* of the (employee's) employment," and so long as the employee does not "exceed his or her authority or act improperly."[*] This assumption of responsibility extends to CAD, even when design firm owners and supervisors do not understand completely the process for which they assume responsibility. Such liability may also extend to consultants and independent contractors.

Liability for *not* using computers. Professionals may be held liable for *not* using computer software that could either produce results more accurately and/or faster. Legally, this liability falls under the category of "wrongful *non*use" and can apply to CAD work as well as to specifications writing, spreadsheets, and databases. As computers continue to pervade the A/D community, designers who choose *not* to use computers will have to work harder to justify such a decision.

Liability for failure to verify a computer software's accuracy or suitability. Professionals may be held liable for using software without first verifying that it produces accurate results or is suitable for the uses to which it is put. Thus, a designer could be held responsible for not doing some field verifications, for example, or for not checking a string of dimensions in an AutoCAD file or the results of a formula used in a spreadsheet program.

---

[*]Harrison Streeter, *Professional Liability of Architects and Engineers* (NY: John Wiley & Sons, 1988), p. 72.

Liability for improper use of computer software. Using a new software package without fully understanding how it works or without undergoing proper training can produce incorrect or suboptimal results. Designers may be held liable for using software unsuited for a particular task or for spending excess time (and client fees) using the software because they didn't understand how to operate it properly.

Liability for a computer's errors. Although computer hardware and software may not be prone to *human* errors, they are prone to programming flaws, and reliance on them can sometimes cause tragic errors in judgment. Although the ultimate liability for these errors may rest with the software manufacturers, designers may be forced to deal in the interim with problems arising from software and hardware problems. They may also be liable for using a flawed program if these flaws were known in the industry, if the users ignored instructions or warnings provided by the software manufacturer, or if they did not immediately report problems they experienced to the manufacturer.

Liability for coordination failures. Used effectively, AutoCAD and other computer software are powerful project coordination and quality-control tools (see Chapters 5 and 6). But if CAD files are distributed without proper monitoring, the potential increases for errors and inconsistencies to crop up as a result of discrepancies in base floor plans and other shared data. As designers are usually the party responsible for coordinating the efforts of many consultants, they can be the first to face liability in this regard.

Liability for professional misconduct (malpractice). Professionals are expected to perform services that, while not perfect, exhibit "due diligence and reasonable care." Failure to do so can result in malpractice suits. The care with which designers are expected to review their hand-produced work applies to computer-based work as well. Computers augment but cannot replace professional judgment. Even if a major error in a building design was caused by a computer "bug," for example, a designer is still expected to use professional expertise and common sense in evaluating the results of CAD and questioning them as necessary.

Protecting Yourself against Liability for CAD

Designers who use AutoCAD and other software can take some basic steps to protect themselves against liability and other legal problems. These steps, however, should be taken in consultation with legal and insurance professionals who are familiar with issues pertaining to the A/D community. Without attempting to become computer "techies," designers have a professional obligation to remain current with the latest developments regarding electronic creation and data transfer as they affect architects and interior designers. They must also be aware that computer law is a dynamic business; overnight new government legislation or a U.S. Supreme Court or Justice Department decision could substantially alter the nature of design practice.

Following are some steps you can take to reduce liability:

- Budget and schedule projects to permit time for proper project coordination and quality control. This applies to CAD and non-CAD projects alike. And quality control should be performed by senior staff (if not principals) who are qualified to evaluate the work being issued.

- With any client, contractor, consultant, or other party, write a contract that spells out clearly *what* computer-based work you are to produce, by when, and in which file format (such as *.dwg*, *.dxf*, or *.iges*). Contracts should state explicitly whether you are responsible for testing file translations (if so, work the time for this into your fees) or whether the other party is responsible.

- When clients provide written standards for the format of the CAD files they will receive, append these standards to the contract. Thereafter, any changes to CAD standards that the client requests justify additional fees.

- In applicable contracts, include a paragraph about the status of CAD files once they are turned over to other parties at a project's completion. The paragraph can include the following stipulations:

    1. A finite time frame (usually 90 days) during which you will agree to change or correct any problems with the files.

    2. A disclaimer refusing responsibility for problems arising from future use of the files once the client has accepted them.

    3. A statement that CAD files are provided for the user's convenience only. In terms of what the designer is responsible for, only hard-copy drawings with a signature and stamp should be your responsibility.

    4. A reminder that all conditions, dimensions, and other information in the drawings must be verified. Even when a CAD file reflects as-built information, the information it contains must always be double-checked against reality.

- If a project requires acquisition, in-house customization, or other special uses of computer software, the ownership of the software, device, or work should be clarified in the contract. The AIA form B163: *Standard Form of Agreement between Owner and Architect for Designated Services* (1993 edition) lists such activities under the category of special services.

- Before giving CAD files of completed projects to clients, strip them of information specific to your firm, such as firm name, address, and telephone number. Unless the client requests otherwise, strip the files also of dimensions, architectural targets, notes, and other elements of the "execution" of your design. The CAD file disclaimer you issue in contracts, as described above, can be replicated on preprinted diskette labels.

Consider providing CAD "as-builts" on CD-ROM rather than on diskettes. Clients can read your files and copy them to their computer, but they can never overwrite or alter the files on the "record" CD-ROM you've provided.

- Be an informed consumer of computer products and their applications. While senior-level design firm staff need not become software experts, they must understand the purpose, basic features, and suitability of any product for which they authorized purchase. The same goes for hardware.

- Research new computer applications with care before approving their use and purchase. This applies particularly to new or obscure products and free "shareware." The major software producers—Autodesk, Adobe, Microsoft—have enough of an audience that flaws in their products are disseminated to the general public fairly quickly. But smaller producers may not have enough market share that product flaws would receive timely or extensive publication.*

- Develop procedures for hiring, reviewing, and training staff who use computers, especially in regards to more complex applications. Since you are responsible for their work, you want your employees to produce work that you have confidence in.

- Include in your office manual procedures for transferring, distributing, filing, and archiving data in electronic form. This includes procedures for e-mail.

---

*Several professional associations, including the American Bar Association, have begun to develop guidelines for appropriate uses of computers and even specific software packages in their work. Eventually, the American Institute of Architects, Institute of Business Designers, and related professions may produce similar guidelines for the A/E/C community.

- Keep paper copies of every electronic document of value in your system. Make sure that both your electronic and paper filing systems enable you to retrieve any sort of document with ease.
- Revise office standard forms for transmittals, memos, and other communications so that users of these forms are reminded to document any computer files that accompany paper-based communications.
- Never, ever use a computerized facsimile of a registered architectural stamp and signature. Stamps and signatures should always be done by hand on prints and, even then, with discretion. With the low cost today of scanning and with Optical Character Recognition software, you may be tempted to use a computer-based facsimile of your signature for routine correspondence—DON'T!
- Charge explicitly for any CAD-related services that go beyond the industry norms. You should not give your work away for free (doing so automatically downgrades its value), and if you are to assume responsibility for your CAD work, you should be compensated for it.
- Keep abreast of developments in computer software and hardware to the same degree that you do in other areas of professional practice. As computers become more of a mainstream tool in architecture and interior design, an informed practitioner will be presumed to have current and well-rounded knowledge of the computer status quo in design applications.
- As computer use expands in your firm, apply a general principle of tort law: the greater the potential harm of your actions, the more care and prudence you must exhibit.

Insuring Data

Designers might imagine that their creativity and their reputation are their most valuable professional asset. In today's marketplace, however, your computer system is probably your most valuable *tangible* asset. Not only do the hardware and software you own have a fairly well-defined market value, but the configuration and reliability of your computer system directly affect your productivity and your ability to obtain and maintain design projects. Therefore, protecting your computer system is essential. Being adequately insured is a major part of this protection.

Having insurance is essential for sustaining the potential loss that theft, fire, floods, or other calamities may cause if disaster strikes your computer system (see Chapter 9). Having proper coverage will not only allow you to replace your hardware in the case of loss or damage, it could possibly protect you from financial losses that may be incurred if disasters prevent you from completing work on schedule.

Taking the following steps can help you to make the best use of your insurance coverage. Note, too, the suggested ways you can limit the need to use that insurance.

- Get an insurance policy that reimburses you for the market, that is, current replacement value of your hardware. Since hardware becomes obsolete so quickly, you need to be able to purchase what is on today's market, not models that were discontinued several years ago. The replacement value should reflect enhancements, such as the installation of additional computer memory, to your machines after you purchased them.

- Your coverage should include not only computers and other hardware in the office but any off-site equipment, such as laptops used at construction sites. Depending on the terms, you may also have to purchase coverage for any equipment you lease or rent.

- Register new copies of AutoCAD or any other major software immediately. This way, you will more easily be able to get replacements of any disks that are lost or damaged.

- As described earlier, maintain a detailed and updated computerized inventory of all your hardware and software. Send a copy to your insurance agent and keep another in your safe-deposit box or in another off-site fire-resistant storage space.

- Keep current tape or disk backups of your AutoCAD drawing files as well as your AutoCAD installation. That way you can immediately replace your AutoCAD working environment on a new machine if the original is stolen or destroyed.

- See if your insurance policy covers the cost of renting machines and other downtime costs while you deal with the effects of a loss from theft or fire.

- Find out if your insurance carrier will reimburse you for the costs of *data recreation*, that is, the time spent re-creating computer data that was lost or damaged. If so, you must understand what the dollar value of this data to be re-created is.

- Many insurance companies offer discounts on premiums if you provide deterrents to theft and fire. Such deterrents include officewide smoke and theft alarm systems or locks and security devices for computers. Evidence of reliable backup systems and procedures may enable you to purchase insurance coverage for data failure at favorable rates.

11 The Future for AutoCAD and Design

"I have seen the future, and it works."

—Lincoln Steffens

Many of the computer fantasies about which designers might dream are already *technologically* feasible. Computers can automate routine tasks—from setting up title blocks to redlining—they can produce more accurate and potentially better designs, and they can help us completely visualize a space, in terms of dimensions, light, color, and even sound. Except perhaps for offering genuine creativity and originality, computers today can do almost anything.

Technological *feasibility*, however, does not guarantee its *implementation*. History shows that the existence and awareness of a new technology does not automatically lead to its use. Technology must be packaged and marketed in such a way that consumers feel compelled to give up other ways of doing things. Computers with a strong graphic user interface (GUI), for example, had originally been invented by Xerox, but they had to be repackaged as Apple computers before individuals would eagerly put one in their house. CAD has been available for over two decades, but only in recent years has it begun to saturate design practice. And it has done so more slowly than in other professions that use CAD because designers have remained skeptical of CAD values and have not uniformly seen the advantages of changing the status quo. This reluctance, combined with the substantial investment in terms of cost, customization, and training time—in a field with limited options for either—has slowed down the infiltration of CAD in architecture and interior design. Although CAD use increasingly is becoming ubiquitous for routine 2D production work, its use as a 3D and animation tool is still in its infancy. Therefore, with regards to the architecture and interior design professions, you can still expect to see a gap between what is technologically feasible and what is actually used on a widespread basis.

Despite the difficulties of predicting the future for CAD in design, in certain respects the future is already here. Whether they like it or not, design firms already have to cope with a new reality in design practice. And firms that are not aggressively pursuing technology nonetheless need to be aware of the directions that computers in general and AutoCAD in particular seem to be taking.

Over the past several years, a number of general trends in CAD suggest the following (near-term) developments:

- As CAD programs become more user-friendly for the building design professions, the barrier between drafting and design will diminish, perhaps disappear altogether, and design firms will finally see the impact on their bottom line–profitability that they'd hoped for all along. On the other hand, if design firms do not eliminate the CAD ghetto and do not build in time to train and mentor up-and-coming drafters, they may encourage an even greater divide between the two functions. If so, the benefits of CAD as a design tool will be lost.

- If the design professions continue to split into more and more specialties, the role of architects and designers as project coordinators will increase. Designers will use CAD as a tool for project coordination and work flow management.

- The *traditional* professional distinctions between designers, engineers, clients, and contractors will blur. New distinctions will emerge that are based not so much on defined spheres of designing or discipline but instead on new electronic forms of analysis and decision making.

- More power will continue to fall into the hands of "civilian" computer users and away from technical specialists. More people are now "literate" in basic computer skills and are exhibiting a consumer force that demands that software and hardware be easy and fast, rather than mysterious and time-consuming, to install and use. For many routine business-oriented applications, including CAD, there will be less need to acquire machines other than what can be purchased at a regular computer supply store or mail-order house. In design firms, increasingly a designated CAD or Management Information Science (MIS) manager can be a trained designer who happens to lean towards computer management, rather than a technical expert who may or may not be skilled in design. This will be a boon for small to medium-sized design firms, for which a non–design-oriented computer manager is a major expense.

- More work will be done using the 3D and dynamic features of AutoCAD and other software. Animation and 3D will be the starting point of the design process. Final designs will then simply be "flattened out" into 2D for issuing construction documents.

- The traditional means of conveying information for both clients and contractors will evolve. Designers will use 3D and other tools to convey their design intent more meaningfully and accurately. They will rely, for example, more on isometric viewpoints and photorealistic renderings for routine construction details.

- Less and less work will be produced on the standard oversized drawing sheets (such as 42" long x 30" high). Rather, as designers recognize that CAD is a database that permanently stores information and then produces it in an infinite number of scales, they will feel more comfortable routinely issuing smaller-size drawings that can be copied, faxed, and stored more easily.

- Computer software will offer more "intelligence." Rather than merely being passive information repositories, they will be expected to solve problems and be active participants in the design process.

- More tools will be available for various forms of quality control. At present, we rely on fairly mundane quality-control tools, such as spell-checkers, grammar checkers, and database verifications. Increasingly, however, software will be expected to verify data as it is entered, to cross-check and double-check data entry and assumptions, and to automatically look for errors and inconsistencies of all sorts. We might, for example, rely on a computer to tell us—before the local building inspectors do—whether or not our stairways meet code!

- With technologies such as Object Linking and Embedding (OLE), software will increasingly become focused on *projects*, and the issue of software *applications* will become virtually irrelevant. Documents will be created with little conscious attention to whether the tools used are word processors, spreadsheets, CAD packages, or other types of software. Indeed, CAD as a separate and distinct function may disappear altogether. Instead, composing a series of schematic designs or a set of construction documents will be seen as a form of integrated desktop publishing or as a form of multimedia.

- The increase in object-oriented programs and cross-application data exchange will make the issue of file exchange increasingly irrelevant. The ability to read an AutoCAD *.dxf* file, for example, will become less important as capabilities for copying, pasting, and linking data between applications increase.

- The issue of different computer platforms will disappear or else take totally new form. Just five years ago, the decision about going to DOS, Windows, or Macintosh was crucial for most organizations. Now with products like file translation software and DOS and Windows simulators for the Macintosh, platform selection is less critical.

- More standards for software use and documentation will emerge. As governments and other institutions worldwide recognize the importance of having a shared format, organization, and means of tracking electronic data, they will increasingly require standard formats and documentation methods. International communications channels such as the Internet will encourage standardization.

- New generations of computer processors will continue to demonstrate major increases in speed and reliability over previous generations of computers. Designers will have to rethink how they do their work, however, in order to take advantage of machines that can operate at lightning speed. To do so may require giving up a certain control and "hands-on" involvement in various aspects of the design process; instead, designers will have to "delegate" more to the computer.

- "Hand" work will become more appreciated. As in other fields, such as woodworking and textile production, hand-produced drawing will be replaced by CAD in most situations, and by virtue of its rarity it may become more valuable. Indeed, design firms may go from having difficulty hiring CAD-proficient designers to facing challenges finding designers who are skilled with *non*-electronic pens and pencils!

Autodesk has already begun to respond to these developments. Recent steps by the company indicate possibilities for its future direction. Even designers who choose not to use AutoCAD should monitor Autodesk's evolution, since it will continue to shape the CAD market in general.

- With regard to the architecture and interior design professions, AutoCAD is already facing competition from other software. Less expensive and more focused software such as DataCAD and MiniCAD, as well as Autodesk's own AutoCAD LT for Windows, may appear to be more affordable CAD solutions, especially for smaller design firms. Customers may be willing to give up AutoCAD's power and features for cheaper software that meets their more focused needs. For AutoCAD to compete effectively, it will have to either reduce its price (unlikely, given precedent) or change its product in such a way that it can deliver more value to A/D customers.

- One direction that Autodesk already seems to be pursuing is splitting up AutoCAD and other products into interactive "modules," or component software separate packages that users can purchase selectively as they need to, and then combine with other modules later on. Thus, if you use AutoCAD only for 2D construction drafting, you would purchase a 2D drafting "tool pack" and delay purchasing other modules for functions such as 3D or programming until you actually need them. AutoCAD LT for Windows, while not a true "module," is a step in this direction; it is an affordable package of basic drafting 2D and 3D tools; the drawings it produces can then be used with full-fledged AutoCAD. AutoCAD recently introduced AutoCAD OEM, which is intended for use by third-party developers who want to create specific AutoCAD-driven applications that require only limited components of AutoCAD software.

- Autodesk has been encouraging AutoCAD users towards the Windows platform (and may eventually drop the DOS platform). If AutoCAD can relinquish the need to be DOS-compatible, then it can become a "true" Windows program. AutoCAD could then offer more Windows-specific features and more interaction with other Windows applications. As a result AutoCAD might evolve into a leaner program, since AutoCAD features that are redundant with Windows features could be eliminated.

- Windows notwithstanding, AutoCAD will revise its software to take better advantage of new computer programming developments. AutoCAD Release 13 already demonstrates evidence of pursuing these ends because Autodesk has rewritten AutoCAD's *.dwg* file format, improved its modeling of geometric entities, and introduced the AutoCAD Runtime Extension (ARX), an object-oriented development feature.

- Autodesk will continue to expand and redefine its mission. Recently the corporation realigned itself into five market groups, including divisions for A/E/C and facilities management; GIS; and "data" publishing. With products such as WorkCenter and View, Autodesk is moving into the technical document software arena. Recent product introductions as well as acquisitions of software producers in the fields of GIS and mechanical engineering suggest that Autodesk may be moving more aggressively into the vertical application markets that it traditionally has left to third-party developers. Reflecting the previously mentioned trend away from software *applications* towards a more holistic focus on user *projects*, AutoCAD and other Autodesk products will evolve into more generic tools while at the same time offering more features of specific interest to the A/E/C community.

  As with most large corporations, including Microsoft, Autodesk has the market share and the financial muscle to move aggressively in many directions. Time will tell whether its corporate strategies for expansion will allow it to expand its industry dominance.

Over the past decade, major shifts in the design world paradigm have occurred and will continue beyond the year 2000. External pressures such as competition and the need to "speak" electronically with other parties will encourage more and more design firms to use CAD and related software. The evolution of AutoCAD and other programs into more user-friendly design tools will appease the CAD skeptics. And as senior designers gradually are replaced by younger ones who have already learned CAD in design school and on their first drafting job, much of the computer phobia that currently infects the industry will disappear.

Whatever emerges in the dynamic field of CAD, as we head into the next century, designers must never forget that their original mission is the cause of good design and servicing their clients. Computer applications should always be selected and used to this end. While CAD can augment and refine good design, it cannot replace or create genuine design talent itself. Used properly, however, CAD can enable good design to flourish in the twenty-first century; it also may help designers expand the somewhat limited opportunities that architecture and interior design offer at present.

Appendix A

Checklist for Configuring/Customizing AutoCAD in Design Offices

Customization Step	*Comments*
1 Install and Configure DOS and Windows.	
Adjust Windows Virtual Memory to accommodate AutoCAD's requirements for hard disk space and swap files.	Review AutoCAD's requirements in conjunction with other Windows applications to be installed on your computer.
Modify Windows defaults to reflect user/office standards and preferences.	Use the Windows Control Panel.
2 Set up Computer Directories and Standards for File Naming.	
Set up directories for AutoCAD directories, project directories, and office standard templates.	Directory names can relate to project name, standard job phases, and the office "hard copy" filing system.
3 Install AutoCAD.	
Choose between typical, custom, or minimal installation options.	
If appropriate, install and configure network version of AutoCAD.	
4 Customize your systemwide AutoCAD configuration.	
Customize system variables through the **Preferences** Dialog Box.	
default drawing name	should be based on a prototypical AutoCAD drawing file
default plot file name	can reference drawing name/number that plots rather than file name that produced the plot
automatic Save frequency	15 to 20 minutes minimum frequency recommended
full-time CRC validation automatic audit	useful if you frequently read and write a lot of *.dxf* or *.dxb* files

Checklist for Configuring/Customizing AutoCAD in Design Offices, continued

Customization Step	Comments
log-in name	
file locking	essential for shared files on networks
spelling dictionaries	coordinate with other software packages used in the office
default typefaces	should reflect office standard typefaces used in other Windows applications
other AutoCAD **SetVars**	
Customize AutoCAD command tools	
keyboard commands and "hot" keys or key "aliases"	Create keyboard templates to place over keyboard function keys to summarize customized commands.
command line	
menus	
toolbars and "flyout" menus	can create custom tool buttons
digitizer tablet menus	With key commands and toolbars, digitizer tablet menus are redundant for basic AutoCAD commands; use for custom symbol libraries and AutoLISP routines.
mouse buttons	Most CAD users rarely use more than four or so.
dialog boxes	requires skill with AutoLISP routines

5 Customize your AutoCAD drawing environment.

Set drawing units.	normally architectural or (for metric) decimal units
Set drawing unit precision.	Usually 1/16" or 1/8" is adequate, depending on the scale of the drawing.
Set drawing limits.	Normally lower limits are 0,0; upper limits match a typical drawing-sheet size, adjusted by a scale factor.
Set scale factors.	
Model Space versus Paper Space	Determine if you will lay out drawing sheets using Paper Space.
block scales	Also will vary according to individual block; specify general approach up front; can be "true" scale or designed to be scaled up or down upon insertion.
celtscales	
LTScales	
dimscales	
Set linetypes.	Should reflect office standard linetype graphics review in terms of **LTScale** and **Celtscales** and target pen weights; can add text and shapes to linetypes.
linetypes for basic construction lines	
linetypes for column grids, limits of work, and other designations	
Set layer standards.	Should consider all possible applications for CAD drawings; see Chapter 7 for layer-naming guidelines.

Checklist for Configuring/Customizing AutoCAD in Design Offices, continued

Customization Step	Comments

Set type styles.
    styles for drawing titles
    styles for design firm information
    styles for client information
    styles for consultants information
    styles for construction notes and legends
    styles for details and elevations
	should relate to office-standard typefaces and also be selected for legibility and space requirements

Set dimensions.
    styles for linear dimensions
    styles for angular dimensions
    styles for leaders and notes
    styles for alternate units (metric/inches combinations)
	Dimension text should relate to office-standard typefaces but be modified for legibility and fit. Apply custom arrows for leaders and ticks or dots for linear dimensions.

Set hatches.
    hatches for building materials
    hatches for space designations
	Hatches should be tested for appropriate scale factor. Should be on separate layer in order to "hide" as necessary. Can add own custom hatches.

Set multiline styles. multiline styles for wall and partition types	**Mlines** representing office standard wall types can be coded as such.
6 Create prototypical AutoCAD drawings, templates, and libraries.	
AutoCAD prototypical drawings	normally architectural or (for metric) decimal units

cover sheets
        30" × 24"
        42" × 30"
        48" × 36"
        other as required
	can include attributed blocks with fields for typical information: client, design firm, consultants, project name, address, and so on

plan sheets
        30" × 24"
        42" × 30"
        48" × 36"
        other as required
    include title and plot information, layers, legends, and reference grids and guidelines
	Can include attributed blocks for title and plot information; AutoLISP routines will ask you to provide comprehensive design information and then will set up title block accordingly.

elevation sheets
        30" × 24"
        42" × 30"
        48" × 36"
        other as required
    include title and plot information, layers, legends and reference grids and guidelines
	provide a separate layer for grids and guides with lines for typical elevation heights, such as soffits, ceilings, and HP fixtures

Checklist for Configuring/Customizing AutoCAD in Design Offices, continued

Customization Step	Comments

detail sheets
    30" × 24"
    42" × 30"
    48" × 36"
    individual 8 1/2" × 11" or 8 1/2" × 14"
    other as required
	Use Paper Space and External References (**Xrefs**) to facilitate layout of sheets with details at multiple scales; include grid and guidelines for layouts.

other standard sheets
    A0 sheets for legends, abbreviations and symbols
    doors, hardware, and windows
    code analysis drawings
    building envelope studies
sketch sheets
    8 1/2" × 11"
    8 1/2" × 14"
    11" × 17"
    other as required
	Paper Space–based forms allow you to "zoom" in on portions of existing drawings at your preferred scale.

include title and plot information, layers, reference grids, and guidelines	

presentation materials
    programming and diagrammatic templates
    finish color boards
    3D presentations and renderings
    slide presentations
    program statements
	Use pretested palettes for colors, materials, lights, and other elements of rendering and color presentations.

AutoCAD block libraries and attendant AutoLISP routines and Scripts	
partition plan legends and symbols	Include AutoLISP routines for partition types.
reflecting ceiling plan legends and symbols	Include AutoLISP routines for laying out test ceiling grids.
power and signal plan legends and symbols	Include attributes that indicate whether floor, wall, or ceiling mounted, number of circuits.
finish plan legends and symbols	can link attributed finish targets to finish legends and schedules
plumbing plan legends and symbols	include HP standard layouts
mechanical plan legends and symbols	
revision legends and symbols	Use AutoLISP routines to draw clouds with revision references attached.
site and landscaping legends and symbols	Use AutoLISP routines to calculate growth rates of plant materials and to lay out HP-compliant parking lots.
furniture and equipment legends and symbols	Use attributed blocks to link items to pricing databases.
handicapped person graphics	See Chapter 7 for typical standard libraries.

standard layouts for enlarged plans
    elevators
    entryways and lobbies
    HP accessible layouts
    stairwells
    toilet room	

Checklist for Configuring/Customizing AutoCAD in Design Offices (continued)

Customization Step	Comments

elevation symbols
        doors and windows
        elevators
        equipment
        exterior facades
        HP accessible fixtures
        stairways
        toilet room fixtures and accessories	

Detail libraries
    door details
    millwork details
    partition details
    roof details
    site details
    window details
    other details	

7 Customize output devices and other peripherals.	

Plotters
    Specify pen settings for all typical output.
    paper sizes
    plotting windows and views
    network queues
    default plot file names
    other plotting parameters
	See Chapter 8.

Printers	Customize as for plotters.
Modems	Test for baud rate; create automated address books for frequently dialed numbers.
Scanners	Test for optimal dpi and color settings and file format conversions.
Slide viewers	
Develop form for tracking plots for reimbursables	See Chapter 8.
Research and establish relationship with plotting/reprographics service.	See Chapter 8.
Research, test, and stock inventories of appropriate media, pens, and ink.	See Chapter 8.
Collect and distribute information on plotting schedules and turnaround time.	See Chapter 8.
8 Customize special AutoCAD applications and links to other software.	
Scripts	
AutoLISP routines	
ADS routines	

Checklist for Configuring/Customizing AutoCAD in Design Offices (continued)

Customization Step	Comments
3D and animation rendering views and palettes	See Chapter 7.
Programmable dialog boxes	

Database and OLE Links
    budgets
    client programs
    schedules for doors, windows, finishes, and so on
    specifications
	See Chapters 6, 7, and 10.

File conversion routines	
9 Address miscellaneous CAD-related tasks.	
Develop standard marketing materials and information for RFPs concerning CAD.	See Chapters 6, 7, and 10.
Create an annual budgeting form for CAD-related income and expenses.	See Chapter 9.
Develop standard contracts or clauses for CAD-related services; review with lawyers.	See Chapters 9 and 10.
Create project management forms that incorporate CAD-related fees, staff, and other issues.	See Chapters 5 and 9.
Create CAD standards manual.	See Chapter 9.
Organize a CAD Technical Committee.	See Chapter 9.
Create in-house training curriculum.	See Chapter 9.
Compile sources for additional training in CAD.	See Chapter 9.
Develop procedures for interviewing, evaluating, hiring, and promoting employees with CAD skills.	See Chapter 9.
Create written job descriptions for CAD-skilled staff.	See Chapter 9.
Reconfigure office workstations and work areas to accommodate new equipment.	See Chapter 9.
Review office security systems.	See Chapter 9.
Create and maintain computer inventory forms and computer licenses.	See Chapter 10.
Develop and document computer maintenance and backup procedures.	See Chapter 10.
Create documentation and procedures for distributing computer files to colleagues and clients.	Modify and create special transmittal forms, diskette labels, and so on, with appropriate disclaimers; see Chapter 10.
Develop a procedure and forms for archiving CAD files.	See Chapter 10.

Checklist for Configuring/Customizing AutoCAD in Design Offices (continued)

Customization Step	Comments
Review insurance coverage.	See Chapters 9 and 10.
Develop a system for importing AutoCAD to desktop publishing and marketing files.	See Chapters 7 and 10.
Review and document preferred file-transfer methods for moving CAD files between AutoCAD and other programs.	See Chapter 10.

AutoCAD Drawing Scale/Scale Factor Matrix

Plotted Drawing Scale	Standard Scale Factor[a]	Linetype Scale[b]	Drawing Limits for Selected Standard Drawing Sizes[c] (Drawing Limits shown refer to upper X,Y coordinates and assume lower X,Y coordinates are 0,0)					Sample AutoCAD Text Conversion Heights for Plotted Text	
			8 1/2" W × 11" H (Vertical or "Portrait")	17" W × 11" H (Horizontal or "Landscape")	36" W × 24" H	42" W × 30" H	48" W × 36" H	3/32" H[a]	1/8" H (9 pt.)[d]
1" = 1" (Full scale)	1	0.5	8.5" × 11"	17" × 11"	3'-0" × 2'-0"	3'-6" × 2'-6"	4'-0" × 3'-0"	0.09375	0.125
6" = 1'-0"	2	1	1'-5" × 1'-10"	2'-10" × 1'-10"	6'-0" × 4'-0"	7'-0" × 5'-0"	8'-0" × 6'-0"	0.1875	0.25
3" = 1'-0"	4	2	2'-10" × 3'-10"	5'-8" × 3'-8"	12'-0" × 8'-0"	14'-0" × 10'-0"	16'-0" × 12'-0"	0.3750	0.50
2" = 1'-0"	6	3	4'-3" × 5'-6"	8'-6" × 5'-6"	18'-0" × 12'-0"	21'-0" × 15'-0"	24'-0" × 18'-0"	0.5625	0.75
1 1/2" = 1'-0"	8	4	5'-8" × 7'-4"	11'-4" × 7'-4"	24'-0" × 16'-0"	28'-0" × 20'-0"	32'-0" × 24'-0"	0.7500	1.00
1" = 1'-0"	12	6	8'-6" × 11'-0"	17'-0" × 11'-0"	36'-0" × 24'-0"	42'-0" × 30'-0"	48'-0" × 36'-0"	1.1250	1.50
3/4" = 1'-0"	16	8	11'-4" × 14'-8"	22'-8" × 14'-8"	48'-0" × 32'-0"	56'-0" × 40'-0"	64'-0" × 48'-0"	1.50	2.00
1/2" = 1'-0"	24	12	17'-0" × 22'-0"	34'-0" × 22'-0"	72'-0" × 48'-0"	84'-0" × 60'-0"	96'-0" × 72'-0"	2.25	3.00
3/8" = 1'-0"	32	16	22'-8" × 29'-4"	45'-4" × 29'-4"	96'-0" × 64'-0"	112'-0" × 80'-0"	128'-0" × 96'-0"	3.00	4.00
1/4" = 1'-0"	48	24	34'-0" × 44'-0"	68'-0" × 44'-0"	144'-0" × 96'-0"	168'-0" × 120'-0"	192'-0" × 144'-0"	4.50	6.00
3/16" = 1'-0"	64	32	45'-4" × 58'-8"	90'-8" × 58'-8"	192'-0" × 128'-0"	224'-0" × 160'-0"	256'-0" × 192'-0"	6.00	8.00
1/8" = 1'-0"	96	48	68'-0" × 88'-0"	136'-0" × 88'-8"	288'-0" × 192'-0"	336'-0" × 240'-0"	384'-0" × 288'-0"	9.00	12.00
3/32" = 1'-0"	128	64	90'-8" × 117'-4"	181'-4" × 117'-4"	384'-0" × 256'-0"	448'-0" × 320'-0"	512'-0" × 384'-0"	12.00	16.00
1/16" = 1'-0"	192	96	136'-0" × 176'-0"	272'-0" × 176'-0"	576'-0" × 384'-0"	672'-0" × 480'-0"	768'-0" × 576'-0"	18.00	24.00
1/32" = 1'-0"	384	192	272'-0" × 352'-0"	544'-0" × 352'-0"	1152'-0" × 768'-0"	1344'-0" × 960'-0"	1536'-0" × 1152'-0"	36.00	48.00

[a] If you work in Paper Space, your drawing sheets can be set in a consistent paper scale factor of 1" = 1". You then specify scale factors to link objects created in Model Space to your Paper Space sheet. For example, a floor plan created in Model Space which is designed to plot at 1/4" scale would be linked to 1" Paper Space by a scale factor of 48.

[b] Scales for linetypes, dimensions, hatches, and other graphic elements can match the standard drawing scale factor. However, most architects and designers find the results unsatisfactory and will adjust them as necessary. As an example, the linetype scale factor shown here is set to half the drawing scale. Always test various scales before selecting one and then document it and use it consistently.

[c] To arrive at the upper limits of a drawing sheet size which is not shown here, multiply the paper width and height each times the appropriate scale factor times 1/12 to arrive at the upper X- and Y-coordinates respectively. For a 48" × 30" drawing sheet plotted at 1/8" scale, for example, the upper limits would be 384' (48 × 96 × 1/12) for the X-coordinate and 240' (30 × 96 × 1/12) for the Y-coordinate.

[d] To translate AutoCAD text, which reads in feet and inches when architectural units are used, to standard font heights expressed in points, remember that 1 point equals 1/72" or .01388". The easiest way to relate AutoCAD text heights to points is by specifying AutoCAD heights in decimal inches, such as 0.1388" for 10 pt. text, 0.1666" for 12 pt. text, and so forth.

Appendix B: AutoCAD Resources

Listed below is information on resources for AutoCAD and related topics. This information was, to the best of the author's knowledge, accurate at the time this book went to press. It is always wise, however, to call and confirm numbers and addresses. Please note that the addresses and telephone numbers listed here pertain only to the United States. For resources in other countries, please contact the Autodesk dealer in that country.

One of the most valuable resources on many aspects of AutoCAD is *The AutoCAD Resource Guide*, which ships with all licensed copies of AutoCAD software. Updated quarterly, the *Guide* lists all Autodesk products as well as products produced by third-party developers, including products specific to architecture and interior design. It also describes various training tools, locations of Autodesk Training Centers, and AutoCAD user groups. You can order additional copies from your local Autodesk dealer or by calling (800) 964-6432.

General Autodesk Information

Corporate Headquarters

Autodesk, Inc.
111 McInnis Parkway
San Rafael, CA 94903
USA

General Telephone Number: (415) 507-5000
General Fax Number: (415) 507-5100
Autodesk Information Line: (415) 507-5927

Sales and Product Information

For the names of Authorized Autodesk Dealers in your area and for product brochures and demonstration software: (800) 964-6432

For Autodesk Multimedia: (415) 507-5666

Technical Support

- Telephone: Contact your local Autodesk Dealer (see above)
- Fax: The Autodesk FAX Information System is available twenty-four hours a day to send you technical support on a wide range of topics. Call (415) 507-5595 (press 1 to hear a list of documents that can be sent to you by fax).
- Global Village Electronic Bulletin Board (BBS) modem line: (415) 507-5921

AutoCAD on the Internet

CompuServe Forums

- For general Autodesk information: GO ADESK
- For information about AutoCAD: GO ACAD
- For information about Autodesk Multimedia: GO AMMEDIA
- For information about Autodesk Retail Products: GO ARETAIL

Internet File Server

- fpt.autodesk.com

World Wide Web (Mosaic)

- http://www.autodesk.com

Books and References

The ultimate source of basic technical information is AutoCAD's own documentation, which is included with all licensed copies of AutoCAD software. This documentation includes general installation and customization instructions, a command reference, and tutorials.

The basic AutoCAD documentation is augmented by many books written by third-party authors. Many of these authors participate in the Autodesk Registered Author/Publisher Program, through which they receive special technical support and access to information about developments with AutoCAD and other related products.

Recent publications on AutoCAD are listed in *The AutoCAD Resource Guide*, mentioned above. In addition, many major publishers of computer references publish books about AutoCAD, and most are available at any bookstore that has a comprehensive selection of computer books. A number of bookstores and mail-order services also carry CAD references, including several that focus specifically on architectural and interior design applications.

The AIA Bookstore
1735 New York Avenue, N.W.
Washington, D.C. 20006-5292
(202) 626-7475

The Architects & Designers Book Service
3000 Cindel Drive
Delran, New Jersey 08370-0001
(800) 854-3233

CAD News Bookstore
1580 Center Drive
Santa Fe, NM 87505-9746
(800) 526-2665

Charrette Corporation
31 Olympia Avenue
P.O. Box 4010
Woburn, MA 01888-4010
(800) 367-3729

Periodicals on AutoCAD and Related Topics

A/E/C Systems Computer Solutions: The Computer Applications and Management Journal for Design and Construction. Published bimonthly by A/E/C/ Systems, Inc., P.O. Box 310318, Newington, CT 06131-0318 Tel: (203) 666-1326. $36.00 for a one-year subscription.

AutoCAD Tech Journal. Published quarterly by Miller Freeman, Inc., 600 Harrison Street, San Francisco, CA 94107, Tel: (800) 289-0484. $59.80 for a one-year subscription.

CADalyst. Published monthly by Advanstar Communications, Inc., 859 Williamette Street, Eugene, OR 97401, Tel: (800) 346-0085. $39.00 for a one-year subscription.

CADENCE. Published monthly by Miller Freeman, Inc., 600 Harrison Street, San Francisco, CA 94107, Tel: (800) 289-0484. $39.95 for a one-year subscription.

Inside AutoCAD: Tips and Techniques for Users of AutoCAD. Published monthly by the COBB Group, 9420 Binsen Parkway, Suite 300, Louisville, KY 40220, Tel: (800) 223-8720. $109.00 for a one-year subscription.

Organizations

North American AutoCAD User Group (NAAUG)
P.O. Box 3394
San Rafael, CA 94912-3394
Tel: (800) 964-6432

NAAUG is a national organization of AutoCAD user groups that, although it operates independently from Autodesk, is recognized by Autodesk. It publishes a regular newsletter and sponsors both local and national meetings and user groups. It also provides a channel for independent evaluations of Autodesk products. See *The AutoCAD Resource Guide* for user groups in your area.

In addition to NAAUG, any local chapter of an A/D professional organization may also sponsor CAD user groups or conduct occasional seminars on CAD-related issues (often for Professional Continuing Education credits). The publication *A/E/C Systems Computer Solutions* is a useful resource that provides comprehensive information about CAD-related activities in building-related industries. Its annual *Software Buyer's Guide* and the *A/E/C Conference Planner* are troves of information. Organizations that may have specific information on the use of computers in the design professions include:

The American Institute of Architects (AIA)
1735 New York Avenue, N.W.
Washington, D.C. 20006-5209
Tel: (202) 626-7000

CADD Management Institute
12 South Sixth Street, Suite 914
Minneapolis, MN 55402
Tel: (612) 333-9222

Construction Specifications Institute (CSI)
601 Madison Street
Alexandria, VA 22314-1791
Tel: (703) 684-0300

Interior Design Educator's Council (IDEC)
P.O. Box 3433
Chicago, IL 60654-0433
Tel: (312) 527-0517

International Facilities Management Association [IFMA]
1 East Greenway Plaza, 11th Floor
Houston, TX 77046-0194
Tel: (800) 359-4362

National Institute of Building Sciences (NIBS)
1201 L Street N.W., Suite 400
Washington, D.C. 20005-4024
Tel: (202) 289-7800

Note: NIBS sponsors the CADD Council, a forum that addresses the use of CAD and other software in various facilities.

Society for the Marketing of Professional Services (SMPS)
99 Canal Center Plaza, Suite 320
Alexandria, VA 22314
Tel: (703) 549-6117

Training, Education, and Placement Services

Autodesk sponsors Autodesk Training Centers (ATCs) at a number of colleges, universities, CAD institutes, and other institutions throughout the United States and abroad. For a current list of locations, see *The AutoCAD Resource Guide* or contact Autodesk at (415) 507-5000. ATCs designated as Premier ATCs offer courses in advanced topics, such as AutoLISP routines and multimedia applications.

In addition to ATCs, most design and technical schools offer introductory courses in AutoCAD, often with a focus on architectural and interior design applications. Many continuing education programs offer courses in CAD as well.

To parallel the growth of consultants and temporary employees in the design industry, a number of placement firms that specialize in placements for people with CAD skills have emerged. Many of these firms also offer courses in CAD, and some now qualify as ATCs as well. A summary of placement firms that cater to the A/E industries can be downloaded to members of the AIA Internet service, AIAOnline. For information call (800) 864-7753.

Continuing Education and Professional Development

A number of organizations sponsor conferences, trade shows, and seminars that are intended to provide ongoing professional education in various aspects of computers in professional practice. Some events are targeted purely to computer applications in general; others are designed for practicing designers who need to keep up with technological trends in their field. For some programs, you may receive a Continuing Education Unit credit (CEU) from AIA, IIDA, or other organizations to which you belong.

For CAD applications in particular, two events are particularly worth noting.

Autodesk University

This is an annual conference and exhibition, usually held in the fall, for users of Autodesk products. It is targeted to all CAD-using disciplines and is especially suitable for CAD users who want to take their skills to a higher level or who are interested in special applications of Autodesk products. It is also an effective way to see demonstrations of many third-party AutoCAD products and to meet their developers. Contact Autodesk at (415) 507-5000 for details.

The A/E/C Systems Show

This is an annual conference and exhibition, usually held in June, for computer users in the fields of design and construction. The event includes seminars on incorporating CAD and related tools in various aspects of A/E/C management. It features all major CAD systems, although AutoCAD and related products are well represented. Call (800) 527-7943 for additional information.

In addition to these events, the major professional organizations (see above) also feature seminars at their annual events; these are often listed in periodicals and local chapter mailings. Typical events include the Boston Society of Architects Build Boston Convention and the NEOCON show at Chicago's Merchandise Mart.

Glossary

This glossary provides brief definitions of selected AutoCAD-related terms that appear in this book. These terms are defined from a managerial rather than a technical viewpoint. For more elaborate and technical definitions, see the AutoCAD documentation that ships with each licensed copy of AutoCAD as well as third-party technical references, many of which are listed in *The AutoCAD Resource Guide* (see the section on "AutoCAD Resources"). Please note that some of the terms here might be defined somewhat differently when applied to other software packages.

ACI—AutoCAD Color Index. AutoCAD's own palette of 256 colors. The palette can be viewed through the **DdColor** command.

ADI—Autodesk Device Interface. specifications for printers, plotters, and other peripherals that will guarantee that these peripherals will respond to AutoCAD commands for output

ADS—AutoCAD Development System. AutoCAD's own variant on the C programming language. ADS programs are similar to AutoLISP routines (see below) but are more complex and powerful.

Aec—Architecture/Engineering/Construction. In computers, Aec is often linked together in terms of computer applications of interest to these building-related professions.

AME—Advanced Modeling Extension. AutoCAD's tools for modeling mathematically complex 3D solids and regions

ASCII—American Standard Code for Information Exchange. An internationally recognized file-format standard that assigns numeric values to letters, numbers, and punctuation marks. AutoCAD files in ASCII format can be read and edited in a word-processing program.

ASE—AutoCAD SQL Extension. an AutoCAD application that links graphic information in AutoCAD drawings to nongraphic information in external database programs

Attributes. data fields that can be attached to AutoCAD blocks and can store alphanumeric information. Attributes can be linked to external databases.

AutoLISP Routines. AutoCAD's own programming language variant on the popular common LISP compiled programming language. AutoCAD software relies on many built-in AutoLISP routines, and you can develop your own custom AutoLISP routines.

Baud Rate. This measures the rate at which a telephone line transmits data, such as a fax, an electronic mail message, or files. Transmission speed may also be expressed in terms of bits (or characters) transmitted per second (bps). Typical baud rates range from 300 to 28800 with most fax/modem devices today offering rates of 1200, 2400, 9600, and 14400.

Blocks. AutoCAD's term for user-defined drawing entities that can be used repeatedly throughout drawings. They are equivalent to *cells* or *symbols* in other CAD programs.

CAD—Computer-Aided Design (and Drafting). refers to both the process and the various software packages available for drafting and designing on computers. Both CAD and CADD are used, although these days CAD is more prevalent.

CAD-CAM—Computer-Aided Design and Manufacturing. refers to both the process and the various software packages available for not only designing but also fabricating manufactured parts

CAD Usage Rate/Utilization Factor. A ratio that measures how heavily CAD is used in an organization. It is calculated by dividing the number of hours CAD is used by the number of hours a firm determines it is available to be used. The higher the ratio the heavier the CAD use. The ratio can be applied to peripherals, such as printers and plotters, as well as to other software.

CAFM—Computer-Aided Facilities Management. software that links CAD drawings to external databases so that organizations can better manage their facilities and the people and equipment in them

CAL—The AutoCAD Calculator. AutoCAD's built-in mathematical and geometric calculation tool, accessed by the command **Cal**

Centralized Network Operating Systems. Centralized network systems, or server-centric systems, rely on a specific dedicated file server as the central controlling spoke of the office network. The file server in this setup normally controls the distribution and backup of files and is the computer to which peripherals, such as plotters, printers, and fax modems, are attached. Centralized networks are preferable for large network systems (ten or more machines), when complex SQL databases or multitasking software packages are used, or when you want to take advantage of more high-powered features of AutoCAD, such as simultaneous use of **Xrefs** or links to SQL databases.

CIM—Computer-Intergrated Manufacturing. refers to the process of using computers for many or all stages of product manufacturing

Clients and Servers. terms that refer to the relationship between computers or between computer files. A computer file server stores files on a computer hard drive and makes them available to other computers (or clients). A server software file is a source file that makes itself available for use by other computer files (or clients).

Computer Phobia. a common syndrome, computer phobia is a form of behavior that is manifested by general avoidance of anything to do with computers and a demonstrable aversion to their use and output. A common behavioral pattern among non-CAD literate members of design firms.

ConDoc. a documentation formatting system developed by, and for, architects. It offers a uniform, systematic approach to organizing construction drawing names and layout, notation, and specifications. Although it has not been uniformly accepted by the American architecture profession, many architects use it and have been inspired by it to develop their own standards for construction documents.

Coordinate Systems. Drawing on basic trigonometry, AutoCAD defaults to the World Coordinate System (WCS) of X, Y, and Z on orthogonal axes. Within AutoCAD, users may apply their own coordinate systems (UCS), and objects may have their own coordinate systems (OSC).

Dialog Boxes. a common software device to enable users to view and select multiple command options simultaneously, thereby allowing users to consider more options simultaneously and to avoid the tedious process of scrolling through individual command lines. AutoCAD commands that use dialog boxes start with the prefix "**Dd**."

DOS—Disk Operating System. an operating system that has run on most IBM-compatible personal computers for over the past decade. Currently it is marketed by Microsoft, but it may eventually be replaced by Microsoft Windows.

.*dwg* File Format—Drawing File. AutoCAD's own file format for drawings. Files must recognize and use the *.dwg* extension to be opened by AutoCAD.

.dxf format—Drawing Interchange Format. a file format developed by Autodesk for facilitating drawing exchange between different versions of AutoCAD and between AutoCAD and other CAD software. Generally the standard drawing exchange format used within the A/E/C communities.

External References (Xref command). AutoCAD's own take on OLE, External References link together separate AutoCAD files to form a composite drawing.

FEA—Finite Element Analysis. a computer-generated model of a machine or component that simulates the materials and conditions an object will undergo in use

File extensions. standard terminology for an optional suffix at the end of file names that helps identify what application created the file or what applications the file can be used with. File extensions have up to three letters, and in DOS and Windows 3.11, the extension is separated from the file name with a ".". An AutoCAD drawing file has the file extension *.dwg*. A text document normally has a *.doc* or *.txt* extension.

File server. standard terminology for a computer, usually on a network, that stores and distributes computer files and applications for other computers to access or retrieve. File servers usually serve as the primary machine for hooking up a group of computers to printers and other peripherals.

Fonts. computer files consisting of alphanumeric characters. Each file defines the basic appearance of type styles. AutoCAD provides its own fonts and can also read Microsoft Windows TrueType and Adobe PostScript fonts.

GDT—Geometric Dimensioning and Tolerancing. AutoCAD's built-in feature to provide drafters with ANSI- and ISO-compliant graphic standards for mechanical symbols and notations.

GIS—Geographic Information Systems. a CAD application that stores graphic information about sites and geological surveys

GUI—Graphical User Interface. the visual experience that a software program provides. Microsoft Windows and Apple Macintosh are known for their user-friendly GUI; DOS generally is not.

.iges—Initial Graphics Exchange Specification. an ANSI-standard file format for the exchange of CAD drawing files between different CAD packages. Unlike the *.dxf* file exchange format, *.iges* was not developed by AutoCAD but rather by an independent committee. AutoCAD sells an add-on, the IGES Translator, to enable AutoCAD files to convert to and from *.iges* format files.

Macros. a single command that automates a collection of other commands and is used in many different software packages. Macros are conceptually similar to AutoLISP routines and Script files; all are intended to speed command routines and to build consistency in the sequence of commands.

Model Space versus Paper Space. Model space is the AutoCAD working drafting/design environment for developing and editing objects in 2D and 3D. Model space draws at "true" or full (1"= 1") scale. Paper space is a page-layout working environment that allows you to place objects drawn in model space on a layout sheet for plotting; the objects can be presented in a range of scales and views. You can switch between the two space modes through the AutoCAD **View** menu.

Nested Block. user-defined AutoCAD blocks containing other blocks, including other nested blocks. No limits exist on the number of blocks and nested layers. Nested blocks help conserve memory and facilitate management of graphic entities.

NURBS—Non-Uniform Rational B-Spline. a type of drawn curve that fits a smooth line to a series of points. New to Release 13, NURBS are smoother or more accurate than AutoCAD's previous features for drawing curves.

OLE—Object Linking and Embedding. refers to a Microsoft Windows feature that enables different software and different files to link data and update automatically.

OOP—Object Oriented Programming. a general trend in Windows and Macintosh computer software that organizes programs by objects rather than procedures. OOP facilitates the use of a single object in multiple programs.

Open Architecture. refers to AutoCAD's ability to permit easy customization of its basic features

Optical Character Recognition (OCR). This widely available software feature "reads" blocks of text and then converts them to editable text files. It works well for text that is presented in a text-only format, but often has difficulty converting text that is embedded in graphics or is presented in a highly decorative typeface. It can be used with text files received as computer files through fax software, for scanned images, and for text found in paint programs.

Output Devices. refers to printers, plotters, and other machines that output files into hard copy or—these days—e-mail and other forms of results

Paper Space. See **Model Space**.

Peer-to-Peer Network Systems. Peer-to-peer network systems link various computers and peripherals together without dedicating one computer as a permanent file server. They are the most cost-effective option for small networks where the primary goal is to exchange files and e-mail and to share a few printers or plotters. For larger networks and high-powered multitasking applications and where many users must share a small number of peripherals, peer-to-peer networks get bogged down; upgrading to, and running concurrently with, a centralized network (see above) is preferable.

Peripherals. devices that are attached to computers to perform various functions. Some are *input* devices, such as mice, digitizer pads, and scanners; others are *output* devices, such as plotters and printers.

PLT Format—Plot File Format. AutoCAD's own file format for drawing files that are to be plotted by output devices such as printers and plotters

PostScript. a file format that saves information in Encapsulated Postscript (*.eps*) file format. The *.eps* files are used to generate special graphic fills and smooth typefaces and to work with PostScript-compatible printers and plotters. Available in AutoCAD through the **Psout** command.

Queuing. This refers to the ability of computers and other intelligent machines to enable computer users to line up—as if at a British bus queue—and wait for their turn at printers, plotters, fax/modems, and other devices that are attached to a network. Various network and printer manager utilities are available to permit and to control queues as necessary.

RAM—Random Access Memory. temporary memory for storage of files that you are working on. The more RAM you have, the more memory you have for operating.

Running *inside* and *outside* AutoCAD. refers to the relationship between AutoCAD and other software. Programs running inside AutoCAD need AutoCAD in order to be used themselves (and are often called "add-ons"). Programs running outside AutoCAD may read or write AutoCAD files but don't need AutoCAD to operate.

Script Files. text files with the *.scr* file extension that "script" or automate a series of AutoCAD commands, such as those needed to create plot files or slide shows. Analogous to macros (see above)

Solids. 3D representations of a mathematically defined, internally consistent object, such as a cone, cube, or torus.

SQL—Structured Query Language. a common database language format that is used by relational database programs. Conforms to ANSI X3.135-1989 standards and is imbedded in AutoCAD's ASE feature (see above).

Swap Files. Swap files refer to space on computer hard drives that AutoCAD and other Windows software must "appropriate" as temporary storage space, or "virtual memory," while the software performs various operations. Because AutoCAD alone requires swap file space of at least 40MB, and other applications need swap file space as well, users should have a strong incentive to keep their hard drives as empty as possible (especially when attempting to "multitask"). Swap files can be temporary—for example, they might disappear when an application is closed—or permanent.

TDM—Technical Document Management. refers to software and other tools that provide a comprehensive environment for management of both electronic and paper-based documents. Autodesk's various TDM products, such as Autodesk WorkCenter, is intended to facilititate management of CAD-generated drawings as well as other types of files.

Third-Party Software Developer. independent software developers, publishers, and other parties that develop applications for AutoCAD and other Autodesk products but do not work for Autodesk

TrueType Fonts. a type of computer font—or typeface—that uses a single font file to manage fonts on both the computer screen and printers. TrueType technology is available to both Macintosh and Windows operating systems, and AutoCAD Release 13 for Windows recognizes TrueType fonts.

Utilities. Utilities are software applications that protect and enhance a computer's overall performance and accessibility. Typical utilities include antivirus software, file managers, typeface managers, and memory "doublers." If you think of software applications, such as CAD, text processors, and databases, as rooms in a building, then computer utilities would be the telephone, electrical, and gas utilities.

Virtual Memory. See **Swap Files**.

Virtual Reality. a 3D, dynamic (or animated) view of the world as simulated by computer. Often used synonymously with the term Cyberspace.

Visualization. in computer terminology, refers to the process of using computer software to plan, construct, and analyze 3D and animated forms.

Windows. the popular "Mac-like" PC-based operating *environment* (not an operating system) developed by Microsoft to make DOS more user-friendly. The latest version, Windows 95, eliminates standard DOS and becomes more of an operating system rather than a DOS overlay.

WYSIWYG—"What You See Is What You Get". refers to a software's ability to have its GUI display on a monitor what will actually appear on a printout or plot

Suggested Reading

Angell, David and Brent Heslop. *The Elements of E-mail Style: Communicate Effectively via Electronic Mail.* Reading, Mass.: Addison-Wesley Publishing Company, 1994.

The American Institute of Architects. *AIA Contract Documents.* Washington, D.C.: American Institute of Architects (various dates).

Autodesk, Inc. *The AutoCAD Release 13 Software Documentation Set.* (7 volumes plus electronic documentation). San Rafael, Calif.: Autodesk, Inc. 1994.

———. *The AutoCAD Resource Guide.* San Rafael, Calif.: Autodesk, Inc. (Quarterly publication).

———. *1995 Annual Report.* San Rafael, Calif.: Autodesk, Inc. 1995.

Baker, Robin. *Designing the Future: The Computer in Architecture and Design.* New York: Thames and Hudson, 1993.

Bertol, Daniela. *Visualizing with CAD: An AutoCAD Exploration of Geometric and Architectural Forms.* New York: Springer-Verlag, 1994.

Dib, Albert. *Forms and Agreements for Architects, Engineers & Contractors.* Vol. 1. Deerfield, Ill.: Clark, Boardman & Callaghan, 1994.

Fishman, Stephen. *The Copyright Handbook: How to Protect and Use Written Works,* 2nd ed. Berkeley, Calif.: Nolo Press, 1994.

Fitch, James Marston. *Historic Preservation: Curatorial Management of the Built World.* Charlottesville, Va.: University Press of Virginia, 1990.

Gerlach, Gary M. *Transition to CADD: A Practical Guide for Architects, Engineers, and Designers.* New York: McGraw-Hill Book Company, 1987.

Guzey, Onkal K., AIA, and James N. Freehof, AIA. *CONDOC: The New System for Formatting and Integrating Construction Documentation* (1991).

Grabowski, Ralph. *The Successful CAD Manager's Handbook.* Albany, N.Y.: Delmar Publishers, Inc., 1994.

Hamer, Jeffrey M. *Facility Management Systems*. New York: Van Nostrand, Reinhold, 1988.

Haviland, David, Hon. AIA, ed. *The Architect's Handbook of Professional Practice*, 12th ed., Vols. 1–4. Washington, D.C.: The American Institute of Architects Press, 1994.

Heuer, Charles E., AIA, Esq. *Means Legal Reference for Design and Construction*. Kingston, Mass.: R. S. Means Company, Inc., 1989.

Kay, David C., and John R. Levine. *Graphics File Formats*. 2nd ed. New York: Windcrest/McGraw-Hill, 1995.

Katz, Genevieve. *AutoCAD for the Mac Visual Guide*. Alameda, Calif.: Sybex, Inc., 1994.

Landy, Gene K., *The Software Developer's and Marketer's Legal Companion: Protect Your Software and Your Software Business*. Reading, Mass.: Addison-Wesley Publishing Company, 1993.

McCullough, Malcolm, William J. Mitchell, and Patrick Purcell, eds. *The Electronic Design Studio: Architectural Knowledge and Media in the Computer Era*. Cambridge, Mass.: MIT Press, 1990.

Neeley, Dennis, and Bob Callori. *CAD and the Practice of Architecture: ASG Solutions*. Santa Fe, N.M.: OnWord Press, 1993.

Orr, Joel. *How to Choose a CAD System*. Sausalito, Calif.: Autodesk.

Ramsey/Sleeper. Hoke, John Ray, Jr., editor-in-chief, FAIA. *Architectural Graphic Standards*. New York: John Wiley & Sons, 1994.

Robert McNeel & Associates. *AutoCAD Release 13: Upgrade Guide*. Seattle, Wash.: Robert McNeel and Associates, 1994–95.

Sabo, Werner, AIA, Esq. *Legal Guide to AIA Documents*. 3rd ed. New York: Wiley Law Publications, John Wiley & Sons, Inc., 1992.

Schilling, Terrence G., and Patricia M. Schilling. *Intelligent Drawings: Managing CAD and Information Systems in the Design Office*. New York: McGraw-Hill, 1987.

Schley, Michael K., ed. *CAD Layer Guidelines: Recommended Designations for Architecture, Engineering, and Facility Management Computer-Aided Design*. Washington, D.C.: The American Institute of Architects Press, 1990.

Shu, Evan H., Ruth Neeman, and Geoffrey Moore Langdon. *CADD and the Small Firm '94: A Resourcebook*. Version 6.0. Boston: The Boston Society of Architects, 1994.

Walker, John. *The Autodesk File: Bits of History, Words of Experience*. 3rd ed. Thousand Oaks, Calif.: New Riders Publishing, 1989.

Wurman, Richard Saul. *Information Anxiety*. New York: Doubleday & Sons, 1989.

Index

Acad.bat, 123
Acad.lin, 102
Acad.mnu, 95
Acad.pgp, 93
Accelerators, graphic, 53
Accessories, ergonomic, 66
Accuracy
　of AutoCAD, 21
　from CAD, 16
　dimensioning, 225–26
　of files, disputes over, 436
　of plotter, 294
Add-ons. *See also* Third-party software
　approximate automated cross-referencing, 180
　electronic product catalogues, 167
　interactive design routines created with, 166
　for materials and textures, 267
　for modeling and animations, 260–61
　for redlining, 179
　for simulating light, 265
　symbol libraries, 216–18
Adjacent buildings, presenting design in context of, 174
Administration, 311–75
　financial, 313–39
　　affordability of AutoCAD, 21–22, 335–39
　　budgeting, 313–15
　　consultants, 334
　　copyright ownership, 334

　　designer's liabilities, 334
　　economies of scale, 320–21
　　expenses, 316–20, 325
　　fees for services, 326–33
　　income, 315–16
　　returns on AutoCAD, 322–26
　　sources of budgeting information, 321–22
　personnel, 340–62
　　burnout, 361–62
　　CAD ghettos, 359–61
　　CAD managers, 351–56
　　CAD performance and CAD parity, 349–51
　　interviewing and hiring, 341–46, 353–56, 440
　　job descriptions and responsibilities, 340–41, 354
　　mentoring and apprenticing, 356–57
　　standards manual, 357–58
　　technical committees, 358–59
　　training, 346–49
　routine, 362–75
　　controlled obsolescence, 367–70
　　memory and speed management, 362–67
　　security, 370–73
　　space planning, 373–75
A/E/E Systems Conference and Exhibit, 51
Affordability of AutoCAD, 21–22, 335–39

AIA Master Systems, 180, 181, 182*n*
AIA-sponsored layers, 244–45
American Institute of Architects (AIA), 145, 396
　AIA Document B163, 148
　AIA Master Systems, 180, 181, 182*n*
　AIA Online, 185*n*, 396
American Institute of Steel Construction (AISC), 168
American National Standards Institute (ANSI), 228, 240
American Society of Mechanical Engineers (ASME), 240
Americans with Disabilities Act (ADA), 155, 194–96
Animated presentations, 150
Animated shadow studies, 174
Animation, 49, 445
　add-ons for, 260–61
Animator Studio, 49
Antivirus software, 58, 59, 92
Apple Computer, 432
Application files, 87
Applications, researching, 440
Appload command, 120, 366
Apprenticing, 356–57
Approval of design, 173–74
Arc command, 178
ARCHIBUS, 209
ARCHIBUS/FM 10, 205
Architectural overlays, 51
Architectural Registration Exam (ARE), 344

Architectural ruler scale, 101
Architectural Works Copyright Protection Act of 1990, 432
Architecture, AutoCAD in, 6–8
Archiving data, 427–31
　drawing files, 427–28
　file used to create area takeoffs, 202–3
　problems with, 428–29
　resolving challenges to, 429–31
Area command, 198, 202
Area takeoffs, 197–203, 328
　databases and, 203
　standard methods, 199–200
　suggested procedure for, 201–3
Array command, 171
ARX, 447
As-builts, 190, 328, 435, 440
ASCII files, 409, 430
ASME, 240
Associative dimensioning features, 222, 225
ATC, 348
Attributed block, 109
Audio/sound devices, 62
Audit
　of errors, 406
　software, 421–22
Authorization code, 89
AutoCAD, 1–8. *See also* Release 13
　accuracy of, 21
　affordability of, 21–22, 335–39
　in architecture and interior design, 6–8
　built-in options, 247
　competitors of, 6
　customizability of, 20
　future for, 443–48

　history of, 2–6
　industry dominance of, 1
　installing, 88–89
　open architecture of, 20, 22
　reasons for using, 19–23
　streamlining setup, 338
　support network for, 20
AutoCAD Color Index (ACI), 210–12, 215
AutoCAD Customization Guide, 97
AutoCAD Development Systems (ADS), 121, 212
AutoCAD LT for Windows, 5, 48–49, 188, 336–37, 447
AutoCAD Release 13 Customization Guide, 279
AutoCAD Release 13 Installation Guide, 31, 239, 293, 302
AutoCAD Resource Guide, The, 31, 50, 67, 68, 79, 216, 239, 261, 309
AutoCAD Return on Investment (ROI) Calculator, 322–25
AutoCAD Runtime Extension (ARX), 447
AutoCAD Training Centers (ATC), 348
Autodesk, 3–4, 250
　new product development, 5
　other software from, 336–37 (*see also specific software*)
　product and marketing strategies, 4–5
Autodesk Device Interface (ADI) drivers, 31, 283
Autodesk Federal Bulletin, 229
Autodesk Multimedia Partner Catalog, The, 267
Autodesk Registered Developer Program, 50

Autodesk University, 348
AutoLISP routines, 119–21
　for ADA compliance, 195
　automating CAD exchange process using, 183
　automating layer setup by writing, 248
　to convert to metric, 279
　feasibility studies using, 155–56
　interactive design routines created with, 166
　translation from English to other languages, 240
Automatic Audit option, 91–92
Automatic Saves, 91
Automation, 16
AutoSketch, 49
AutoSurf, 49, 259
AutoVision, 49, 173, 250, 261, 265, 267

Backgrounds in physical models, 250–51
Backing up data, 372–73, 423–27
　media options for, 123, 424–25, 430
　procedures for, 425–26
　protecting backups, 426–27
　software for, 59
.bak files, 365
Baud rate, 64, 394
Benefits, 344
Berne Convention, 241
Bid process, 184–86
Big-font file format, 239
Billing documentation, 61
Bitmapped images, 398
Blipmode option, 366
Block(s), 109–14
　attributed, 109

474   INDEX

data attributes attached to, 153
documenting, 112–13
evaluation of, 343
exploded, 114
libraries of, 112, 194–96, 261
nested, 112
Block command, 110
Block redefinition capabilities, 158–59
.bmp file extension, 221
Boundary command, 202
Brochures, 150
Budgeting, 313–15
   for customization, 80–82
   for data upgrades, 418
   sources of information on, 321–22
Bulk discounts, 34
Burnout, 361–62
Business Software Alliance, 421
Bytes per second (BPS), 394

CAD. *See* Computer-Aided Design (CAD)
CAD Drafter/Designer, 341
CAD ghettos, 359–61
CAD operators, 341
CAFM. *See* Computer-Aided Facilities Management (CAFM)
Cal command, 172
Calculator, AutoCAD Return on Investment (ROI), 322–25
CAM, 208–9, 251
Cameras, digital, 62–63
Canadian Inventory of Historic Buildings, 230
Cartoon sets, 134–35
Cash purchase of equipment, 318–19, 368

CD-recorders (CD-Rs), 34, 392
CD-ROM
   as-builts on, 440
   circulating CAD-based drawings on, 185
   data input from, 391–93
CD-ROM drives, 34, 62
Centralized network file server, 136, 137
Change command, 175
Check sets, 140–41, 176–78, 304
CIM, 187
Client markets, marketing AutoCAD by, 148–49
Clients
   AutoCAD use and, 21
   CAD-related tools used by, 157
   development of, 147–52
   layer-naming system using names of, 247
   OLE, 412
   promotional needs of, 192
Clipboard utility, 409–10, 412
Code compliance, 194–96
Code-compliance diagrams, 174
Color(s), 209–15
   color palette, 210–12, 214–15
   customization of, 121
   layer-naming system using, 247–48
   of light, 264, 265
   on-screen versus plotting, 213–14
   plotter options and, 289–92
   uses for color output, 210
Color boards, finish, 173
Color laser copiers, 308
Color plotting capabilities, 294
Color printers, 308

Command line, 93, 94
Commands, keyboard, 93. *See also specific commands and functions*
Commerce Business Daily, 227
Commercial symbol libraries, 216–18
Communications
   managing team, 139–40
   office, 37–38
Compatibility
   with consultants, 183
   software, 227–28
Complementary software, 46–61
   collecting information about, 67–69
   other Autodesk software, 48–49
   product features to consider, 48
   selecting, 69–71
   third-party add-ons, 49–54
   typefaces, 53, 61
   utilities, 58–61, 319–20, 364, 372, 387
   Windows applications, 54–58
Complexity, project, 131
CompuServe, 394
Computer-Aided Design (CAD), 1–2. *See also* Modeling
   availability of, 128
   decision making about, 9–24
   appropriate usage, 17–23
   decision tree for, 12–13
   foregoing CAD, 13–15
   opting for CAD, 15–17
   as part of design firm strategy, 9–11
   design training and, 116

INDEX 475

dissatisfaction with, 6–7
in the field, 188
graphic standards and, 81
history of, 2–6
impact on design, debate over, 162
limitations of, 6
publications related to, 296
service centers, 268–69
skepticism about, 161, 162
standards for, 129
successful use of, 7–8
utilization ratio, 325–26
Computer-Aided Design and Manufacturing (CAD/CAM), 2
Computer-Aided Facilities Management (CAFM), 52, 157, 189, 203–9
benefits of, 204
design practice and, 205–6
in-house development of, 208
selecting CAFM system, 206–9
software, 52, 159
Computer-aided modeling and manufacturing (CAM), 208–9, 251
Computer committee, 352
Computer/drafter ratio, 129
Computer-Integrated Facilities Management (CIFM), 209
Computer-Integrated Manufacturing (CIM), 187
Computer inventories, 422–23
Computer platforms, 26. *See also specific personal computer platforms*
Computer regulations, international, 241

Computer systems
configuration of, 136–37
liabilities involving, 437, 438
obsolete, 368–70
speed of, 131
Computer trade shows, 68–69
Computer workstations, personalizing, 86
ConDoc, 181–82, 246, 385–86
Config menu, 115
Configuration, 25–72
complementary software, 46–61
collecting information about, 67–69
other Autodesk software, 48–49
product features to consider, 48
selecting, 69–71
third-party add-ons, 49–54
typefaces, 53, 61
utilities, 58–61, 319–20, 364, 372, 387
Windows applications, 54–58
decision tree for, 25–26
hardware selection, 31–35
purchasing guidelines, 33–35
networks, 35–46
advantages of, 35–38
AutoCAD configured for, 42–44
centralized, 43–44
configuring, 40–42
costs of, 39
disadvantages of, 38
office culture and, 44–46
peer-to-peer, 38, 43
of overall operating systems, 84–86

peripherals and other hardware, 61–67
audio/sound devices, 62
CD-ROM drives, 34, 62
digital cameras and videos, 62–63
digitizer pads on tablets, 63, 95
ergonomic accessories, 66
input devices, 63, 338
Macintosh versus Windows, 62
modems, 64, 300, 393–95
power protectors, 66
printers, plotters, and other devices, 64–65, 283, 308
scanners, 65, 309
platforms, 26–31
DOS, 27–28
Macintosh, 27, 30
multi-platform office, 30–31
personal computers, 26
Windows, 27, 28–29
workstations, 26
systemwide, 89–90
training and evaluation tools, 67
of Windows, 84–86
Configure menu, 298
Configure Operating Parameters option, 91
Construction administration, 186–90
Construction documents, 236
Construction signs, 189
Construction Specifications Institute (CSI), 181, 240–41, 246

Consultant(s)
  AutoCAD use and, 21
  contracts with, 334–35
  coordination with, 182–84
  for customization, 78–80
  financial administration
    and, 334–35
  as manager, 352
  training from, 348
  working in 3D, 268
Contract(s)
  with consultants, 334–35
  design, 331–32
  liability protection in,
    439–40
  maintenance, 317
  service, 295–96
Contractors, CAD use by, 188
Conversion, 122, 402–11
  file-exchange
    documentation, 410–11
  file formats and, 403–11
  raster-to-vector, 399
  software for, 59
Coordination, 439
  with consultants, 182–84
  in document production,
    174–76
  liability for failures in, 438
  managing team, 139–40
  of moves, 189
  project, 36
  using CAD, 17
Copiers, color laser, 308
Copyright
  on databases, disputes over,
    434
  ownership of, 334
  protection provided by, 241,
    431–34
  violations of, 420–21
Copyright Act of 1976, 432

Corps of Engineers Guide
  Specifications (CEGS), 228
Cost(s)
  of AutoCAD system, 335
  of networks, 39
  opportunity, 320, 324
  per plot, 298
  per workstation, 14–15, 320
  of upgrading, 414, 418
Cost estimation for ADA or
  code compliance
  renovation, 195
Cost estimation software,
  179–80
Craftsmanship, CAD, 350
CRC validation, 91, 92
Credentials of job candidate,
  343–44
Cross-referencing, 180–82
CSI, 181, 240–41, 246
Culture, networks and, 44–46
Custom installation, 88
Customization of AutoCAD,
  20, 73–124
  AutoLISP routines, 119–21
  color, 121
  configuring overall
    computer operating
    system, 84–86
  database and OLE links, 122
  defaults for AutoCAD
    commands, 93–98
  drawing environment,
    98–107
    dimension styles, 106–7
    drawing limits, 99–101, 109
    drawing units, 99
    hatches, 107
    layer standards, 103–5
    linetypes, 102–3
    multiline styles, 107
    scale factors, 101–2
    typestyles, 105–6

  evaluating, 342–43
  expenses, 317–18
  external consultants for,
    78–80
  feedback and subsequent
    revisions, 124
  file conversion routines, 122
  general guidelines for, 74–75
  implementing, 123–24
  in-house, 76, 82–83
  installing AutoCAD, 88–89
  office standard versus
    individualized
    customization, 98
  other Windows applications
    and, 83
  output devices, 115–18
  performers of, 124
  programmable dialog boxes,
    121
  protecting, 123
  prototypical drawings,
    templates, and libraries,
    107–15, 387
  scheduling and budgeting
    for, 80–82
  scripts, 119
  setting up computer directo-
    ries and file-naming
    standards, 86–88
  steps in, 83
  systemwide AutoCAD
    configuration, 89–90
  third-party software
    add-ons, 76–78
  3D parameters, 121–22
  underlying system
    operating variables,
    91–93
  upgrades and, 415
Cutting machines, 308–9
Cyclic Redundancy Check
  (CRC), 91, 92

Damage, system, 372
Databases, 54–56, 122
  area takeoffs and, 203
  copyright disputes over, 434
  creation of, 153–55
  GIS, 227, 233, 251, 401–2, 437
Data management, 377–442
  archiving data, 427–31
    drawing files, 427–28
    problems with, 428–29
    resolving challenges to, 429–31
    used to create area takeoffs, 202–3
  backing up data, 372–73, 423–27
    media options for, 123, 424–25, 430
    procedures for, 425–26
    protecting backups, 426–27
    software for, 59
  copyright protection, 431–34
  data conversion, 122, 402–11
    file exchange documentation, 410–11
    file formats, 403–11
    raster-to-vector, 399
    software for, 59
  data input, 390–402
    from files, 391–95
    from public domain, 401–2
    from scanning, 395–401
  data linking and embedding, 411–14
  directories, 379–90
    maintaining, 365
    organization of, 381
    setting up, 86–88
    files, 379–90
      External References (Xrefs), 387
      file access, 381–83
      file naming approaches, 86–88, 383–86
      plot files, 387–88
      prototypical drawings and utilities, 387
      special AutoCAD file managers, 388–90
  insuring data, 441–42
  liability and legal disputes over data, 434–41
  licences for data, 418–23
    computer inventories and, 422–23
    employees and, 421–22
  upgrading data, 414–18
    planning, 416–17
    scheduling and budgeting, 418
Data recreation, 442
Ddattext command, 153
Ddim command, 170
Ddinsert command, 162
Ddlmodes, 97
Ddmodify, 97, 170
Ddunits command, 99, 277
Dealers, AutoCAD, 69, 348
Defaults, AutoCAD, 93–98
Delegating tasks, 136–39
Deliverables, 332–33
Depreciation schedules, 368
Design contracts, 331–32
Design process, 145–92
  benefits of, 146–47
  bid process, 184–86
  construction administration, 186–90
  design development, 160–73
    design checklists, 172–73
    evolution from schematics to developed designs, 161–65
    finish color boards, 173
    geometry and math, 169–73
    interactive design routines, 166–67, 172
    models and mock-ups, 168–69
    products and components, 167–68
  document production, 174–84
    check sets, 176–78
    coordination and quality control, 174–76
    coordination with consultants, 182–84
    cross-referencing, 180–82
    links to specifications and cost estimation software, 179–80
    redlines, 178–79
  marketing and client development, 147–52
  postconstruction activities, follow-up and evaluation, 190–91
  predesign services, 152–56
  promotion, 192
  review and approval stage, 173–74
  schematic design, 156–60
  traditional approach to using CAD, 145, 146
Design publications, 68
Design training, 116
Desktop Photogrammetry, 233
Desktop publishing (DTP), 56–57, 218–21
  within AutoCAD, 219–20
  file exchange, 220–21
  slide presentations, 220

Detail, in 3D modeling, 262–63
Detail libraries, 51–52, 114–15
Device and Default Selection box, 116
Diagrams, 156–58
Dialog boxes, 97
  Alternate Units, 277, 278
  Browse/Search, 389
  Color, 265
  Dimension Primary Units, 225–26
  Dimension Styles, 106
  interactive, for ADA compliance, 195
  interactive design routines created with, 166
  Layer Control, 104
  Lights, 211, 263, 264
  Linetype, 102, 103
  Load Multiline Styles, 108
  Materials Library, 265, 266
  Menu Customization, 94–95
  Modify Standard Material, 267
  Paper Size, 117–18
  Plot Device and Default, 299
  Preferences, 90, 277
  Preferences Misc, 366
  programmable, 121
  Scenes, 264
  Select File, 388
  Set Layer Filters, 243
  Units Control, 99, 100
Dialog Control Language (DCL), 97, 121
Dictionaries, spelling, 92
Digital cameras and videos, 62–63
Digitizers, 63, 95, 338

Dimensioning, field conditions and, 222–26
Dimensioning accuracy, 225–26
Dimensioning toolbar, 97
Dimension styles, 106–7
Direct imaging plotters, 291
Directories, 379–90
  maintaining, 365
  organization of, 381
  setting up, 86–88
Diskettes, data input from, 391
Dist command, 195, 406
Dlgcolor command, 97
Documentation
  AutoCAD, 419
  of construction, 186
  of customization, 123
  file-exchange, 410–11
  for historic preservation, 230–32
  management and storage of, 152
  about memory management, 363
  of standards for CAD layer names, 249–50
  time and billing, 61
Document management, 329
  software for, 49, 52
  technical (total), 157, 329, 378–79, 389, 390
Document production. *See under* Design process
Donating secondhand equipment, 370
DOS platform, 27–28
Downtime
  strategies against, 371
  upgrades and, 415

Drafter/computer ratio, 129
Drafting
  by hand, 127, 130
  simultaneous, 36
Draft quality plots, 132
Drawing environment, 98–107
  dimension styles, 106–7
  drawing limits, 99–101, 109
  drawing units, 99
  hatches, 107
  layer standards, 103–5
  linetypes, 102–3
  multiline styles, 107
  scale factors, 101–2
  typestyles, 105–6
Drawing exchange format. *See* .dxf files
Drawing file. *See* .dwg files
Drawing file-exchange format, 337
Drawing information, merging, 156
Drawing limits, 99–101, 109
Drawing matrix, 135–36
Drawing name, default, 91
Drawings
  multiple, 270–73
  prototypical, 108–14, 248, 387
Drawing sheets, 294, 445
Drawing titles, file names based on, 384–85
Drawing type, dividing work by, 138
Drawing units, 99
Dtext command, 202
DTP. *See* Desktop publishing (DTP)
Dual dimensioning units, 277–79

.dwg files, 20, 220, 233, 332, 403–5, 447
  archiving, 427–28
  maintaining, 365–67
  name of, 110
Dxbin command, 91
.dxf files, 20, 52, 220, 221, 240, 332, 337, 405–8
Dxfin command, 91, 405, 406
Dxfout command, 405

Economies of scale, 320–21
Editing information, 158–59
Efficiency from CAD, 16
Electrophotographic (laser/LED) plotters, 291
Electrostatic plotters, 292
Ellipse command, 161
e-mail, 46, 397–98
Employees. *See also* Personnel administration
  liability for mistakes of, 437
  licenses for data and, 421–22
  revenues and expenses per, 325
  selling computers to, 370
  temporary, 345
.eps file extension, 220
Equipment
  destruction of, 371
  leasing of, 319, 368
  purchase of, 33–35, 293–96, 339
  secondhand, 369
Ergonomic accessories, 66
Ergonomic space planning, 373–75
Errors
  audit of, 406
  liability for, 437, 438
European Economic Community (EEC), 241

Evaluation
  of design process, 190–91
  personnel, 342–44
  of project, 191
  tools for, 67
Exchange process, automation of CAD, 183
Expandability, 33, 367–68
Expenses, CAD related, 316–20, 325
Explode command, 110
Exploded blocks, 114
Export Data command, 409
Extended warranties, 295–96
External References (Xrefs), 136, 138, 159, 387
  for historic preservation project, 236–38
  to link multiple drawings, 272–73
  to link multiple floors, 270
  multiple phases and, 275

Fabrication, 186–87
  full-scale details of historical elements for, 236
Fabricators, use of CAD by, 187
Facilities Contract User's Manual, 229
Facilities management, 52, 190, 328
Feasibility studies, 155–56
FEA software, 187
Federal Paperwork Reduction Act (1995), 379
Fees for services, 326–33
Field conditions and dimensioning, 222–26

File(s), 379–90
  access to, 381–83
  application, 87
  bad, 364
  control of, 37
  conversion of (*see* Conversion)
  as copyrightable materials, 432–33
  data input from, 391–95
    on CD-ROM, 391–93
    on diskettes, 391
    by modem, 393–95
    on networks, 395
    on tapes, 393
  disputes over accuracy of, 436
  exchanges of, 183
    with desktop publishers, 220–21
    media of, 333
    preferences in other countries, 240
    under Release 13, 46
  firm policy about sharing, 434
  font, 93
  formats of, 220–21 (*see also specific formats*)
    data conversion and, 403–10
    preferred, 332–33
    standards for, 439
    upgrades to, 430
  guidelines on contents of, 333
  plot, 387–88
  project drawing, 87
  read-only, 382
  relationship between hard copy and, 377–78

requirements of plotting services, 299
software versus user, 420
swap, 43
transfer capabilities of plotting services, 299–300
translation software, 52
unauthorized use of, 436
File compression software, 60
File/document management software, 59
File-exchange format, 337
File locking, 89, 92
File management, 365–67
File naming approaches, 86–88, 383–86
File sharing, electronic, 402
File transfers on networks, 36
Fill option, 366
Film plots, 132, 298
Filter command, 104
Financial administration, 313–39
  affordability of AutoCAD, 21–22, 335–39
  budgeting, 313–15
    for customization, 80–82
    for data upgrades, 418
    sources of information on, 321–22
  consultants, 334–35
  copyright ownership, 334
  designer's liabilities, 334
  economies of scale, 320–21
  expenses, 316–20, 325
  fees for services, 326–33
  income, 315–16
  returns on AutoCAD, 322–26
  sources of budgeting information, 321–22

Financing, 318–19
Financing strategy of design firms, 11
Finish color boards, 173
Finite Element Analysis (FEA) software, 187
Flat–file databases, 54
Flat/lump sum fees, 330
Floating licenses, 36, 43, 44
Floor(s)
  dividing work by, 138
  multiple, 273–74
Floppies, 391
Flowcharts, 156–58
Fly-out menus, 97
Fonts, 61, 65, 105, 239
Full-size plots, minimizing use of, 304

General managers, CAD managers and, 354
General Services Agency (GSA), 229
Geographical flexibility with CAD, 17
Geographical markets, marketing AutoCAD by, 149
Geographic Information Systems (GIS) databases, 227, 233, 251, 401–2, 437
Geometric Dimensioning and Tolerancing (GDT) capabilities, 187, 222
Geometry in design development, 169–73
Ghettos, CAD, 359–61
Government work, 226–29
Graphic accelerators, 53

Graphical information, merging different sets of, 156
Graphical User Interface (GUI), 28, 86
Graphics
  quality and consistency of, 16
  standards for, 81, 228, 240
  use of special, 327
Graphics and desktop publishing, 56–57
Grids, 99, 109
Group command, 111
Growth, with CAD, 17
Guidelines, 109

HABS, 227, 229
Hand drafting
  comparing CAD to, 127
  hybrid production using CAD and, 130
Hand-lettering typestyles, 105
Hard copy
  archives, 429
  relationship between files and, 377–78
Hard drives
  optimizing, 364
  partitioning, 87–88, 364
Hardware expenses, 316
Hardware lock, 43–44, 89, 239
Hardware selection, 31–35
  purchasing guidelines, 33–35
Hatch command, 107, 202
Hatches, 107
Hewlett-Packard Graphics Language (HPGL) and HPGL/2 plotting languages, 293

Hiring. *See* Interviewing and hiring
Historic American Buildings Survey (HABS), 227, 229
Historic preservation, 229–38
　analysis of, 235
　construction documents and specifications, 236–38
　documentation for, 230–32
　inputting historical data in AutoCAD, 232–34
History, fee estimation based on, 331
HVAC, 319–20
Hybrid printer/plotters, 292
Hybrid production, 130

Icons, 96
.iges files, 240, 332, 403, 408
IGES Translator v 5.1, 408
Imaging plotters, direct, 291
Imperial drawings to metric, converting, 279
Import Data command, 409
Income, CAD-related, 315–16
Industrial designer, use of CAD by, 187
Industry standard guidelines, electronic versions of, 168
Industry trends, fee estimation based on, 331
Information, manufacturers, 52
Information superhighway, 394, 396–98
In-house plotters
　choosing between external plotting services and, 283–86
　equipment selection guidelines, 293–96

In-house training programs, 349
Injury to workers, 374
Inkjet plotters, 290–91, 301
Input, data, 390–402
　from files, 391–95
　from public domain, 401–2
　from scanning, 395–401
Input devices, 63, 338
Insurance, 373
　of data, 441–42
Interactive design routines, 166–67, 172
Interior design, AutoCAD in, 6–8
Internal Revenue Service, 80
International AutoCAD, 238–41
International Standards Organization (ISO), 149, 240
　layer standards, 246–47
　9000 Series, 228
Internet, 185, 228, 369, 396–98
Internetwork Protocol Exchange (IPX), 44
Interviewing and hiring, 341–46
　of AutoCAD-experienced employees, 21
　of CAD managers, 353–56
　evaluation, 342–44
　existing versus new staff, 341–42
　firm policy for, 11
　procedures for, 440
　salaries and benefits, 344
　service bureaus, 345–46
　temporary staff, 345

Inventories
　computer, 422–23
　graphic building, 232
　software, 422–23
ISO. *See* International Standards Organization (ISO)
Isometric models, 259
Isoplane command, 259

Japanese Industrial Standard (JIS), 240
Job descriptions and responsibilities, 340–41, 354
Job titles, 341

Keyboard characters for international AutoCAD, 239
Keyboard commands, 93

Laptop computer, 35
Large-format scanners, 309
Large screen monitors, 309
Laser copiers, color, 308
Laser/LED plotters, 291
Lasers, CAM, 251
Layer classes, 243
Layer-naming systems, 242–50
　standards for, 103–5
　documenting, 249–50
　implementing, 248–49
　typical approaches to, 244–48
Layers, 159
　AutoCAD layer characteristics, 242–44
　dividing work by, 137–38
　fees set by quantity of, 330

Leasing equipment, 319, 368
Legal disputes, 434–41
  over databases, 434
  possible sources of, 435–38
Liability(ies)
  for data, 434–41
  of designer, 334
  protection against, 438–41
Library(ies)
  block, 112, 194–96, 261
  detail, 51–52, 114–15
  marketing, 151–52
  product, 167
  of sample materials, 266
  slide, 115, 151–52
  of standard light types and scenes, 264–65
  symbol, 51, 216–18
  templates, 195–96
  Windows-based, 58
License(s)
  for data, 418–23
    computer inventories and, 422–23
    employees and, 421–22
  floating, 36, 43, 44
  multiuser, 419
  policy for, 78
Light
  simulation of, 265
  3D models with, 263–65
Linetypes, 102–3
Linetypes command, 366
List command, 170, 406
Lock, hardware, 43–44, 89, 239
Locking, file, 89, 92
Log-in name, default, 92
Logs of project archives, 430
Ltscale command, 101, 103

Ltype option, 102
Lump sum fees, 330

Macintosh systems, 22–23, 27, 30
  file names, 384
  hardware selection for, 31
Magazines about AutoCAD, 68
Mail-order houses, 69
Maintenance
  memory and speed management, 363
  of plotters, 304
  systemwide, 36
Maintenance contracts, 317
Malpractice, liability for, 438
Management, networks and, 44–46
Managers
  CAD, 351–56
    assigning responsibilities, 351–52
    establishing need for, 352–53
    hiring, 353–56
  project, 125–43, 178, 256
    CAD knowledge needed by, 142
    central role of, 125, 126
    management of team communications and coordination, 139–40
    project scope refinement by, 134–36
    quality control implementation by, 140–41
    scheduling by, 127–32
    scheduling output, 141–43

    staffing decisions by, 132–34
    task delegation by, 136–39
Manufacturers' symbol libraries, 216–18
Marketing, 147–52
Matchline for multiple files, 272
Materials
  CAD-generated images of, 173
  interaction of light with, 264
  3D models with, 265–67
Materials command, 266
Math, in design development, 169–73
Matrix
  to document office-standard layers, 249–50
  drawing, 135–36
  file-exchange format, 410–11
Media
  backup, 424–25, 430
  for file exchange, 333
  liability for, 334
  plot, 300–301
  stability of, 429
Memory, 362–67
  amount installed, 33–34
  AutoCAD memory management, 365–67
  evaluation of, 343
  in plotter, 295
  system memory, 364–65
Memory enhancers, 60
Memory managers, 350
Mentoring, 356–57
Menus, 93–94, 366
Meshes, surface, 259
Metric system, 228, 276–79
Microsoft, 432

Microsoft Windows, 27, 28–29
  applications, 54–58
    collecting information
      about, 67–69
    customization of
      AutoCAD and, 83
    databases, 54–56
    graphics and desktop
      publishing, 56–57
    libraries, 58
    open, 364
    presentation software, 57
    project management and
      scheduling tools, 57–58,
      133
    quality control tools
      provided in, 175–76
    selecting, 69–71
    spreadsheets, 56
    type size, 106
    word-processing software,
      56
  AutoCAD installation for, 88
  Autodesk's push toward,
    447
  configuring, 84–86
  Windows 3.1, 28
  Windows 95, 29, 84, 85
    file names in, 384
    plug and play technology,
      316
  Windows for Workgroups,
    28–29
  Windows NT, 29, 84
Minimal installation, 88
Mirror command, 171
Misconduct, liability for, 438
.mli files, 265, 266
Mline command, 107, 167, 366
Mock-ups, 169
Modeling, 250–69
  physical models, 250–51
  surface, 49

3D, 185, 251–69, 445
  add-ons for, 260–61
  advantages of, 252
  guidelines, 262–63
  internal versus external,
    268–69
  with light, 263–65
  limitations of 3D
    AutoCAD, 260
  with materials, 265–67
  options, 259–61, 269
  planning, 256–58
  presentation options,
    267–68
  tips on, 269
  2D compared to, 255–56
Models for design
  development, 168–69
Model space, 102
Modems, 64, 300, 393–95
Modules, 447
Monitoring, network, 46
Monitors, 34, 309
Mouse, 95
Move coordination, 189
Mslide command, 150, 152, 220
Mtext command, 202
Multiline styles, 107
Multimedia programs, 56
Multiple drawings, 270–73
  separate CAD file approach
    to, 272
  single CAD file approach to,
    270–72
  Xrefs to link or embed,
    272–73
Multiple floors, 273–74
Multiple phases, 274–75
Multiple schemes, presenting,
  160
Mvsetup command, 99

National Council of
  Architectural Registration
  Boards (NCARB), 344
National Institute of Building
  Sciences (NIBS), 246
National Park Service, 227,
  229, 233, 235
Natural terrain, models of,
  251
Nested blocks, 112
NetBIOS, 44
Network management tools,
  60
Networks, 35–46
  advantages of, 35–38
  AutoCAD configured for,
    42–44
  AutoCAD installed on,
    88–89
  centralized, 43–44
  configuring, 40–42
  costs of, 39
  data input from, 395
  disadvantages of, 38
  logging on and off, 364
  office culture and, 44–46
  peer-to-peer, 38, 43
Nonuniform rational
  B-splines (NURBS), 161
Number, layers by, 245

Object Linking and
  Embedding (OLE), 122,
  155, 180, 411–14
Object snaps (osnaps), 172,
  350
Obsolescence, 316
  controlled, 367–70
  plotter, 293–94
Occupational Safety and
  Health Administration
  (OSHA), 149

Office planning, 319
Office standards, 37
    for color palettes, 214–15
    modification of, 129
Office tours, 151
OLE. *See* Object Linking and
    Embedding (OLE)
Olelinks command, 412
On-screen colors, 213–14
Open architecture, 20, 22
Open command, 178
Operating systems, 27–29, 30.
    See also specific platforms
    configuring, 84–86
Operating variables, system,
    91–93
Opportunity costs, 320, 324
Optical Character Recognition
    (OCR), 64, 401
Optimization software, 59
Options, built-in, 247
Orthophotos, 232–33
OSHA, 149
Osnap command, 171
Output, color, 213–14
Output devices, 281–309
    customizing, 115–18
    other devices, 308–9
    overall CAD strategy and,
        281, 282
    plot media, options for,
        300–301
    plotters, 64–65, 283–92
        break-even point for
            purchasing, 286–87
        configuring, 301–3
        customizing, 118
        internal versus external,
            283–86, 293–300
        options, 287–92, 301
        reasons not to purchase,
            285–86

reasons to purchase,
    284–85
selection guidelines,
    293–96
vendor selection
    guidelines, 296–300
printers, 64–65, 68
    color, 308
    software for, 53
scheduling, 141–43
Output size, flexibility in, 295
Overlays, 51, 103, 175
    for ADA and code
        compliance, 195
    tablet, 115
Ownership of work, disputes
    over, 435

Pantone Matching System,
    210
Paper space, 102
Paperwork Reduction Act of
    1995, 228, 397
Partitioning hard drives,
    87–88, 364
Partners, CAD managers and,
    354
"Patches," 414–15, 420
.pcp files, 118, 298–99
Peer-to-peer networks, 38, 43
Pen Assignments, 117
Pencil plotters, 289
Pen plotters, 288–89
Pen settings
    layer-naming system using,
        247–48
    tracking plotter, 302–3
Performance, personnel,
    349–51
Periodicals, budgeting
    information from, 321

Peripherals, 61–67. *See also
    specific peripherals*
    compatible, 31–32
    Macintosh versus Windows,
        62
    sharing, on networks, 38
Personal computer platform,
    26
Personal Information
    Managers (PIMs), 58
Person-hours, fees by, 329
Personnel administration,
    340–62
    burnout, 361–62
    CAD ghettos, 359–61
    CAD managers, 351–56
        assigning responsibilities,
            351–52
        establishing need for,
            352–53
        hiring, 353–56
    CAD performance and CAD
        parity, 349–51
    interviewing and hiring,
        341–46, 353–56
        of AutoCAD-experienced
            employees, 21
        evaluation, 342–44
        existing versus new staff,
            341–42
        firm policy for, 11
        procedures for, 440
        salaries and benefits, 344
        service bureaus, 345–46
        temporary staff, 345
    job descriptions and respon-
        sibilities, 340–41, 354
    mentoring and
        apprenticing, 356–57

networks and, 44–46
standards manual, 357–58
technical committees, 358–59
training, 346–49
Photogrammetry (stereophotogrammetry), 232–33
Photographic image managers, 60
Physical models, 250–51
.pict files, 221, 410
PIMs, 58
Placement agencies, 345
Platforms, 26–31. *See also* specific operating systems and computers
  computer, 26
  operating systems, 27–29, 30
Plot files, 387–88
  naming of, 91, 304
Plot information, 110
Plot log, 305, 306
Plot media, options for, 300–301
Plot spooling, 295
Plotters, 64–65, 283–92
  configurations, 298–99, 301–3
  customizing, 118
  direct imaging, 291
  drivers for, 293
  electrophotographic (laser/LED), 291
  electrostatic, 292
  hybrid printer/plotters, 292
  inkjet, 290–91, 301
  internal versus external, 283–86, 293–300

  pen, 288–89
  pencil, 289
  plot media, 300–301
  purchasing, 284–87
    break-even point for, 286–87
    reasons against, 285–86
    reasons for, 284–85
    selection guidelines, 293–96
  thermal imaging, 291
Plotting, 132
  file formats, 295
  quality of, 132
  reimbursables, 297, 304–5, 306
  schedules, 38, 132, 141–42, 305–8
  software, 53
  time-out value for, 302
  tips for, 303–4
Plotting services, 141, 283–86
  vendor selection guidelines, 297–300
.plt files, 221, 365
Plug and play technology, 62, 316
Pointing devices, 95
Point lights, 264
Polygon tool, 170
Portable computer, 35
Portfolio, review of, 342
Postconstruction activities, 190–91
Postevaluation study, 191
PostScript, 61, 65, 105
Power protectors, 66
Precedents for similar CAD projects, 130–31
Predesign services, 152–56

Premier ATCs, 348
Presentation boards, 150
Presentation options for 3D models, 267–68
Presentation software, 57
Preservation. *See* Historic preservation
Printed output. *See* Output devices
Printers, 64–65, 283
  color, 308
  software, 53
Proactive approach to CAD, 126
Product bundles, 34
Product fairs, 296
Production. *See* Output devices
Productivity, enhancing, 16
Product libraries, 167
Products
  commercial symbol libraries of, 216–18
  design development and, 167–68
Professional organizations
  budgeting information from, 322
  fee estimation based on, 331
Profit center, plotter as, 284–85
Programming, 153–55
Project(s)
  complexity of, 131
  coordination of, 36
  evaluation of, 191
  file organization by, 380
Project management and scheduling tools, 57–58, 133

Project manager(s), 125–43, 178, 256
  CAD knowledge needed by, 142
  central role of, 125, 126
  management of team communications and coordination, 139–40
  project scope refinement by, 134–36
  quality control implementation by, 140–41
  scheduling by, 127–32
  scheduling output, 141–43
  staffing decisions by, 132–34
  task delegation by, 136–39
Project name and number, file names based on, 384–85
Project stage or task, dividing work by, 138
Project team, physical setup of, 139–40
Promotion, 192
Proposals, AutoCAD files used in, 150
Prototypical drawings, 108–14, 248, 387
Publications
  CAD-related, 296
  drawings from AutoCAD in marketing, 150
Public domain data, 401–2
  disputes over, 436–37
Pulldown menus, 93–94
Punch lists, 189–90
Purge command, 173, 248, 365

Qsave command, 91
Qtext, 366
Quality control, 439
  in document production, 174–76
  implementing, 140–41
  Windows-based tools for, 175–76

Radiosity, 265
Raster-driven plots, 290
Raster-to-vector software, 399
Ray tracing, 265
Read-only files, 382
Realism in 3D modeling, 262–63
Redefine command, 110, 158, 162
Redlines, 178–79
Redundant Array of Inexpensive/Independent Disks (RAID), 426
Reimbursables, 297, 304–5, 306, 333
Relational databases, 54
Release 13, 5, 22–23, 28
  ADI drivers for major models of, 302
  computer speed and, 131
  curved shapes with, 161
  database creation in, 155
  file exchange capabilities in, 46, 404
  file management and, 87
  file preview function in, 388
  Geometric Dimensioning and Tolerancing (GDT) feature, 187, 222
  OLE in, 413
  on-screen working environment, 94
  options for command input, 93
  plotting drivers and plotter configuration files, 293
  system requirements of, 84–85
  typestyle options in, 105
  upgrade cost, 414
  Windows-compatible SQL drivers in, 55
Render command, 212
Rendering, 49
Render toolbar, 263, 265
Renting equipment, 319, 368
Reprographics houses, 346, 401
Resolution of monitors, 34
Resume, review of, 342
Return on investment, time frame for earning, 314
Returns on AutoCAD, 322–26
Revenues per employee, 325
Reverse-read plots, 294
Review and approval, design, 173–74
Rich-text format, 409
Rmat command, 266
Rolling upgrade, 417
Rotate command, 171
.rtf files, 409
Ruler scale, architectural, 101

Salaries, 344
Salvage values, 368
Sample materials library, 266
Saveasr12 command, 404
Saves, Automatic, 91
Scale
  in 3D modeling, 262
  in 3D models, 252
Scale economies, 320–21
Scale factors, 101–2
Scanners, 65
  large-format, 309
Scanning, 395–401
Scenario testing, 16

Scheduling, 127–32
  for customization, 80–82
  for data upgrades, 418
  of output, 141–43
  plotting, 38, 132, 141–42, 305–8
  software for, 57–58
Schematic design, 156–60
  evolution to developed designs from, 161–65
Scope of project, refining, 134–36
Screen capture utilities, 60–61
Screen savers, 151
Script command, 150
Scripts, 119, 183, 248
Search commands, 172–73
Secondhand equipment, 369
Security, 370–73
Separate files approach to multiple floors, 274
Server
  installation on, 89
  OLE, 412
Service bureaus, 268–69, 318, 345–46
Service contracts, 295–96
Shade command, 264
Shop drawings, 186
Side menus, 93–94
Signage, 275–76
Signatures, 441
Simultaneous drafting, 36
.sld files, 220
Slidelib.exe, 115, 152, 220
Slide libraries, 115, 151–52
Slide presentations, 150, 220
Sociology, network, 40
Softdesk, 51, 261

Software. *See also* Complementary software; Third-party software
  advances in, data archiving and, 429
  antivirus, 58, 59, 92
  audit of, 421–22
  CAD ghetto concept and, 361
  CAFM, 52, 159
  cost estimation, 179–80
  document management, 49, 52
  intelligent, 445
  inventory of, 422–23
  liabilities involving, 437–38
  for plot queuing and management, 295
  plotting, 53
  project management, 133
  project-oriented, 445–46
  scheduling, 57–58
  supplementary, 327–28
Software expenses, 317
Software files, 420
Software stores, 69
Solid models, 260
Sound devices, 62
Space planning, 373–75
Speakers, 62
Special services, marketing AutoCAD by, 148
Special studies, 328
Specialty color output devices, 308
Specifications
  in historic preservations, 236–38
  linking document production to, 179–80
  linking signage drawings to, 276

Speed management, 362–67. *See also* Memory
  evaluation of, 343
Spell command, 175
Spelling dictionaries, 92
Spline command, 161
Spotlights, 264
Spreadsheets, 56
SQL. *See* Structured Query Language (SQL)
Square footage, fees set by, 330
Staff. *See also* Personnel administration
  expenses related to, 318
  for projects, 132–34
  selling computers to, 370
  temporary, 345
Stamps, 441
Standard Form of Agreement between Owner and Architect for Designated Services, 145, 148, 332, 440
Standards
  availability of, 129
  for file formats, 439
  for forms, 228
  for graphics, 81, 228, 240
  layer, 103–5
  layer-naming, 103–5
    documenting, 249–50
    implementing, 248–49
  office, 37
  for color palettes, 214–15
  modification of, 129
  in other countries, 240–41
  use of special, 327
  work environment, 374–75
Standards manual, 357–58
Stereophotogrammetry, 232–33

Storage formats, 429
Strategy, output management and, 281, 282
Structured Query Language (SQL), 55, 122, 154–55
  approximate automated cross-referencing, 180
  area takeoffs and, 203
  drivers, 55
  to highlight discrepancies with external database, 175
  links to external databases for historic preservation, 236
Subdirectories, maintaining, 365
Suites, program, 70
Supplementary software, 327–28
Supplies, 319
Support network, 20
Surface models or meshes, 259
Surfaces
  interaction of light with, 264
  modeling of, 49
  in physical models, AutoCAD for, 250–51
Surge protectors, 66
Surveying tools, 233
Swap file, 43
Swap space, 89
Symbol libraries, 51, 216–18
System backups, 372–73
System damage, 372
System monitoring and evaluation, 37
System operating variables, 91–93

Tablet overlays, 115
Tape backups, 123, 423, 425–26
Tapes, data input from, 393
Task delegation, 136–39
Team communications and coordination, managing, 139–40
Technical committees, 358–59
Technical expertise of plotting services, 300
Technical school, donating obsolete equipment to, 370
Technical (total) document management (TDM), 157, 329, 378–79, 389, 390
Temp agencies, 345
Templates, 107–15
  AutoCAD signage, 276
  design checklist through, 172
  feasibility studies using, 155–56
  libraries of, 195–96
  to reconcile differences between the ADA and building codes, 195
  for use with typical projects, 133
Temporary staff, 345
Test preview of 3D model, 263
Tests, 343
Text lengths in international versions of AutoCAD, 239
Text management tools, 53
Theft prevention, 371–72
Thermal imaging plotters, 291
Third-party software, 49–54
  for customization, 76–78
  training with, 347
  upgrades and, 415

3D AutoCAD, 269
3D modeling, 185, 251–69, 445
  add-ons for, 260–61
  advantages of, 252
  guidelines, 262–63
  internal versus external, 268–69
  with light, 263–65
  limitations of 3D AutoCAD, 260
  with materials, 265–67
  options, 259–61, 269
  planning, 256–58
  presentation options, 267–68
  tips on, 269
  2D compared to, 255–56
  working with software other than 3D AutoCAD, 260–61
3D parameters, 121–22
3D presentations, 150–51
3D shapes, 169, 170
3D shape tools, 260
3D Studio (Autodesk), 49, 150, 173, 250, 261, 265, 267
.tiff files, 220, 221
Time and billing documentation, 61
Time-out value for plotting, 302
Tolerancing capabilities, 222
Toolbars, 95–97
Topology, network, 40
Total Document Management (TDM), 157, 329, 378–79, 389, 390
Tracking schemes, 159–60
Trade shows, 68–69, 321

Training, 67, 143, 346–49
  design, 116
  expenses associated with, 318
  of job candidate, 343–44
  procedures for, 440
Translation software, 59
Transmission protocol, 64
Trojan horses, 58
TrueType fonts, 61, 65, 105
Turnaround time of plotting services, 298
Tutorials, 247
2D AutoCAD, 255–56, 259
2D shapes, 169, 170
.txt files, 409
Typefaces, 53, 61, 93
Type size in Windows applications, 106
Type styles, 105–6
Typical installation, 88

Unicode character encoding capability, 239
United Computer Exchange, 369
U.S. Department of Defense, 228, 408
Universal Copyright Convention, 241

Universities, AutoCAD training from, 348
Upgrading
  of data, 414–18
  planning, 416–17
  scheduling and budgeting, 418
  selective, 339
  training after, 349
Usage factor, 315, 325–26
  plotter, 284, 285
User groups, local, 69
Users, network, 36
Utilities, 58–61, 319–20, 364, 372
  prototypical, 387
Utilization ratio. *See* Usage factor

VDT workers, injuries to, 374
Vectorizing, 398–400
Vendors
  budgeting information from, 321
  selection guidelines, 296–300
Video cards, 34
Video projectors, 309

Videos, digital, 62–63
Viewer participation in 3D models, 252
Viewres, 366
Views, 366
Visualization, 250
Visualization studies, 328
Visualization tools, 53
Vslide command, 150, 220

Wblock command, 272, 433
Windows. *See* Microsoft Windows
Wire frame models, 259
.wmf files, 220, 221, 410
Word-processing software, 56
WorkCenter (Autodesk), 141, 179, 389, 390
Workers, injury to, 374
"Works" program, 70
Workstation(s), 26
  cost per, 14–15, 320
  personalizing, 86
  space planning, 375
  uniform, 339
Writable CD drives (CD-Rs), 392

Xrefs, 115, 366, 413